AF179645

AESS Interdisciplinary Environmental Studies and Sciences Series

Series Editor
Wil Burns, Forum for Climate Engineering Assessment, School of International Service, American University, Washington, DC, USA

Environmental professionals and scholars need resources that can help them to resolve interdisciplinary issues intrinsic to environmental management, governance, and research. The AESS branded book series draws upon a range of disciplinary fields pertinent to addressing environmental issues, including the physical and biological sciences, social sciences, engineering, economics, sustainability planning, and public policy. The rising importance of the interdisciplinary approach is evident in the growth of interdisciplinary academic environmental programs, such Environmental Studies and Sciences (ES&S), and related 'sustainability studies.'

The growth of interdisciplinary environmental education and professions, however, has yet to be accompanied by the complementary development of a vigorous and relevant interdisciplinary environmental literature. This series addresses this by publishing books and monographs grounded in interdisciplinary approaches to issues. It supports teaching and experiential learning in ES&S and sustainability studies programs, as well as those engaged in professional environmental occupations in both public and private sectors.

The series is designed to foster development of publications with clear and creative integration of the physical and biological sciences with other disciplines in the quest to address serious environmental problems. We will seek to subject submitted manuscripts to rigorous peer review by academics and professionals who share our interdisciplinary perspectives. The series will also be managed by an Editorial board of national and internationally recognized environmental academics and practitioners from a broad array of environmentally relevant disciplines who also embrace an interdisciplinary orientation.

More information about this series at http://www.springer.com/series/13637

Wil Burns • David Dana
Simon James Nicholson
Editors

Climate Geoengineering: Science, Law and Governance

Springer

AESS
Association for Environmental Studies and Sciences

Editors
Wil Burns
Co-Director, Institute for Carbon Removal Law & Policy, American University
Visiting Professor, Environmental Policy and Culture program, Northwestern University
Evanston, IL, USA

David Dana
Northwestern University
Chicago, IL, USA

Simon James Nicholson
American University
Washington DC, WA, USA

ISSN 2509-9787 ISSN 2509-9795 (electronic)
AESS Interdisciplinary Environmental Studies and Sciences Series
ISBN 978-3-030-72371-2 ISBN 978-3-030-72372-9 (eBook)
https://doi.org/10.1007/978-3-030-72372-9

© Springer Nature Switzerland AG 2021
This work is subject to copyright. All rights are reserved by the Publisher, whether the whole or part of the material is concerned, specifically the rights of translation, reprinting, reuse of illustrations, recitation, broadcasting, reproduction on microfilms or in any other physical way, and transmission or information storage and retrieval, electronic adaptation, computer software, or by similar or dissimilar methodology now known or hereafter developed.
The use of general descriptive names, registered names, trademarks, service marks, etc. in this publication does not imply, even in the absence of a specific statement, that such names are exempt from the relevant protective laws and regulations and therefore free for general use.
The publisher, the authors, and the editors are safe to assume that the advice and information in this book are believed to be true and accurate at the date of publication. Neither the publisher nor the authors or the editors give a warranty, expressed or implied, with respect to the material contained herein or for any errors or omissions that may have been made. The publisher remains neutral with regard to jurisdictional claims in published maps and institutional affiliations.

This Springer imprint is published by the registered company Springer Nature Switzerland AG
The registered company address is: Gewerbestrasse 11, 6330 Cham, Switzerland

Contents

Introduction

Wil Burns, David Dana, and Simon James Nicholson

The Paris Agreement's entry into force in November of 2016 was hailed as a hallmark achievement by the world community in addressing what many believe is the greatest existential global threat of this century and beyond, climate change.[1] In

This book grew out of a conference held under the auspices of the Northwestern University Center on Law, Business and Economics, formerly the Searle Center on Law, Regulation, and Economic Growth. We would like to thank the Center at the outset for bringing together many of the contributors to this volume to workshop some of the topics in this book.

[1] 1 Fiona Harvey, *Paris Climate Change Agreement Enters into Force,* The Guardian, Nov. 3, 2016, https://www.theguardian.com/environment/2016/nov/04/paris-climate-change-agreement-enters-into-force, site visited on Feb. 15, 2017; Natalya D. Gallo, et al., *Ocean commitments under the Paris Agreement,* 7 Nature Climate Change 833, 833 (2017). The Agreement has been ratified 189 Parties to the United Nations Framework Convention on Climate Change to date, UNFCCC, Paris Agreement – Status of Ratification, https://unfccc.int/process/the-paris-agreement/status-of-ratification, encompassing more than 97% of the world's greenhouse gas emissions, World Resources Institute, Paris Agreement Tracker, https://cait.wri.org/source/ratification/#?lang=en. Of course, a blow has been dealt to the Agreement by the decision of the largest industrial greenhouse emitters, the United States, to withdraw from Paris, which will take place on November 4, 2020, Michael M. Pompeo, Press Statement, U.S. Department of States, On the U.S. Withdrawal from the Paris Agreement, November 4, 2019, https://www.state.gov/on-the-u-s-withdrawal-from-the-paris-agreement/; US formally starts withdrawal from Paris climate accord, Euractiv, Nov. 4, 2019, https://www.euractiv.com/section/climate-environment/news/us-formally-starts-withdrawal-from-paris-climate-accord/

W. Burns (✉)
Visiting Professor, Environmental Policy and Culture Program,
Northwestern University, Evanston, IL, USA
e-mail: william.burns@northwestern.edu

D. Dana
Pritzker School of Law, Northwestern University, Evanston, IL, USA

S. J. Nicholson
School of International Service, Washington, DC, USA

© Springer Nature Switzerland AG 2021

W. Burns et al. (eds.), *Climate Geoengineering: Science, Law and Governance*,
AESS Interdisciplinary Environmental Studies and Sciences Series,
https://doi.org/10.1007/978-3-030-72372-9_1

seeking to avoid some of the most serious potential climatic impacts for human institutions and ecosystems, the Agreement, *inter alia,* calls for "[h]olding the increase in the global average temperature to well below 2 °C above pre-industrial levels and to pursue efforts to limit the temperature increase to 1.5 °C above pre-industrial levels, recognizing that this would significantly reduce the risks and impacts of climate change."[2] To operationalize this goal, the Agreement also calls upon the parties to "aim to reach global peaking of greenhouse gas emissions as soon as possible and to undertake rapid reductions thereafter in accordance with best available science, so as to achieve a balance between anthropogenic emissions by sources and removals by sinks of greenhouse gases in the second half of this century."[3]

However, the Nationally Determined Contributions (NDCs)[4] made by the Parties to Paris to date put us on track to exhaust the remaining "carbon budget" to hold temperatures to below 1.5 °C by 2030, and 2.0 °C within 35–41 years.[5] Indeed, the current NDCs place us on an emissions trajectory by which temperatures will reach 2.6–3.7 °C above pre-industrial levels by 2100,[6] and continue to increase for centuries beyond due to the substantial inertia of the climate system.[7] It's daunting to note that the policy ambitions in the NDCs would have to triple to put them in line with the Paris Agreement's 2 °C goal.[8] Moreover, a recent study concluded that only seven of 25 major emitting States are meeting their tepid pledges, potentially leading to temperature increases of as much a 4.4 °C above pre-industrial levels by

[2] The Paris Agreement, FCCC/CP/2015/L.9, Conference of the Parties, 21st Session (2015), at art. 2(1)(a).

[3] *Id.* at art. 4(1).

[4] *Id.* at art. 3.

[5] Samer Fawzy, et al., *Strategies for mitigation of climate change: a review,* 18 Envtl. Chemistry Letters 2069, 2072 (2020); Andrew Freedman & Chris Mooney, *Earth's carbon dioxide levels hit record high, despite coronavirus-related emissions drop*, Washington Post, June 4, 2020, https://www.washingtonpost.com/weather/2020/06/04/carbon-dioxide-record-2020/; Phillip Goodwin, et al., *Pathways to 1.5 °C and 2 °C warming based on observational and geological constraints*, 11 Nature Geosci. 102, 104 (2018).

[6] Mathias Fridahl & Mariliis Lehtveer, *Bioenergy with carbon capture and storage (BECCS): Global potential, investment preferences, and deployment barriers,* 42 Energy Res. & Soc. Sci. 155, 155 (2018); Joeri Rogelj, et al., *Paris Agreement Climate Proposals Need a Boost to Keep Warming Well Below 2 °C,* 534 Nature 631, 634 (2016); Climate Action Tracker, Paris Agreement: Stage Set to Ramp up Climate Action, Dec. 12, 2015, http://climateactiontracker.org/news/257/Paris-Agreement-stage-set-to-ramp-up-climate-action.html, site visited on Feb. 15, 2017; World Resources Institute, Why are INDC Studies Reaching Different Temperature Estimates?, http://www.wri.org/blog/2015/11/insider-why-are-indc-studies-reaching-different-temperature-estimates, site visited on Feb. 15, 2017.

[7] Peter U. Clark, et al., *Consequences of Twenty-First Century Policy for Multi-Millennial Climate and Sea-Level Change,* 6 Nature Climate Change 360, 361 (2016).

[8] Bipartisan Policy Center, *Investing in Climate Innovation: The Environmental Case for Direct Air Capture of Carbon Dioxide*, May 2020, at 7, https://bipartisanpolicy.org/report/investing-in-climate-innovation-the-environmental-case-for-direct-air-capture-of-carbon-dioxide/

2100.[9] Temperature increases that exceed the Paris temperature target by this magnitude would have extremely serious consequences for human institutions and natural ecosystems.[10]

The sobering reality of the disconnect between the resolve of the world community to effectively address climate change, and what actually needs to be done, has led to increasing impetus for consideration of a suite of approaches collectively known as "climate geoengineering," or "climate engineering." Indeed, the feckless response of the world community to climate change has transformed climate geoengineering from a fringe concept to a potentially mainstream policy option.[11]

Climate geoengineering is defined broadly by the UK's Royal Society as "the deliberate large-scale manipulation of the planetary environment to counteract anthropogenic climate change."[12] Climate geoengineering technologies are usually divided into two broad categories, solar radiation management approaches (SRM) and carbon dioxide removal approaches (CDR).[13]

SRM options could be used to reduce the amount of solar radiation absorbed by the Earth (pegged at approximately 235 W m^{-2} currently[14]) by an amount sufficient

---

[9] Noah Sachs, *The Paris Agreement in the 2020s: Breakdown or Breakup?*, 46(1) Eco. L.Q. 865, 893 (2019); Some Progress Since Paris, But Not Enough, as Governments Amble Towards 3 °C of Warming,
ClimateActionTracker(Dec.11,2018),https://climateactiontracker.org/publications/warmingprojections-global-update-dec-2018/. *See also,* Kevin Anderson, et al., *A factor of two: how the mitigation plans of 'climate progressive' nations fall far short of Paris compliant pathways*, 20(10) Climate Pol'y 1290–1304 (2020).

[10] Climate Change 2014, Synthesis Report, *Summary for Policymakers*, UNFCCC (2014),
at 18–19, http://www.ipcc.ch/pdf/assessment-report/ar5/syr/AR5_SYR_FINAL_SPM.pdf, site visited on Jan. 16, 2016; *Report of the Conference of the Parties on its nineteenth session*, held in Warsaw from 11
to 23 Nov. 2013, Further Advancing the Durban Platform, UNFCCC (Jan. 31, 2014), at CP/2013/10,
¶ 2(b); *INDCs as Communicated by Parties*, UNFCCC, http://www4.unfccc.int/submissions/
indc/Submission%20Pages/submissions.aspx, site last visited Dec. 29, 2016; V. Ramanathan & Y. Feng, *On Avoiding Dangerous Anthropogenic Interference with the Climate System: Formidable Challenges Ahead*, 105(3) PNAS 14245, 14,245 (2008).

[11] Netherlands, Norway, Norway, Sweden, Non-Paper on Carbon Capture and Storage, Klima-, Energi- og Forsyningsudvalget 2020–21 KEF Alm.del - Bilag 87 Offentligt (2020); Robin Gregory, Terre Satterfield & Ariel Hasell, *Using Decision Pathway Surveys to Inform Climate Engineering Policy Choices,* 113 PNAS 560, 560 (2016); Shinichiro Asayama, *Catastrophism Toward 'Opening Up' or 'Closing Down'? Going Beyond the Apocalyptic Future and Geoengineering,* 63(1) Current Sociology 89, 90 (2015). For a history of geoengineering over the past fifty years, *see* Wil Burns & Simon Nicholson, *Governing Climate Geoengineering, in* New Earth Politics 345–50 (Simon Nicholson & Sikina Jinnah eds., 2016).

[12] The Royal Society, *Geoengineering the Climate: Science, Governance and Uncertainty* (2009), at 11. http://royalsociety.org/Geoengineering-the-climate/, site visited on Jan. 16, 2017.

[13] William C.G. Burns, *Geoengineering the Climate: An Overview of Solar Radiation Management Options,* 46 Tulsa L. Rev. 283, 286 (2012).

[14] J.T. Kiehl & Kevin E. Trenberth, *Earth's Annual Global Mean Energy Budget,* 78(2) Bull. Am. Meteorological Soc'y 197, 198 (1997), http://climateknowledge.org/figures/Rood_Climate_

to offset the increased trapping of infrared radiation by rising levels of greenhouse gases.[15] Alternatively, SRM options could be deployed at a smaller scale to offset only a proportion of projected warming.[16] By contrast, carbon dioxide removal options seek to remove and sequester carbon dioxide from the atmosphere, either by enhancing natural sinks for carbon, or deploying chemical engineering to remove carbon dioxide from the atmosphere.[17] This, in turn, can increase the amount of long-wave radiation emitted by Earth back to space, reducing radiative forcing, thus, exerting a cooling effect.[18]

Examples of SRM approaches include stratospheric aerosol injection (SAI), which seeks to enhance planetary albedo (and thus negative forcing) through the injection of a gas such as sulfur dioxide, or another gas that will ultimately react chemically in the stratosphere to form sulfate aerosols[19]; marine cloud brightening (MCB), which seeks to increase the albedo of maritime clouds through seeding with seawater droplets,[20] and space-based methods seeking to reduce the amount of solar radiation reaching the Earth by positioning sun-shields in space to reflect or deflect radiation.[21]

CDR technologies include bioenergy and carbon capture and storage (BECCS), a process by which biomass is converted to heat, electricity, or liquid or gas fuels, coupled with carbon dioxide capture and sequestration (CCS),[22] ocean iron fertilization, a process for dispersing iron in iron-deficient regions of the world's oceans regions to stimulate phytoplankton production, thus potentially enhancing carbon

---

Change_AOSS480_Documents/Kiehl_Trenberth_Radiative_Balance_BAMS_1997.pdf, site visited on Feb. 2, 2017.

[15] Samer Fawzy, et al., *Strategies for mitigation of climate change: a review,* 18 Envtl. Chemistry Letters 2069, 2086 (2020).

[16] David W. Keith & Peter J. Irvine, *Solar geoengineering could substantially reduce climate risks – A Research hypothesis for the next decade,* 4 Earth's Future 549, 552 (2016).

[17] Timothy Lenton, *The Global Potential for Carbon Dioxide Removal,* Geoengineering of the Climate System 53 (Roy Harrison & Ron Hester eds., 2014).

[18] T.M. Lenton & N.E. Vaughan, *The Radiative Forcing Potential of Different Climate Geoengineering Options,* 9 Atmos. Chem. Phys. 5539, 5540 (2009).

[19] Sean Low & Matthias Honegger, *A Precautionary Assessment of Systematic Projections and Promises From Sunlight Reflection and Carbon Removal Model Modeling,* Risk Analysis 1, 1 (2020) Peter J. Irvine, et al., *An Overview of the Earth System Science of Solar Geoengineering,* WIREs Climate Change, doi: 10.1002/2 cc.423 (2016), at 7.

[20] John Latham, et al., *Global Temperature Stabilization via Controlled Albedo Enhancement of Low-Level Maritime Clouds,* 366 Phil. Transactions Royal Soc'y 3969, 3970 (2008); Keith Bower, et al., *Computations Assessment of a Proposed Technique for Global Warming Mitigation via Albedo-Enhancement of Marine Stratocumulus Clouds,* 82(1–2) Atmospheric Res. 328, 329 (2006).

[21] Takanobu Kosugi, *Role of Sunshades in Space as a Climate Control Option,* 67 Acta Astronautica 241, 242 (2010).

[22] Joris Kornneeff, et al., *Global Potential for Biomass and Carbon Dioxide Capture, Transport and Storage up to 2050,* 11 Int'l J. Greenhouse Gas Control 117, 119 (2012); U.S. Environmental Protection Agency, Carbon Dioxide Capture and Sequestration, http://www3.epa.gov/climatechange/ccs/#CO2Capture, site visited on Feb. 17, 2017.

dioxide uptake,[23] increasing ocean alkalinity, and thus carbon dioxide uptake, by adding substances such as lime or olivine to oceans or in coastal regions,[24] direct air capture (DAC), a process to extract carbon dioxide from ambient air in a closed-loop industrial process,[25] terrestrial enhanced mineral weathering, a process to accelerate the uptake of carbon dioxide from the atmosphere by magnesium and calcium-rich rocks,[26] and afforestation and reforestation initiatives,[27] and efforts to increase sequestration of carbon dioxide in soils.[28]

Field research on most climate geoengineering options is currently either at an early stage, or has not even begun. However, preliminary research indicates that both SRM and CDR approaches could potentially help to ameliorate warming and the climatic impacts of burgeoning greenhouse gas emissions. For example, recent studies have concluded that large-scale deployment of SRM approaches could begin to return temperatures to pre-industrial levels within a few years of deployment,[29] and potentially restore temperatures to pre-industrial conditions by the end of this century.[30] The vast majority of mitigation scenarios developed in integrated assessment models under which temperatures are kept to 2 °C or below contemplate extensive deployment of CDR technologies during the course of this century,[31] with

---

[23] Matthew Hubbard, *Barometer Rising: The Cartagena Protocol on Biosafety as a Model for Holistic International Regulation of Ocean Fertilization Projects and Other Forms of Geoengineering,* 40 WM. & MARY ENVTL. L. & POL'Y REV. 591, 598 (2016); Christine Bertram, *Ocean Iron Fertilization in the Context of the Kyoto Protocol and the Post-Kyoto Process,* 8 ENERGY POL'Y 1130, 1130 (2010).

[24] Wil Burns & Charles R. Corbett, *Antacids for the Sea? Artificial Ocean Alkalinization and Climate Change*, 3 ONE EARTH 154–56 (2020); Andrew Lenton, et al., *Assessing carbon dioxide removal through global and regional ocean alkalization under high and low emission pathways,*.9 EARTH SYS. DYNAMICS 339–257 (2018).

[25] Robert Socolow, et al., *Direct Air Capture of CO_2 with Chemicals* (2011), American Physical Society, at 7–9, https://www.aps.org/policy/reports/assessments/upload/dac2011.pdf, site visited on Feb. 14, 2017; R. Stuart Haszeldine, *Can CCS and NETs Enable the Continued Use of Fossil Carbon Fuels after CoP21?,* 32(2) OXFORD REV. ECON. POL'Y 304, 310 (2016).

[26] David J. Beerling, et al., *Potential for large-scale CO2 removal via enhanced rock weathering with croplands*, 583 NATURE 242–62 (2020); P. Renforth, et al., *The dissolution of olivine added to soil: Implications for enhanced weathering*, 61 APPLIED GEOCHEMISTRY 109–118 (2015).

[27] Jean Francois-Bastin, et al., *The global tree restoration potential,* 365 SCI. 76–79 (2019); Matthew E. Fagin, et al., *How Feasible are global forest restoration goals?,* 13(3) CONSERVATION LETTERS 1–8 (2020), https://doi.org/10.1111/conl.12700

[28] Xiongxiong Bai, et al., *Responses of soil carbon sequestration to climate-smart agriculture practices: A meta-analysis*, 25 GLOBAL CHANGE BIO. 2591–2606 (2019).

[29] The Royal Society, *supra* note 12, at 34.

[30] David P. Keller, Ellias Y. Feng & Andreas Oschlies, *Potential Climate Engineering Effectiveness and Side Effects During a High Carbon Dioxide-Emission Scenario,* NATURE COMM., Feb. 25, 2014, DOI: https://doi.org/10.1038/ncomms4304, at 5-6, http://www.nature.com/ncomms/2014/140225/ncomms4304/pdf/ncomms4304.pdf, site visited on Feb. 14, 2017.

[31] Intergovernmental Panel on Climate Change, Fifth Assessment Report, Working Group III, Ch. 6, *Assessing Transformation Pathways,* at 93; Giulia Realmonte, et al., *An inter-model assessment of the role of direct air capture in deep mitigation pathways,* 10 NATURE COMMUNICATIONS, 3277 (2019), at 3; Etsushi Kato & Yoshiki Yamagata, *BECCS Capability of Dedicated Bioenergy Crops*

bioenergy and carbon capture and storage cited as the primary option.[32] Global climate models project that the globe may need removal of between 700–1000 $GtCO_2$ between 2011–2100 to stabilize temperatures at either 1.5C or 2.0C above pre-industrial levels.[33] There has been increasing recognition by policymakers to incorporate that into their climate planning processes. For example, the European Commission's proposals for a long-term EU climate strategy envisions economy-wide net-negative emissions in the second half of this century.[34]

However, climate geoengineering approaches may also pose serious risks to society and ecosystems. For example, the SRM option of stratospheric aerosol injection (as well as marine cloud brightening) could alter global hydrological cycles, potentially modifying the Asian and African monsoons, "impacting the food supply to billions of people,"[35] and visiting "humanitarian disasters" upon such regions.[36] Large-scale deployment of SAI geoengineering options could also delay recovery of the ozone layer for 30–70 years or more,[37] increase sulfuric acid deposition in the troposphere, with potential negative implications for both terrestrial and aquatic ecosystems,[38] and potentially lead to an increase in summer heat extremes in high-northern latitude regions during the boreal summer.[39] Moreover, SRM

---

under a Future Land-Use Scenario Targeting Net Negative Carbon Emissions, 2 EARTH'S FUTURE 421, 421 (2014).

[32] Fridahl & Lehtveer, *supra* note 6, at 155; T. Gasser, et al., *Negative Emissions Physically Needed to Keep Global Warming Below 2 °C,* 6 NATURE COMM., Art. No. 7958 (2015), at 5; *See also* José Roberto Moreira, et al., *BECCS Potential in Brazil: Achieving Negative Emissions in Ethanol and Electricity Production Based on Sugar Cane Bagasse and Other Residues,* 179 APPLIED ENERGY 55, 56 (2016) (BECCS "will play a vital role in reaching the required level of emission reductions in the future"); Sabine Fuss, *Betting on Negative Emissions,* 4 NATURE CLIMATE CHANGE 850, 850 (2014).

[33] James Mulligan, et al., *Technological Carbon Removal in the United States* 5 (Sept. 2018), https://www.wri.org/publication/tech-carbon-removal-usa

[34] Wilfried Rickels, et al., *The Future of (Negative) Emissions Trading in the European Union,* Kiel Working Paper, No. 2164 (2020), at 5, https://www.ifw-kiel.de/experts/ifw/wilfried-rickels/the-future-of-negative-emissions-trading-in-the-european-union-15070/

[35] The Royal Society, *supra* note 12, at 31. *See also* Charles C. Gertler, *Weakening of the Extratropical Storm Tracks in Solar Geoengineering Scenarios*, 47 GEOPHYSICAL RES. LETTERS 1–9, e2020GL087348 (2020).

[36] Holly Jean Buck, *Geoengineering: Re-Making Climate for Profit or Humanitarian Intervention?,* 43(1) DEV. & CHANGE 253, 255 (2011).

[37] Simone Tilmes, Rolf Müller & Ross Salawitch, *The Sensitivity of Polar Ozone Depletion to Proposed Geoengineering Schemes,* 320 SCI. 1201, 1204 (2008). *See also* Khara D. Grieger, et al., *Emerging risk governance for stratospheric aerosol injection as a climate management technology*, 39 ENV'T SYSTEMS & DECISIONS 371, 372 (2019).

[38] Grieger et al., *supra* note 37, at 2; MIT, *The Unintended Consequences of Sulfate Aerosols in the Troposphere and Lower Stratosphere,* Department of Civil Engineering (2009), at 11, https://zero-geoengineering.com/2016/unintended-consequences-sulfate-aerosols-troposphere-lower-stratosphere/

[39] Katherine Dagon & Daniel P. Shrag, *Regional Climate Variability Under Model Simulations of Solar Geoengineering,* 122 J. GEOPHYSICAL RES., ATMOSPHERES 12,106, 12, 112 (2017).

deployment could result in serious geopolitical tensions or conflict should it be pursued unilaterally.[40]

On the CDR side of the equation, many of the contemplated technologies and techniques could also pose serious risks, especially at large-scales of deployment. For example, ocean iron fertilization could result in shifts in community composition that could threaten the integrity of ocean ecosystems.[41] Large-scale deployment of bioenergy and carbon capture and storage could divert large swathes of land from food production, imperiling food security for vulnerable populations,[42] It could also result in large land grabs,[43] and threaten biodiversity.[44] Enhanced mineral weathering could pose risks to agricultural applications by releasing potentially toxic levels of chromium and nickel,[45] could pose potential threats to human health through inhalation of ultrafine particles,[46] and might adversely impact ocean environments through substantially altering biogeochemical cycles.[47]

---

[40] Anna Lou Abatayo, et al., *Solar geoengineering may lead to excessive cooling and high strategic uncertainty*, PNAS LATEST ARTICLES (2020), at 5, https://www.pnas.org/content/early/2020/05/28/1916637117, The Royal Society, *Solar radiation management: the governance of research* 16 (2011), https://royalsociety.org/topics-policy/projects/solar-radiation-governance/report/

[41] R.S. Lampitt, et al., *Ocean Fertilization: A Potential Means of Geoengineering?*, 366 PHIL. TRANS. R. SOC'Y 3919, 3925 (2008).

[42] Pete Smith, et al., *Biophysical and Economic Limits to Negative CO_2 Emissions*, 6 NATURE CLIMATE CHANGE 42, 46 (2016). *See also* Phil Williamson, *Scrutinize CO_2 Removal Methods*, 530 NATURE 153, 154 (2016); Markus Bonsch, et al., *Trade-offs Between Land and Water Requirements for Large-Scale Bioenergy Production*, 8 GCB BIOENERGY 11, 11 (2014).

[43] Lorenzo Catula, Nat Dyer & Sonja Vermeulen, *Fuelling Exclusion? The Biofuels Boom and Poor People's Access to Land*, International Institute for the Environment and Development and Food and Agriculture Organization, at 14, http://pubs.iied.org/pdfs/12551IIED.pdf, site visited on Feb. 15, 2017.

[44] Andrew Wiltshire & T. Davies-Barnard, *Planetary Limits to BECCS Negative Emissions*, AVOID2, Mar. 2015, at 15, http://avoid-net-uk.cc.ic.ac.uk/wp-content/uploads/delightful-downloads/2015/07/Planetary-limits-to-BECCS-negative-emissions-AVOID-2_WPD2a_v1.1.pdf, site visited on Jan. 14, 2017.

[45] Mike E. Kelland, et al., *Increased yield and CO2 sequestration potential with the C4 cereal Sorghum bicolor cultivated in basaltic rock dust-amended agricultural soil*, 26 GLOBAL CHANGE BIO. 3658, 3659 (2020).

[46] Romany M. Webb, *The Law of Enhanced Weathering for Carbon Dioxide Removal*, Columbia Law School, Sabin Center for Climate Change Law (2020), at 31, https://climate.law.columbia.edu/sites/default/files/content/Webb%20-%20The%20Law%20of%20Enhanced%20Weathering%20for%20CO2%20Removal%20-%20Sept.%202020.pdf, site visited on January 6, 2021.

[47] Jens Hartmann, et al., *Enhanced Chemical Weathering as a Geoengineering Strategic to Reduce Atmospheric Carbon Dioxide, Supply Nutrients, and Mitigate Ocean Acidification*, 51 REV. GEOPHYS. 113, 113 (2013).

At the same time, business as usual scenarios for climate change also pose grave threats to the world community.[48] This emphasizes the need for a full-throated assessment of society's options, including pertinent metrics to evaluate trade-offs. This includes focusing on critical issues of law and governance, including the management risks, which is the focus of many of the chapters in this book.

As a number of commentators have noted, climate geoengineering could pose a wide variety of thorny global legal and governance issues over the course of the next few decades.[49] Common across the spectrum of CDR and SRM approaches is the need to manage the emergence of high-risk/high-reward technological options. This entails steering between the potential hazards associated with different use scenarios for the various approaches, while at the same time fostering needed research and innovation. In addition, it makes little sense to look at climate geoengineering in a vacuum. No climate geoengineering option offers a single-shot fix for climate change, but rather at best will be some small component of humanity's overall efforts to ameliorate and adapt to climate disruption. This means that evaluation and governance of climate geoengineering approaches must happen alongside the full suite of available and potential climate change response options.

At the same time, the interaction of climate geoengineering approaches with other means to address climate change gives rise to one of the most-discussed risks associated with climate geoengineering. This is the "moral hazard" or "mitigation deterrence" risk – basically, that climate geoengineering might serve as a willful or inadvertent distraction from work to stem greenhouse gas emissions. For some, this risk is serious enough that even contemplation of climate geoengineering options is a bad idea. Others have argued that climate geoengineering is inherently ungovernable or that climate geoengineering will serve to entrench the social and economic dynamics that have given rise to climate change. For people in this "no climate geoengineering" camp, governance of climate geoengineering means imposing moratoria or exceedingly strict limitations on research.

"No climate geoengineering" is one pole along the climate geoengineering governance spectrum. A range of other positions exist, concerned to varying degrees with enabling climate geoengineering research and potential deployment and guarding against various attendant risks. Such perspectives share a faith that climate

[48] Intergovernmental Panel on Climate Change, Summary for Policymakers, *Climate Change 2013: The Physical Science Basis. Contribution of Working Group I to the Fifth Assessment Report of the Intergovernmental Panel on Climate Change* 1–28 (2013), http://www.ipcc.ch/pdf/assessment-report/ar5/wg1/WG1AR5_SPM_FINAL.pdf, site visited on May 7, 2018; NASA, *The Consequences of Climate Change,* Vital Signs of the Planet, https://climate.nasa.gov/effects/, site visited on May 7, 2018.

[49] Peter J. Irvine & David W. Keith, *Halving warming with stratospheric aerosol geoengineering moderates policy-relevant climate hazards,* 15 Envt'l Research Letters (202), 044011, at 4; Mason Inman, *Planning for Plan B,* Nature Reports Climate Change, Dec. 17, 2009, http://www.nature.com/climate/2010/1001/full/climate.2010.135.html, site visited on Jan. 19, 2017; Scott Barrett, *Solar Geoengineering's Brave New World: Thoughts on the Governance of an Unprecedented Technology,* 8(2) Rev. Envt'l Econ. & Pol'y 249, 266 (2014).

geoengineering approaches *can* be governed using institutions and instruments currently available or newly developed.

For this broad "climate geoengineering ought to be explored" coalition, a new consensus is emerging that the label "climate geoengineering" has outstayed its welcome. Instead, it makes more sense, in governance terms, to distinguish SRM from CDR, since these two distinct buckets of climate change response would operate, if developed and used at scale, in quite different ways with quite different risks. Moreover, there is a need to distinguish *within* each of the broad buckets. Stratospheric aerosol injection and marine cloud brightening are both forms of SRM that would require the dispersal of materials into the atmosphere. SAI, though, would have a variegated global impact if utilized at any meaningful scale, while MCB might be utilized to have a localized or regionalized set of primary impacts. Moreover, the two different approaches call for different kinds of materials dispersed into different layers of the atmosphere, with the materials falling out at different rates. These kinds of distinctions matter for the kinds of specific forms of oversight and management that will need to be developed.

Many of the chapters in this book address issues of this nature, as well as others pertinent to governance of these emerging technologies, including the role of human rights and the interface of domestic and international law.

Douglas MacMartin, Peter Irvine, Ben Kravitz, and Joshua Horton initially consider how the decisions that need to be made around the deployment of solar geoengineering ought to influence and impact the design of governance arrangements. Solar geoengineering, they contend, is not just a yes or no proposition. Instead, there is a variety of parameters that could each be manipulated in ways that would allow some tailoring of a solar geoengineering intervention. Ideally, governance arrangements would need to be attentive to and responsive to these parameters and, moreover, to the fact that the parameters would need to be tweaked through time. Time, though, becomes its own governance challenge. This is because while some feedback from a geoengineering climate system would reveal itself quickly, other forms of feedback would only become clear over many years. This suggests forms of governance that can take swift action if more or less geoengineering is suddenly required, but that can also demonstrate extraordinary patience.

Kimberly Gray situates the debate over geoengineering in the larger context of the planetary climate system and the array of options available to address anthropogenic climate change. First, Gray emphasizes that the earth is a complex system, and that highly engineered solutions – solutions that flow from what she dubs "an engineering mentality" to the natural world - quite often founder in the face of that complexity. Gray then reviews the options to address climate change and concludes that mitigation is technically feasible, but it seems not so politically. Specifically, Gray explains that there "are an enormous number of mitigation actions and a plethora of synergies and cascading benefits to be exploited among mitigation, ecologically based CDR and adaptation endeavors." Gray views geoengineering of direct air capture, solar radiation management or glacier containment as quintessential expressions of the engineering mentality she critiques – a mentality that assumes away the complexities of the natural world. As Gray explains, these would be the "'mother of all engineering projects' for three basic reasons – the massive scale at which the technologies must be deployed, the need to integrate these actions with mitigation, adaptation and other CDR measures and the necessity of maintaining flexible designs since they will have to be adjusted as we learn how the climate system responds." In the end, Gray's analysis is a plea that political leaders embrace mitigation and adaptation so that we are not forced to embrace geoengineering, which she characterizes as "emergency, life-support engineering at a technological, economic and political scale that defies both logic and perhaps, feasibility."

Lisa Ferrari and Elizabeth Chalecki provide, for geoengineering, a moral framework based on just war theory. Their concern is with what they call "commons-based geoengineering" – that is, geoengineering responses to climate change that can or would need to be deployed in the global commons, thereby ensuring trans-boundary impacts and potentially global environmental change. Wide-ranging geo-engineering interventions have much in common, they contend, with the conduct of warfare, such that the ethical strictures and legal codes that have grown to shape and constrain war may have lessons for those contemplating geoengineering. The upshot is translation of a set of just war criteria into a parallel set of "just geoengineering" criteria, to guide States in the ethical consideration of large-scale geoengineering options.

Brian Citro and Patrick Smith engage the important question of what normative framework should be employed to specify nation's obligations with respect to geo-engineering. Citro and Smith argue that normative framework that has dominated public policy discussions regarding geoengineering to date has explicitly or implicitly been cost-benefit analysis, which they critique as inadequate. As an alternative, Citro and Smith propose a human rights framework, according to which nations would have both procedural and substantive human rights obligations. In a human rights framework, nations would focus on vulnerable or marginalized groups, prioritize nondiscrimination, require that affected communities participate in decision-making processes that impact their lives, and assign duties, accountability, and remedies for human rights violations. Citro and Smith acknowledge that sometimes human rights obligations might contradict one another in the geoengineering context, and that, especially as to substantive obligations, there may be a lack of clarity as to the content of the obligations in the first place. Building on this acknowledgement, Citro and Smith sketch out an approach by which competing human rights claims could be prioritized.

Ademola Jegede's chapter looks at CDR in the South African context. In particular, he is interested in the opportunities and challenges CDR options present for the fulfillment of human rights in South Africa, and, in turn, how international and domestic human rights laws might provide guidance for the use (or not) have CDR options. Jegede's assessment is mixed. The chapter shows that there are significant uncertainties when it comes to whether and how South Africa could make use of CDR options. In addition, there are uncertainties and limitations when it comes to the kinds of guidance that can be found in human rights instruments and principles. This is because a given CDR option can at once promote and undermine a variety of different rights. The message of the chapter is that great care must be taken if CDR is to be compatible with the rights and needs of South Africa's citizens.

One question about geoengineering governance is whether and how devices or approaches used in other governance regime might be adapted to geoengineering. Pursuing this question, Anthony Chavez's chapter addresses two ways governments can incentivize the development of geoengineering technologies. Noting that none of the current technologies with respect to Carbon Direct Removal (CDR) now seem feasible at the scale needed to have a major impact on carbon levels, Chavez argues that governments must incentivize improvements in current CDR technologies, as well as the creation of new technologies. Chavez explores the potential of two legal devices that various governments in the United States and Europe have used to promote the development of renewable energy technologies – renewable portfolio standards (RPSs) and feed-in-tariffs (FITs). Both RPSs and FITs make financially feasible the deployment of energy sources that otherwise might not be adopted. Chavez reviews the advantages and disadvantages of these devices generally and of extending them to CDR; he concludes that, in the CDR context, RPSs and FITs would work best if they were adopted and implemented together. One advantage of Chavez's recommended approach would be that it would not necessarily require governments to predetermine which CDR technologies deserve the most investment, but rather would support a multitude of approaches in different jurisdictions, just as RPSs in the United States have supported a very wide range of renewable energy technologies. Chavez thus seeks to harness what we have learned from policies regarding renewable energy development for fashioning the optimal geoengineering law and policy.

David Dana writes of what he calls the "question of weakened resolve." His concern is with whether and how contemplation or development of geoengineering options might weaken efforts to reduce greenhouse gas emissions. Dana argues that the question of weakened resolve is of most importance among elite decisionmakers, since it is elite action of various stripes that will ultimately determine whether and how geoengineering impacts emissions abatement activities. The chapter reviews existing socio-psychological work and finds it lacking, for its focus on publics rather than elites and for its being confined to the United States and a small number of countries in Europe. Dana calls for more and broader efforts to understand the question of weakened resolve. The chapter closes with ideas about how to conduct research and how to guide public and elite consideration of geoengineering to lessen the likelihood of weakened resolve.

Tara Rhighetti's chapter focuses on one particular form of geoengineering – carbon storage, and in particular storage in connection with the burning of oil for energy. Rhigetti argues that there are substantial opportunities for the expanded use of carbon storage as a response to the threat of climate change. Indeed, Rhighetti notes that carbon storage in connection with energy production is already taking place, albeit on a very limited scale. Rhigetti argues that next generation technologies can make storage a more feasible strategy, and that such technologies need to be encouraged. However, for carbon storage to be achieve the desired scale, legal and regulatory modifications are required. State level property law needs to address the question of subsurface trespass and other boundary issues raised by storage. Expanded carbon storage also requires that state law address issues of long term environmental and tort liability and transition of ongoing monitoring responsibilities. The potential of storage can be further achieved through implementation of programs that discourage or prohibit the use of natural CO_2, and which create incentives for the transportation and use of CO_2 from anthropogenic and direct air capture sources.

Soheil Shayegh, Garth Heutel, and Juan Moreno-Cruz take a modeling approach to the study of international cooperation regarding climate policy when solar geoengineering is a policy option available to nations. Their chapter utilizes two different types of models. First, the authors use an analytical theoretical model to show how the equilibrium levels of emissions abatement and geoengineering are affected by the level of cooperation between countries. This model indicates that cooperation between countries leads to lower emissions and more geoengineering. To quantify these results, the authors modified a numerical integrated assessment model, DICE, to include solar geoengineering and cooperation among nations. Their simulation results show that the effect of cooperation on policy depends crucially on whether damages from geoengineering are local or global. With local damages, more cooperation leads to more geoengineering, but the opposite is true for global damages.

Finally, Kalyani Robbins' chapter engages the important question of how the precautionary principle – rhetorically at least, a cornerstone of international environmental law and discourse – should be understood in the context of geoengineering. Robbins reviews the literature regarding the precautionary principle, which, as she explains, boils down to the idea that a cautious course should be preferred to human actions that carry with them a substantial uncertainty of disastrous consequences. Given the uncertainties regarding harms associated with known forms of geoengineering, the precautionary principle would seem to counsel against the deployment of geoengineering. On the other hand, given the enormous harms associated with anthropogenic climate change, the precautionary principle arguably favors the deployment of geoengineering. Thus, geoengineering presents an instance of what Robbins dubs dueling precautions. Robbins explains that, in assessing which of two precautionary courses is in fact the most precautionary, the relative certainty of harms matters. Thus, now, Robbins suggests, a precautionary approach arguably disfavors deployment of geoengineering. But as nations continue to fail to adopt climate change mitigation, to the point where mitigation no longer seems possible in time to avoid disastrous climate change scenarios, a precautionary approach would support – indeed perhaps mandate – the deployment of geoengineering.

Climate geoengineering is a dynamic and highly variegated field. This book does not seek to capture all of the facets of the current debates, but it our hope that it highlights some of the emerging issues that society, including the legal community must grapple with as we determine what role, if any, these approaches will play in addressing climate change.

References

1. Bai, X., et al.: Responses of soil carbon sequestration to climate-smart agriculture practices: A meta-analysis. Global Change Bio. **25**, 2591–2606 (2019)
2. Buck, H.J.: Geoengineering: Re-making climate for profit or humanitarian intervention? Dev. & Change. **43**(1), 253–255 (2011)
3. Burns, W.C.G.: Geoengineering the climate: An overview of solar radiation management options. Tulsa L. Rev. **46**, 283–286 (2012)
4. Clark, P.U., et al.: Consequences of twenty-first century policy for multi-millennial climate and sea-level change. Nat. Climate Change. **6**, 360–361 (2016)
5. Fawzy, S., et al.: Strategies for mitigation of climate change: a review. Eniron. Chem. Lett. **18**, 2069–2086 (2020)
6. Hartmann, J., et al.: Enhanced chemical weathering as a geoengineering strategic to reduce atmospheric carbon dioxide, supply nutrients, and mitigate ocean acidification. Rev. Geophys. **51**, 113 (2013)
7. Keith, D.W., Irvine, P.J.: Solar geoengineering could substantially reduce climate risks – A Research hypothesis for the next decade. Earth's Future. **4**, 549–552 (2016)
8. Kelland, M.E., et al.: Increased yield and CO2 sequestration potential with the C4 cereal Sorghum bicolor cultivated in basaltic rock dust-amended agricultural soil. Global Change Bio. **26**, 3658–3659 (2020)
9. Kosugi, T.: Role of sunshades in space as a climate control option. Acta Astronautica. **67**, 241–242 (2010)
10. Lampitt, R.S., et al.: Ocean fertilization: A potential means of geoengineering? Phil. Trans. R. Soc'y. **366**, 3919–3925 (2008)
11. Lenton, T.M., Vaughan, N.E.: The radiative forcing potential of different climate geoengineering options. Atmos. Chem. Phys. **9**, 5539–5540 (2009)

Characteristics of a Solar Geoengineering Deployment: Considerations for Governance

Douglas G. MacMartin, Peter J. Irvine, Ben Kravitz, and Joshua B. Horton

Consideration of solar geoengineering as a potential response to climate change will demand complex decisions. These include not only the choice to deploy or not, but decisions regarding how to deploy, and ongoing decision making throughout deployment. However, relatively little attention has been paid to envisioning what a solar geoengineering deployment would look like in order to clarify what types of decisions would need to be made. We examine the science of geoengineering to ask how it might influence governance considerations, while consciously refraining from making specific recommendations. The focus here is on a hypothetical deployment (and beyond) rather than research governance. Geoengineering can be designed to trade off different outcomes, requiring an explicit specification of multivariate goals. Thus, we initially consider the complexity surrounding a decision to deploy. Next, we discuss the on-going decisions that would be needed across multiple time-scales. Some decisions are inherently slow, limited by detection and attribution of climate effects in the presence of natural variability. However, there is also a need for decisions that are inherently fast relative to political time-scales: effectively managing some uncertainties would require frequent adjustments to the

D. G. MacMartin (✉)
Sibley School of Mechanical and Aerospace Engineering, Cornell University, Ithaca, NY, USA
e-mail: dgm224@cornell.edu

P. J. Irvine
Earth Sciences, University College London, London, UK

B. Kravitz
Department of Earth and Atmospheric Sciences, Indiana University, Bloomington, IN, USA

Atmospheric Sciences and Global Change Division, Pacific Northwest National Laboratory, Richland, WA, USA

J. B. Horton
Harvard Kennedy School, Cambridge, MA, USA

© Springer Nature Switzerland AG 2021

W. Burns et al. (eds.), *Climate Geoengineering: Science, Law and Governance*, AESS Interdisciplinary Environmental Studies and Sciences Series,
https://doi.org/10.1007/978-3-030-72372-9_2

geoengineered forcing in response to observations. We believe that this exercise can lead to greater clarity in terms of future governance needs by articulating key characteristics of a hypothetical deployment scenario.[1]

1 Introduction

There is increasing awareness of the substantial gap between the amount of mitigation needed to avoid dangerous anthropogenic climate change and current mitigation commitments.[2] Solar geoengineering approaches have the potential to provide an additional option for managing the risks of climate change as illustrated qualitatively in Fig. 1,[3] with the most frequently discussed option being the addition of aerosols to the stratosphere to reflect some sunlight back to space.[4] Not enough is currently known to support informed decisions regarding deployment of such approaches,[5] but preliminary climate modeling suggests that solar geoengineering in addition to mitigation is likely to reduce many climate risks.[6]

Deployment of solar geoengineering would have global effects, leading to the question of how one might govern use of these technologies.[7] The international

[1] This chapter is based on D.G. MacMartin et al.: *Technical characteristics of a solar geoengineering deployment and implications for governance,* Climate Policy (2019).

[2] E.g., J. Rogelj et al.: *Paris Agreement climate proposals need a boost to keep warming well below* 2 °C, 534 Nature 534 (2016), see also IPCC: *Global warming of 1.5C, an IPCC special report on the impacts of global warming of 1.5C above pre-industrial levels and related global greenhouse gas emission pathways, in the context of strengthening the global response to the threat of climate change, sustainable development, and efforts to eradicate poverty* (2018)

[3] From D. G. MacMartin, K. L. Ricke and D. W. Keith: *Solar geoengineering as part of an overall strategy for meeting the 1.5 °C Paris target*, phil. Trans. Royal Soc. A (2018), see also T. M. L. Wigley: *A combined mitigation/geoengineering approach to climate stabilization*, Science 314 (2006); J. C. S. Long and J. G. Shepherd: *The strategic value of geoengineering research*, Global Environmental Change 1 (2014). For more quantitative assessments of overshoot scenarios see K. L. Ricke, R. J. Millar and D. G. MacMartin: Constraints on global temperature target overshoot, Scientific Reports 7 (2017) and MacMartin, Ricke and Keith (2018).

[4] P. J. Crutzen: *Albedo Enhancement by Stratospheric Sulfur Injections: A Contribution to Resolve a Policy Dilemma?*, Climatic Change 77 (2006); National Academy of Sciences: Climate Intervention: Reflecting Sunlight to Cool Earth, 2015.

[5] See for example D. G. MacMartin et al.: *Geoengineering with stratospheric aerosols: what don't we know after a decade of research*?, Earth's Future 4 (2016).

[6] E.g., D. W. Keith and P. J. Irvine: *Solar geoengineering could substantially reduce climate risks – A research hypothesis for the next decade*, Earth's Future 4 (2016).

[7] See e.g., J. Reynolds: *Solar geoengineering to reduce climate change: A review of governance proposals,* Proc. Royal Soc. A, 475 (2019), E. A. Parson: *Climate engineering in global climate governance: Implications for participation and linkage*, Transnational Environmental Law (2013); E. A. Parson and L. N. Ernst: *International governance of climate engineering*, Theoretical Inquiries in Law 14 (2013); Steve Rayner et al.: *The Oxford Principles*, Climatic Change 121 (2013); D. Bodansky: *The who, what, and wherefore of geoengineering governance*, Climatic Change 121 (2013); Scott Barrett: *Solar Geoengineering's Brave New World: Thoughts on the*

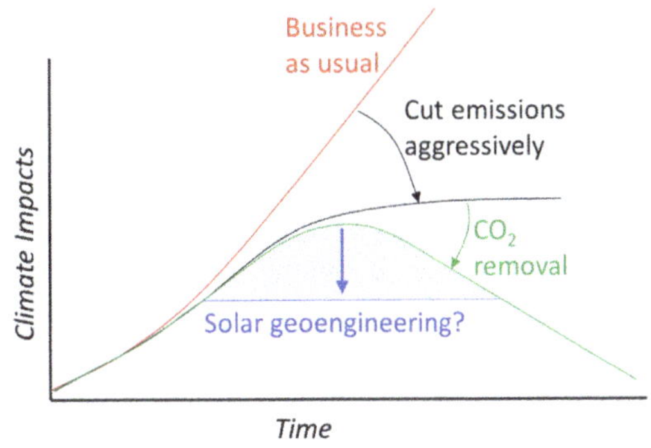

Fig. 1 Reducing greenhouse gas emissions, combined with future large-scale atmospheric CO_2 removal, may lead to long-term climate stabilization with some overshoot of desired temperature targets. There is a plausible role for temporary and limited solar geoengineering as part of an overall strategy to reduce climate risks during the overshoot period. This graph (from MacMartin et al. 2018) represents climate impacts conceptually, not quantitatively

community has agreed upon a limit of 1.5–2 °C rise in global mean temperature above preindustrial levels,[8] but 1.5 °C could be surpassed within the next 1–2 decades.[9] This poses some degree of urgency in terms of developing geoengineering governance mechanisms, while simultaneously continuing scientific research necessary to assess impacts and risks.

To understand what it is that needs governing, an important question to help focus discussion is what decisions need to be made and when? Clearly the most significant choice is simply whether or not to deploy any form of solar geoengineering. However, if a choice is made to deploy, that requires further choices that are neither binary nor static. Different design choices will lead to different projected outcomes. But since outcomes will never exactly match projections, observations made during deployment will then drive subsequent decisions across a wide range of timescales. The nature of these more complex decisions may influence the needs of governance structures. To understand these choices, it is necessary to consider the characteristics of a well-intentioned deployment in greater detail. One might then hope to structure governance that could enable and encourage such an ideal scenario. This is the aim here: to articulate what we know from climate science and engineering that is relevant to defining needs for solar geoengineering governance.

Governance of an Unprecedented Technology, Review of Environmental Economics and Policy 8(2) (2014); J. B. Horton and J. L. Reynolds: *The International Politics of Climate Engineering: A Review and Prospectus for International Relations*, International Studies Review 18 (2016); Jesse L. Reynolds: *Climate Engineering and International Law*, D. A. Farber and M. Peeters (eds.): Forthcoming in Climate Change Law, Elgar Encyclopedia of Environmental Law, vol. 1, 2016.

[8] UNFCCC: *Adoption of the Paris Agreement*, Available at https://unfccc.int/resource/docs/2015/cop21/eng/l09.pdf, 2015.

[9] B. Kirtman et al.: *Near-term Climate Change: Projections and Predictability*, Climate Change 2013: The Physical Science Basis. Contribution of Working Group I to the Fifth Assessment Report of the Intergovernmental Panel on Climate Change, 2013, Fig 11.25.

Much of the initial climate research into solar geoengineering has been exploratory, e.g., how models respond differently to a decrease in sunlight versus an increase in greenhouse gas concentrations. Multiple climate models have simulated an idealized reduction in sunlight,[10] and many climate models have simulated the response to a continuous addition of stratospheric sulfate aerosols; typically in the form of SO_2 that subsequently oxidizes and forms sulfate aerosols.[11] While informative, it would be a mistake to interpret any of these simulations as describing how the climate would respond to solar geoengineering because they test ad hoc strategies rather than intentionally designed ones.

Research is only now engaging with three fundamental questions. First, how could a solar geoengineering deployment be *designed* to achieve some desired outcomes or minimize other effects[12]; this is a necessary precursor to assessing climate impacts. With stratospheric aerosols, for example, not only could one aim for more or less global cooling, but one could put more emphasis at high versus low latitudes, or Northern versus Southern hemispheres. Choices such as these will influence the distribution of benefits and harms. Second, how could a solar geoengineering deployment be *managed* to maintain desired outcomes in the presence of uncertainty in the climate response.[13] No amount of research will reduce uncertainty to zero, and decisions will inevitably be revisited in light of the observed response of the climate to such interventions. That is, some form of adaptive management[14] is essential. However, this introduces a third challenge: how can observed changes be

---

[10] E.g., B. Kravitz et al.: *Climate model response from the Geoengineering Model Intercomparison Project (GeoMIP)*, J. Geophys. Res. 118 (2013). See also P. Irvine et al.: *Halving warming with idealized solar geoengineering moderates key climate hazards*, Nature Climate Change, 9 (2019).

[11] See e.g., G. Pitari et al.: *Stratospheric ozone response to sulfate geoengineering: Results from the Geoengineering Model Intercomparison Project (GeoMIP)*, J. Geophys. Res. A 119(5) (2014).

[12] D. G. MacMartin et al.: *Management of trade-offs in geoengineering through optimal choice of nonuniform radiative forcing*, Nature Climate Change 3 (2013); B. Kravitz et al.: *Geoengineering as a Design Problem*, Earth Systems Dynamics 7 (2016); D. G. MacMartin et al.: *The climate response to stratospheric aerosol geoengineering can be tailored using multiple injection locations*, J. Geophys. Res. A 122 (2017); B. Kravitz et al.: *First simulations of designing stratospheric sulfate aerosol geoengineering to meet multiple simultaneous climate objectives*, J. Geophys. Res. A 122 (2017), MacMartin and Kravitz: *The engineering of climate engineering*, Annual Rev. Control, Robotics & Auton. Systems (2019).

[13] D. G. MacMartin et al.: *Dynamics of the coupled human-climate system resulting from closed-loop control of solar geoengineering*, Clim. Dyn. 43(1–2) (2014); B. Kravitz et al.: *Explicit feedback and the management of uncertainty in meeting climate objectives with solar geoengineering*, Env. Res. Lett. 9(4) (2014); Kravitz et al.: Geoengineering as a Design Problem (*supra* note 16); Kravitz et al.: First simulations of designing stratospheric sulfate aerosol geoengineering to meet multiple simultaneous climate objectives (*supra* note 16).

[14] E.g., C. S. Holling: Adaptive Environmental Assessment and Management, 1978; R. Chris: Systems Thinking for Geoengineering Policy: How to Reduce the Threat of Dangerous Climate Change by Embracing Uncertainty and Failure, 2015.

correctly *attributed* to solar geoengineering[15] in the presence of both natural variability and uncertainty in the response to other anthropogenic forcings?

We consider these three characteristics of deployment, along with their corresponding challenges for decision-making.

1. *Geoengineering is a design problem (Section 2).* Geoengineering can be designed to achieve a range of different possible climates. Given that, what are the goals for deployment? This is more complex than simply manipulating a "global thermostat;" deployment is not a univariate decision.
2. *Some uncertainties can be managed through feedback (Section 3.1).* Climate models don't need to be perfect, as the forcing does not need to be perfectly predicted in advance; it can be adjusted in response to the observed climate – a feedback process. However, this requires frequent updates that cannot be effectuated in a political environment that is usually characterized by extremely slow decision making.
3. *Detection and attribution of regional changes will take decades (Section 3.2).* There will always be unpredictable weather and climate events, and determining causation with confidence will take time. (Conversely, if it is difficult to detect some climate shift, that implies that the shift is small compared to natural variability, and may not be important.) Thus, some decisions involve extreme patience.

These last two propositions, associated with the time-scales of evolving decisions, may appear to be ostensibly contradictory. In reality, there will be a continuum of time-scales associated with different features in the climate response. We explicitly avoid any discussion in Sections 2 and 3 regarding how one might design governance to enable decisions. Section 4 concludes with some brief thoughts tying the nature of decisions explicated in the previous sections to the needs of governance.

2 Spatial and Temporal Goals

Mitigation primarily involves a single decision variable, net greenhouse gas emissions, or equivalently, the atmospheric concentration of greenhouse gases. While mitigation involves trade-offs between economics and climate outcomes, there aren't substantive trade-offs associated directly with climate outcomes: lower emissions yields less climate damage than higher, and as a consequence, a single number such as "2 °C" can stand in as a proxy for a wide collection of impacts. That is not true for solar geoengineering.

First, solar geoengineering does not affect the climate the same way that reduced concentrations of atmospheric greenhouse gases would, leading to potentially

[15] D.G. MacMartin et al.: *Timescale for detecting the climate response to stratospheric aerosol geoengineering*, J. Geophys. Res. A 124 (2019)

disparate regional outcomes,[16] and feeding into the well-known concern over "who gets to set the thermostat." However, reality is more complex. The climate response to geoengineering will depend on how it is deployed. With stratospheric aerosols, for example, one could choose how much to inject at different latitudes to obtain some influence over climate outcomes[17]; this is illustrated in Fig. 2.[18] By injecting aerosols into one or the other hemisphere, one could influence the relative cooling between hemispheres and use this degree of freedom to minimize shifts in the ITCZ that could disrupt tropical precipitation patterns.[19] By injecting aerosols at higher latitudes, one could put more emphasis on cooling higher rather than lower latitudes. The number of independent degrees of freedom that could be achieved is unclear, but is at least these three; introducing seasonal dependent injection rates might allow more options,[20] while other solar geoengineering approaches such as

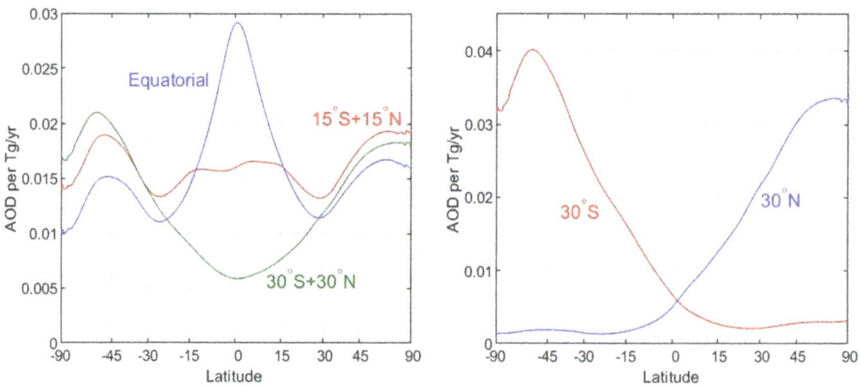

Fig. 2 Illustration of design aspect to geoengineering. The aerosol optical depth (AOD) is shown, scaled for a 1 Tg per year injection of SO_2, calculated in a fully-coupled chemistry-climate model. In the left panel, for equatorial injection (blue), and split equally between either 15°S and 15°N (red) or 30°S and 30°N (green); each leading to different emphasis between low and high latitudes. The right panel shows injection at either 30°S (red) or 30°N (green), yielding different emphasis on each hemisphere. Choosing different combinations of these will result in quite different climate outcomes, allowing some potential to design the deployment to achieve specified goals

[16] Discussed for example in K. L. Ricke, M. Granger Morgan and M. R. Allen: *Regional climate response to solar-radiation management*, Nature Geoscience 3 (2010).

[17] MacMartin et al.: The climate response to stratospheric aerosol geoengineering can be tailored using multiple injection locations (*supra* note 15); Z. Dai, D. Weisenstein and D. W. Keith: *Tailoring meridional and seasonal radiative forcing by sulfate aerosol solar geoengineering*, Geophys. Res. Lett. (2018). See also MacMartin and Kravitz: *The engineering of climate engineering*, (*supra* note 16).

[18] Based on simulations described in S. Tilmes et al.: *Sensitivity of aerosol distribution and climate response to stratospheric SO_2 injection locations*, J. Geophys. Res. A. 122 (2017).

[19] J. M. Haywood et al.: *Asymmetric forcing from stratospheric aerosols impacts Sahelian rainfall*, Nature Climate Change 2013; Kravitz et al.: Geoengineering as a Design Problem (*supra* note 16).

[20] D. Visioni et al., *Seasonal injection strategies for stratospheric aerosol geoengineering*, Geophys. Res. Lett., 46, (2019).

marine cloud brightening[21] might also allow more. It is thus insufficient to agree only on a target for global mean temperature; a decision to deploy must include a clear articulation of the high-level multivariate goals for the deployment.[22]

The U.N. Framework Convention on Climate Change committed nations to avoiding dangerous anthropogenic interference in the climate system. Over time, this qualitative goal was translated into the quantitative goal of limiting warming to well below 2 °C.[23] A similar exercise could arrive at multivariate quantitative goals for solar geoengineering. This could be as simple as specifying the desired global mean temperature, maintaining some minimum amount of Arctic sea ice extent, while minimizing shifts in tropical precipitation. More complex multivariate goals could be defined, with the constraint that the spatial scale of these high-level goals needs to be at least somewhat commensurate with the spatial scale of the available degrees of freedom, and that there is sufficient understanding of the physical relationship between these to use as a basis for design (a non-trivial requirement). The ability to design for multivariate goals could complicate negotiations, in that there are more choices to be made, but could also simplify them, as some concerns that lead to conflicting desires may be partially alleviated.

There will still be fundamental trade-offs, and what constitutes the "ideal" climate is not clear. A plausible goal would be to avoid significant change with respect to some baseline climate state (e.g., the climate at the time geoengineering is commenced), but there will still be trade-offs. A 2 °C world achieved purely through mitigation will not be the same as a 2 °C world achieved through less aggressive mitigation and some amount of geoengineering. However, with multiple degrees of freedom, geoengineering can be designed to make these cases more similar than much of the early research would suggest.[24] Nonetheless, there will still be differences between how geoengineering affects the climate and how other anthropogenic influences affect the climate, due to the different mechanisms of radiative forcing (though it is not clear today how significant these changes might be). Furthermore, the entire climate system is coupled. Even if we understood the system perfectly, it would not be possible to independently adjust every possible climate outcome, neither choosing different effects at spatially proximate locations, nor simultaneously determining temperature and precipitation outcomes at any location, nor eliminating extreme events.

The temporal aspect to the goal also needs to be defined. If solar geoengineering were ever deployed, there are several reasons to only gradually ramp up the forcing

[21] J. Latham: *Control of Global Warming?*, Nature 347 (1990). Spraying salt aerosols into boundary layer clouds is expected, in the right locations and under the right meteorological conditions, to result in more reflective clouds, but the method is currently less well understood than stratospheric aerosol injection.

[22] See Kravitz et al.: Geoengineering as a Design Problem (*supra* note 16).

[23] See Article 2, A*doption of the Paris Agreement (supra* note 12)

[24] Kravitz et al.: Geoengineering as a Design Problem (*supra* note 16); Kravitz et al.: First simulations of designing stratospheric sulfate aerosol geoengineering to meet multiple simultaneous climate objectives (*supra* note 16).

over time rather than immediately demanding a substantial forcing level to cool the planet quickly. This strategy allows possible surprises to be discovered earlier[25] while forcing is still relatively small. Furthermore, rapid changes in forcing can also lead to unnecessary climate impacts, such as a short-term reduction in monsoonal precipitation due to the differential rate of land versus ocean cooling.[26] Thus for example, in the presence of still-rising atmospheric greenhouse gas concentrations, one might choose to maintain conditions as close as possible to the year in which deployment starts, as implied by Fig. 1. Other scenarios include limiting only the rate of change of warming.[27]

A decision to deploy would also need to define the initial strategy to meet these goals (e.g., how much SO_2 to inject per year at which latitudes, and how that is expected to change as a function of time), what the justification is for concluding that that strategy would meet the goals, what the projected impact would be on any climate variable not explicitly specified, and an assessment of (and justification for) confidence in projections.[28] Climate scientists and engineers can in principle provide this type of information, and indicate what is and is not achievable, but the definition of goals is a policy choice.

3 Evolving Decisions

No amount of research will reduce the uncertainty in projected impacts to zero. Uncertainty arises due to a variety of sources.[29] Uncertainty in specific processes, such as aerosol microphysical growth assumptions, or ozone-chemistry reaction rates, might be sufficiently reducible through a combination of better observations after volcanic eruptions[30] and small-scale process-level field experiments.[31] However, an experiment to directly measure the climate response to forcing – how

[25] e.g., D. W. Keith and D. G. MacMartin: *A temporary, moderate and responsive scenario for solar geoengineering*, Nature Climate Change 5 (2015).

[26] See D. G. MacMynowski, H.-J. Shin and K. Caldeira: *The frequency response of temperature and precipitation in a climate model*, Geophys. Res. Lett. 38 (2011) and discussion in A. Robock et al.: *Studying geoengineering with natural and anthropogenic analogs*, Climatic Change 121(3) (2013).

[27] As in D. G. MacMartin, K. Caldeira and D. W. Keith: *Solar geoengineering to limit rates of change*, phil. Trans. Royal Soc. A 372 (2014).

[28] MacMartin and Kravitz, *Mission-driven research for stratospheric aerosol geoengineering*, Proc. Nat. Ac. Sci. (2019).

[29] See for example MacMartin et al.: Geoengineering with stratospheric aerosols: what don't we know after a decade of research? (*supra* note 9).

[30] Robock et al.: Studying geoengineering with natural and anthropogenic analogs (*supra* note 30).

[31] D. W. Keith, R. Duren and D. G. MacMartin: *Field experiments on solar geoengineering: report of a workshop exploring a representative research portfolio*, Phil. Trans. R. Soc. A 372 (2014); J. A. Dykema et al.: *Stratospheric-controlled perturbation experiment: a small-scale experiment to improve understanding of the risks of solar geoengineering*, Phil. Trans. R. Soc. A 372 (2014).

variables such as regional temperature and precipitation might change in response to geoengineering – would require both considerable time and considerable forcing,[32] making such a test practically equivalent to deployment.[33] Indeed, even early deployment would not likely involve sufficient forcing to quickly resolve many uncertainties,[34] as described below. Thus, there will always be some residual level of uncertainty in the climate response at the time of a deployment decision.

If it becomes clear during deployment that some outcome is not what was predicted, a choice will be faced as to whether to modify the strategy for meeting goals (such as increasing or decreasing the amount of SO_2 injected at some latitude), modify the goals themselves (put more or less emphasis on some outcome), or potentially phase-out deployment altogether. The next two sub-sections consider what these decisions over time might look like. These can pose additional challenges for how to structure international governance either by requiring immediate action, or conversely, by requiring a high degree of patience and consequent longevity of institutions.

While the climate system does not provide any clear separation of time-scales, some structure can be imposed based on how decisions might be made, by dividing the problem into those relatively few high-level climate goals that the intervention is designed to meet, and all of the vast number of other climate system variables that affect humans and ecosystems. For example, if geoengineering was intended to maintain global mean temperature at 2 °C, then any sustained period warmer or cooler than that could justify increasing or decreasing the amount of geoengineering; the sign of the effect this would have on temperature is clear from basic physics. However, the impact on precipitation in some country might, at the time of a deployment decision, be uncertain even in sign; this type of effect would need to be monitored, any observed changes determined as to whether they were attributable to the deployment or not, and a decision made as to whether to alter the deployment in response. These two examples yield quite different timescales for decisions.

3.1 Managing Uncertainty through Feedback

No engineered system is perfectly understood. Rather than simply introducing an input and hoping for the best, systems from aircraft flight control to manufacturing plants all rely on feedback: the output is monitored, compared with the desired value, and the inputs slightly adjusted so that over time the output converges to the desired value. One relies on the same fundamental principle every time one drives a car or takes a shower in an unfamiliar place; in an ecosystem context this is known

---

[32] D. G. MacMynowski et al.: *Can we test geoengineering?*, Energy Environ. Sci. 4 (2011).

[33] MacMartin and Kravitz, *Mission-driven research for stratospheric aerosol geoengineering*, (*supra* note 32), Robock et al.: *A test for geoengineering?*, Science 327 (2010).

[34] MacMartin et al.: *Timescale for detecting the climate response to stratospheric aerosol geoengineering* (*supra* note 19).

as adaptive management.[35] In the context of earth system management, Schellnhuber and Kropp[36] term this "geocybernetics". This feedback process compensates for some degree of uncertainty in the strength of the relationship between input and output. Thus, for example, the amount of solar reduction required to offset the warming from some amount of CO_2 varies from model to model.[37] Following Jarvis and Leedal,[38] MacMartin et al.[39] demonstrated the idea of using feedback of the "observed" global mean temperature to adjust the amount of solar reduction in a climate model; Kravitz et al.[40] then demonstrated that this process was sufficiently robust so that even if the feedback algorithm was tuned using simulations from one climate model, it still yielded the desired outcomes in a second. This idea has been extended to manage multiple climate variables simultaneously,[41] and to do so by adjusting the amount of SO_2 injection at multiple latitudes[42] rather than idealized patterns of solar reduction. In each of these cases, there is a clear physical relationship between the input and output; e.g., if you increase the aerosol injection rate you will decrease temperature, if you shift more of the injection to one hemisphere from the other, you will preferentially cool that hemisphere. However, the exact relationship does not need to be known, and thus some amount of uncertainty can be managed. To successfully implement solar geoengineering to achieve some temperature target, for example, we do not need to know either how much radiative forcing is exerted by a given rate of aerosol injection, or how much the climate cools in response – just that increased injection causes increased cooling.

This capability to manage uncertainty requires the ability to constantly make slight adjustments to the system. Anyone who has impatiently tried to adjust a shower temperature knows how difficult the task can be if there is substantial time delay between moving the knob and feeling the resulting change. If one waited for 10 years to see what the effect of geoengineering was on the temperature before making any adjustment, then on average that information is now 5 years old, introducing a substantial time-delay. It is better to make minor adjustments every year, even if the lack of statistical significance means that one might be reacting to climate variability, and indeed, such an algorithm will always react to and modify

[35] Holling: Adaptive Environmental Assessment and Management (*supra* note 18).

[36] H. J. Schellnhuber & J. Kropp: *Geocybernetics: Controlling a Complex Dynamical System Under Uncertainty*, Naturwissenschaften 85 (1998).

[37] That is, the efficacy is uncertain; e.g., D. G. MacMartin, B. Kravitz and P. J. Rasch: *On solar geoengineering and climate uncertainty*, Geophys. Res. Lett. 42 (2015).

[38] A. Jarvis and D. Leedal: *The Geoengineering Model Intercomparison Project (GeoMIP): A control perspective*, Atm. Sci. Lett. 13 (32012).

[39] MacMartin et al.: Dynamics of the coupled human-climate system resulting from closed-loop control of solar geoengineering (*supra* note 17).

[40] Kravitz et al.: Explicit feedback and the management of uncertainty in meeting climate objectives with solar geoengineering (*supra* note 17).

[41] Kravitz et al.: Geoengineering as a Design Problem (*supra* note 16).

[42] Kravitz et al.: First simulations of designing stratospheric sulfate aerosol geoengineering to meet multiple simultaneous climate objectives (*supra* note 16).

climate variability.[43] If such a feedback process were used in a solar geoengineering deployment, the details regarding how much to adjust would be esoteric, although the basic concept is straightforward.

The need for a rapid decision-making capability is not restricted to managing uncertainty. An additional reason would be if a large volcanic eruption occurred during a deployment of stratospheric aerosol geoengineering. One could choose to do nothing different; in this case the decrease in global temperature might still be less than if there were no geoengineering due to nonlinearities in sulfate aerosol microphysics.[44] However, it would be wiser to decrease injection immediately – on a time-scale of weeks – to compensate for the increase in stratospheric sulfate from the eruption. Furthermore, an eruption in one hemisphere will preferentially cool that hemisphere, shifting the intertropical convergence zone (ITCZ) towards the opposite hemisphere, and shifting tropical precipitation with it; this can have significant human consequences such as Sahelian drought.[45] Thus one might want to rapidly increase the injection of aerosols into the opposite hemisphere to counterbalance the effect of the eruption over the ensuing year.

The need for short time-scale decisions clearly has ramifications for governance, as described in Section 4. However, other decisions may present governance challenges at the opposite end of the spectrum due to the long time-scales involved in detection and attribution of changes not predicted at deployment.

3.2 *Detection and Attribution May Take Decades*

The example given earlier for high-level goals included global mean temperature, Arctic sea ice extent, and tropical precipitation. However, the ultimate goals of reducing climate damages are more complicated and multi-dimensional. Prior to deployment there would presumably be a comprehensive multi-model assessment of the predicted impact of geoengineering, not only for high-level goals, but for regional climate shifts, changes in probability of different weather events, and so forth. If models predict that geoengineering will increase the likelihood or magnitude of some particular type of extreme weather event, and if such an event does occur, it is reasonable to (at least fractionally) attribute that event to the deployment; this may be useful in compensation schemes for example.

However, there will always be uncertainty in model predictions, and prediction skill will be more limited for some variables than others. This leads to a challenge: acknowledging model uncertainty requires a willingness to learn through

---

[43] MacMartin et al.: Dynamics of the coupled human-climate system resulting from closed-loop control of solar geoengineering (*supra* note 17).

[44] See, e.g., A. Laakso et al.: *Radiative and climate impacts of a large volcanic eruption during stratospheric sulfur geoengineering*, Atmos. Chem. Phys. 16 (2016).

[45] Haywood et al.: Asymmetric forcing from stratospheric aerosols impacts Sahelian rainfall (*supra* note 23).

observations, while at the same time *not* responding to every weather event or perceived shift in climate that occurs. Learning where model predictions were meaningfully wrong will take time. Furthermore, even the benefits of deployment will not be immediately apparent.

If one learns that the deployment is leading to some undesired and unpredicted shift in regional climate (including changes in the magnitude or frequency of some extreme weather events), one could alter the high-level goals; e.g., allow global mean temperature to increase slightly so that less geoengineered forcing is required, or change the relative emphasis on high vs low-latitudes, or introduce additional goals. Indeed, a possible decision would be to terminate the deployment altogether (ideally through a gradual phase-out as in MacMartin, Caldeira and Keith[46] to avoid a "termination shock"[47]).

The challenge with this collective set of decisions, involving every climate variable at any spatial scale, is that the very concept of "climate" that is at the core of either climate change or climate engineering describes long-term multi-decadal characteristics. Over shorter time-scales there is considerable variability that can mask the response due to geoengineering. For example, despite the duration of anthropogenic greenhouse-gas forced climate change today, while there is no ambiguity regarding the sign of the effect on some metrics like global mean temperature or Arctic sea ice extent, there is still considerable uncertainty in how increased greenhouse gases have affected regional precipitation patterns,[48] and even at the global scale there can be substantial decadal variability in the trend (e.g., the so-called "hiatus" of the early 2000's). Attribution of individual storms or droughts to climate change is improving but remains difficult today,[49] in part because of insufficient statistics on the probability of rare events. There will always be unusual events; in any year one might expect 1% of the world's population to experience a once-in-a-century event. The difficulty of attributing any individual event to geoengineering early in a deployment may be even more challenging than attributing an individual event to climate change is today, simply because the forcing will likely start out smaller. Furthermore, solar geoengineering would be taking place simultaneously with increased greenhouse gas forcing whose detailed impact remains uncertain.

---

[46] MacMartin, Caldeira and Keith: Solar geoengineering to limit rates of change (*supra* note 27).

[47] If deployment was abruptly terminated, the temperature would rise rapidly to roughly the value it would have been had solar geoengineering never been started, with likely severe consequences. See, e.g., Trisos et al.: *Potentially dangerous consequences for biodiversity of solar geoengineering implementation and termination*, Nature Ecology & Evolution, 2 (2018).

[48] Kirtman et al.: Near-term Climate Change: Projections and Predictability (*supra* note 12), Box 11.2.

[49] S. C. Herring et al. (eds.): *Explaining extreme events of 2017 from a climate perspective*, Bulletin Am. Met. Soc. 100, 2019.

As noted earlier, a plausible deployment scenario might be to maintain conditions as close as possible to the year in which deployment starts. Such a scenario was simulated by Kravitz et al.,[50] where the background anthropogenic climate change emissions follow a high-end RCP 8.5 scenario[51] and SO_2 injection is used to maintain 2020 conditions. A few results from that simulation are shown in Fig. 3 below, both at a global scale and for northern India.[52] Geoengineering simulations are typically plotted showing both the no-geoengineering and geoengineered cases as different colored lines, and often averaged over time or over multiple simulations to estimate the forced response. However, if geoengineering were actually used, the alternate universe in which geoengineering was not used will only exist as a hypothetical in models. The actual climate that anyone experiences will continue to be marked by variability and unusual events, and it will only be over decades that some changes will become apparent. For example, in Fig. 3, the change in slope (rate of increase) for global mean temperature is statistically significant with a 95% confidence after 10 years, and similarly for the change in global mean precipitation. It takes 20 years for the change in slope of the temperature over northern India to be statistically significant, and in this single model simulation, the change in annual-mean precipitation over that region is never statistically significant at a 95% confidence, yet averaging over many similar simulations does show that at least in this model, with this deployment strategy, the precipitation is expected to decrease slightly in this region.[53] Changes in many other variables, such as precipitation averaged over only one season, or the frequency of extreme weather events, may be even more difficult to detect in the presence of natural climate variability.

A long time-scale for detection and attribution is not in and of itself a problem. If it is hard to detect a change in some variable, it is hard precisely because the change is small relative to natural variability, and thus that change might not have serious adverse impacts.

However, how should one respond if observations suggest an 80% chance that some variable has changed? Or a 50% chance? Increased certainty will require waiting for more time to pass. Furthermore, with a sufficiently large space of climate variables being monitored, roughly 5% will show unusual changes that appear to be statistically significant at a 95% confidence level. In principle, models can be used to assess the plausibility of a physical connection with the geoengineering deployment, rather than simply relying on analysis of time series. However, the entire

---

[50] Kravitz et al.: First simulations of designing stratospheric sulfate aerosol geoengineering to meet multiple simultaneous climate objectives (*supra* note 16).

[51] M. Meinshausen et al.: *The RCP Greenhouse Gas Concentrations and their Extension from 1765 to 2300*, Climatic Change 109 (2011).

[52] Using the region defined by K. Dagon and D. P. Schrag: *Regional climate variability under model simulations of solar geoengineering*, J. Geophysical Research A 122 (2017).

[53] Cheng et al.: *Soil moisture and other hydrological changes in a stratospheric aerosol geoengineering large ensemble,* J. Geophysical Research A 124 (2019)

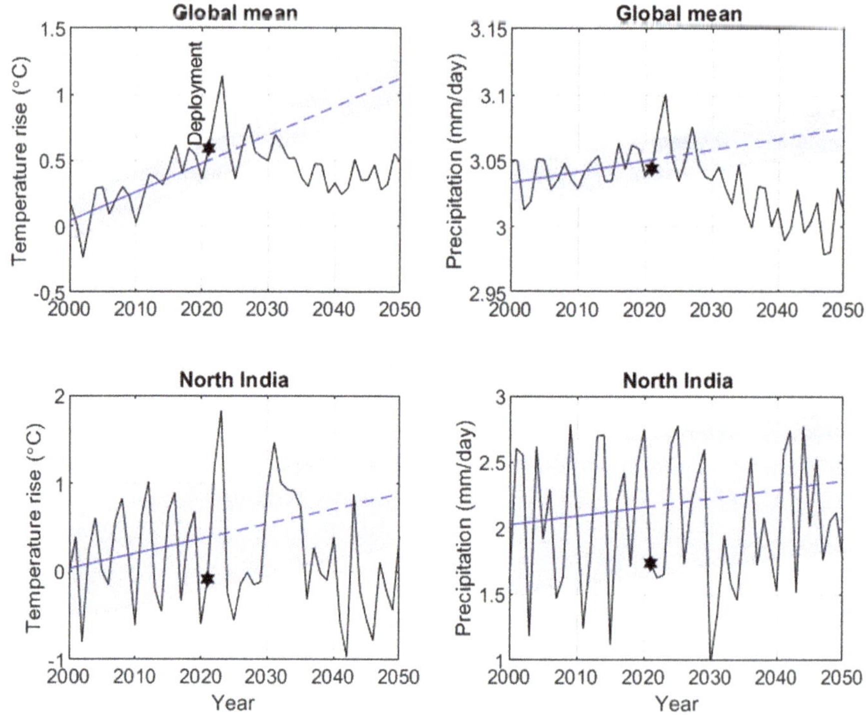

Fig. 3 Annual mean temperature and precipitation change relative to 1975–2020 averaged over the globe and over Northern India in a simulation in which stratospheric aerosol geoengineering was initiated in 2020 with the goal of keeping temperatures at 2020 levels in the presence of increasing greenhouse gas concentrations (see Kravitz et al. (2017) for details). In each plot, the black line shows the simulated trajectory, with the star indicating the start of (low-level) deployment. The blue line and shaded band are the best fit slope to 1975–2020, and ± 1 standard deviation of natural variability about this. The change in slope of global mean temperature and precipitation are statistically significant at the 95th percentile after roughly 10 years (using Welch's unequal-variances t-test); the regional temperature change over northern India takes 20 years to show a statistically significant change in trend, while the change in precipitation over this region is not statistically significant in this simulation even by 2100, but does show a drop when averaged over sufficient ensemble members. Note that the unusual response in year 2023 is due to a model error and should be ignored. Similar plots could be generated for other variables and spatial scales, including frequency of weather events such as the number of Atlantic hurricanes per season

motivation for looking for possible changes in regional climate arises from concern that the models are imperfect, and so it is the difference from predictions that one is most interested in uncovering through observation.

Finally, if the strategy is adjusted in response to some observed change, that resets the clock on attribution, with a similar time-scale required to be sure that the new strategy indeed yields a different result.

4 Implications for Governance

The international community has been able to agree to a target of holding temperatures to substantially below a 2 °C rise in global mean temperature above preindustrial levels. Thus, there is at least precedent for global agreement on climate goals. One of the challenges with reaching agreement on one "global thermostat" for solar geoengineering is that different regions might differentially benefit or have different desired amounts of warming or cooling. While it might seem that if agreeing on one number is hard, and thus agreeing on multiple goals would be harder still, that may not be true if the ability to independently manage multiple goals means that the distribution of benefits and harms is more uniform. Nonetheless, it will not be possible to design a deployment that can achieve every possible goal in every region of the world, and the trade-offs involved will require the ability to agree on more complex choices than simply a number. Deployment goals would be fundamentally political, reflecting not only policy considerations but deeper struggles over the notion and content of an ideal climate, nature vs. artifice, etc. Scientists and engineers can present what is possible and likely or unlikely, but can't (or shouldn't) decide what objectives to pursue.

Once deployed, there will be a variety of decisions that will need to be made over a wide range of time-scales. Both "slow" and "fast" decisions present interesting challenges for governance.

The primary challenge in the former may be to avoid action when it is not warranted by the available evidence. The issue of attributional time-scale puts pressure on organizational lifetimes, with inherent time-scales for solar geoengineering that are not only inter-generational in the overall lifetime of the deployment but at least multi-decadal in the ability to monitor, assess, and modify key decisions about the deployment. While uncertainty about the climate response needs to be accepted, and a culture of adaptive management supported, the long time-scales for attribution also create a need to establish processes that would counter the impulse to constantly change the goals of the deployment in response to the latest climate event; there will always be unusual weather events whether geoengineering is deployed or not.

The shorter time-scales associated with either managing uncertainty or responding to events such as volcanic eruptions are not well matched to political processes; one can't delay because of procedural discussions or political posturing without suffering consequences. Furthermore, political processes may also be ill-suited to these decisions because of the technical knowledge needed to determine the appropriate action. Instead, governance may involve agreeing to the guidelines behind such adjustments and empowering an expert technical body to make them.

Clearly, decisions about feedback and attribution raise critical questions about the role of technocracy in governing a hypothetical geoengineering deployment. While the timeframes for such decisions would vary considerably, both types of decisions would be characterized by a need to insulate decision processes from broader debates about the overall purposes, goals, and objectives of geoengineering. Given the specialized knowledge required for making sound operational decisions and probability estimates based on statistical methods, substantial decision-making authority would need to be delegated to technical experts. These decisions would need to be largely "apolitical" in order to ensure consistency and predictability, in support of the ultimate goal of climate stability. (However, this characterization does not apply to more fundamental decisions about whether to deploy and what goals to pursue, which are primarily political in nature.) Other commentators have argued that such technocratic requirements would necessarily render governance of SRM deployment undemocratic.[54] However, on both short and long time-scales, modern society offers multiple examples of effective technocratic processes successfully embedded within democratic political systems.[55] Electrical grids are managed on a minute scale by trained experts at local utilities and regional system operators under the public oversight of subnational, national, and regional regulatory bodies. Economists at central banks, typically coordinating on an international basis, have wide latitude to set monetary policy to smooth out multi-year business cycles, but they do so within parameters set by the political system, and are ultimately accountable to elected representatives.

To be sure, striking an appropriate balance between expert autonomy and political oversight, particularly on the decadal time-scale required for robust determinations of attribution, will pose serious challenges for any proposal to deploy geoengineering. Just as geoengineering itself is a design problem, so too is geoengineering governance, and solutions will not be easy. However, SRM governance also resembles SRM technology in that it is not binary in character, that is, it is not either democratic or technocratic. Rather, like other forms of global governance, it is likely to entail a mixture of these and other modes of social control, with ample scope for institutional innovation.

In summary, a decision to deploy is more than a simple yes/no, but a responsible deployment decision should also include

- Definition and agreement on quantitative high-level climate goals. This will likely occur in conjunction with the scientific/engineering process of determining the deployment strategy that best meets these goals, evaluating the resulting projected impacts, and explicitly assessing confidence in these projections. Without this definition of goals there is no basis on which to make choices such as where and how much aerosol to inject.

[54] B. Szerszynski et al.: *Why Solar Radiation Management Geoengineering and Democracy Won't Mix*, Environment and Planning A 45(12) (2013).

[55] J. Horton et al.: *Solar geoengineering and democracy*, Global Env. Politics 18 (2018).

- An agreed-upon approach for updating the deployment as a function of time, including observational resources, how the resulting data will be analyzed, how to conduct attribution and how that feeds into adjustments to the original plan (and potentially also compensation), including rules for how to adjust forcing (e.g. SO_2 injection rates) across multiple time-scales, and potentially the formation of an expert body to execute at least some of these rules.

Governance of geoengineering will require international trust, long organizational lifetimes, complex decision-making, and a culture of adaptive management in order to encourage sound decisions about well-intentioned and well-designed climate interventions.

References

1. Cheng, et al.: Soil moisture and other hydrological changes in a stratospheric aerosol geoengineering large ensemble. J. Geo. Res. A. **124** (2019)
2. Crutzen, P.J.: Albedo enhancement by stratospheric sulfur injections: A contribution to resolve a policy dilemma? Clim. Change. **77** (2006)
3. Dagon, K., Schrag, D.P.: Regional climate variability under model simulations of solar geoengineering. J. Geo. Res A. **122** (2017)
4. Herring, S.C., et al. (eds.): Explaining extreme events of 2017 from a climate perspective. Bulletin Am. Met. Soc. **100**, 111 (2019)
5. Horton, J., et al.: Solar geoengineering and democracy. Glob. Environ. Politics. **18** (2018)
6. Jarvis, A., Leedal, D.: The Geoengineering Model Intercomparison Project (GeoMIP): A control perspective. Atmos. Sci. Lett. **13** (2012)
7. Keith, D.W., Irvine, P.J.: Solar geoengineering could substantially reduce climate risks – A research hypothesis for the next decade. Earth's Future. **4** (2016)
8. Keith, D.W., MacMartin, D.G.: A temporary, moderate and responsive scenario for solar geoengineering. Nat. Clim. Change. **5** (2015)
9. Keith, D.W., Duren, R., MacMartin, D.G.: Field experiments on solar geoengineering: report of a workshop exploring a representative research portfolio. Phil. Trans. R. Soc. A. **372** (2014)
10. Kravitz, B., et al.: Climate model response from the Geoengineering Model Intercomparison Project (GeoMIP). J. Geophys. Res. **118** (2013)
11. Laakso, A., et al.: Radiative and climate impacts of a large volcanic eruption during stratospheric sulfur geoengineering. Atmos. Chem. Phys. **16** (2016)
12. MacMartin, Kravitz: Mission-driven research for stratospheric aerosol geoengineering. Proc. Nat. Ac. Sci. **116**, 1089–1094 (2019)
13. MacMartin, D.G., et al.: Management of trade-offs in geoengineering through optimal choice of nonuniform radiative forcing. Nat. Clim. Change. **3** (2013)
14. MacMartin, D.G., et al.: Dynamics of the coupled human-climate system resulting from closed-loop control of solar geoengineering, Clim. Dynamis. **43**(1–2) (2014a)
15. MacMartin, D.G., Caldeira, K., Keith, D.W.: Solar geoengineering to limit rates of change, phil. Trans. Royal Soc. A. **372** (2014b)
16. MacMartin, D.G., Kravitz, B., Rasch, P.J.: On solar geoengineering and climate uncertainty. Geophys. Res. Lett. **42** (2015)
17. MacMartin, D.G., et al.: Geoengineering with stratospheric aerosols: what don't we know after a decade of research? Earth's Future. **4** (2016)
18. MacMartin, D.G., Ricke, K.L., Keith, D.W.: Solar geoengineering as part of an overall strategy for meeting the 1.5 °C Paris target. Philosophic. Trans. R. Soc. A. **376** (2018)

19. MacMartin, D.G., et al.: Technical characteristics of a solar geoengineering deployment and implications for governance, Clim. Policy. **19**, 1325 (2019a)
20. MacMartin, D.G., et al.: Timescale for detecting the climate response to stratospheric aerosol geoengineering. J. Geophys. Res. A. **124** (2019b)
21. MacMynowski, D.G., Shin, H.-J., Caldeira, K.: The frequency response of temperature and precipitation in a climate model. Geophys. Res. Lett. **38** (2011a)
22. MacMynowski, D.G., et al.: Can we test geoengineering? Energ. Environ. Sci. **4** (2011b)
23. Meinshausen, M., et al.: The RCP greenhouse gas concentrations and their extension from 1765 to 2300. Clim. Change. **109** (2011)
24. Pitari, G., et al.: Stratospheric ozone response to sulfate geoengineering: Results from the Geoengineering Model Intercomparison Project (GeoMIP). J. Geophys. Res. A. **119**(5) (2014)
25. Reynolds, J.: Solar geoengineering to reduce climate change: A review of governance proposals. Philosophic. Trans. R. Soc. A. **475** (2019)
26. Ricke, K.L., Granger Morgan, M., Allen, M.R.: Regional climate response to solar-radiation management. Nat. Geosci. **3** (2010)
27. Rogelj, J., et al.: Paris Agreement climate proposals need a boost to keep warming well below 2 °C. Nature. **534**, 534 (2016)
28. Schellnhuber, H.J., Kropp, J.: Geocybernetics: Controlling a complex dynamical system under uncertainty. Naturwissenschaften. **85** (1998)
29. Szerszynski, B., et al.: Why Solar Radiation Management Geoengineering and Democracy Won't Mix. Environ. Plann. A. **45**(12) (2013)
30. Tilmes, S., et al.: Sensitivity of aerosol distribution and climate response to stratospheric SO_2 injection locations. J. Geophys. Res. A. **122** (2017)
31. Wigley, T.M.L.: A combined mitigation/geoengineering approach to climate stabilization. Science. **314**, 452–454 (2006)

Climate Action: The Feasibility of Climate Intervention on a Global Scale

Kimberly A. Gray

1 Introduction

Rarely a week goes by without surpassing yet another threshold marking our rapidly changing climate. Each season atmospheric CO_2 levels set new records.[1] In response to the steady climb in CO_2 concentrations, average surface temperatures continue to climb, as illustrated in Fig. 1.[2] In fact, 17 of the warmest 18 years on record have occurred in this century, the 5 warmest years on record have all occurred since 2010, and 2017 was the 41st consecutive year of temperatures higher than the twentieth century average.[3]

Despite decades of mounting concern and numerous stabs at international accords, countries have failed to curtail their emissions of CO_2 and other greenhouse gases (GHG) and hence, the very complex cascade of global impacts. Many believe that time is running out to reverse or even slow global climate changes, and in the wake of failed efforts to remake our carbon intensive economies and lifestyles, the call for taking drastic measures is upon us. This chapter explains and analyzes from a science and engineering perspective the various options available to us in the current climate intervention toolbox. Special focus is given to

[1] Waldman S. Atmospheric CO_2 sets record high. ClimateWire (05.03.18); https://www.eenews.net/climatewire/2018/05/03/stories/1060080715

[2] Fountain H., Patel J.K., Popovich N. 2017 Was One of the Hottest Years on Record. New York Times; https://www.nytimes.com/interactive/2018/01/18/climate/hottest-year-2017.html

[3] NOAA: 2017 was 3rd warmest year on record for the globe; http://www.noaa.gov/news/noaa-2017-was-3rd-warmest-year-on-record-for-globe

K. A. Gray (✉)
Department of Civil & Environmental Engineering, Northwestern University,
Evanston, IL, USA
e-mail: k-gray@northwestern.edu

© Springer Nature Switzerland AG 2021

W. Burns et al. (eds.), *Climate Geoengineering: Science, Law and Governance*,
AESS Interdisciplinary Environmental Studies and Sciences Series,
https://doi.org/10.1007/978-3-030-72372-9_3

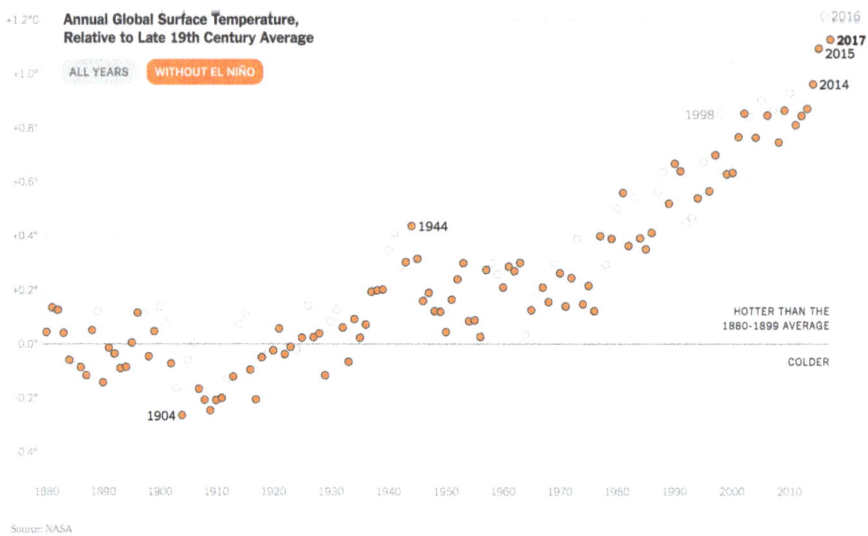

Fig. 1 Annual Global Surface Temperature, relative to a late nineteenth century average[2]

geoengineered strategies aimed at either reducing atmospheric CO_2 concentrations or lessening the impacts of elevated GHG levels. This technical discussion is punctuated with cautionary tales to illustrate the potential challenges, uncertainties, and unexpected consequences of going down the path of engineering climate adjustments at a global scale.

2 Cautionary Tale: 1. The Arctic – Extreme Events Locally and Afar

No where else is warming more evident than in the Arctic, where the rate of warming is twice that anywhere on the planet. In contrast to an average global temperature increase of about 1 °C, Arctic warming is closer to 3 °C.[4] This accelerated warming is disrupting long established cycles of ice melting in the summer and ice formation in the winter, exposing more open water, which increases wave action and mixing, and absorbs much more solar radiation, all of which further enhance melting and delay the onset of ice formation in the fall.[5]

For tens of thousands of years, the Arctic was continuously covered by ice, even in the summer, and water temperatures rarely, if ever, rose above freezing. Over the

[4]Watts J. Arctic warming: scientists alarmed by 'crazy' temperatures. The Guardian (02.27.18). https://www.theguardian.com/environment/2018/feb/27/arctic-warming-scientists-alarmed-by-crazy-temperature-rises

[5]NASA Global Climate Change; Arctic Sea Ice Minimum. https://climate.nasa.gov/vital-signs/arctic-sea-ice/

last 40 years, however, the maximum extent of summer ice melt, which occurs in September, has been steadily and dramatically increasing. In September 2017 Arctic ice cover reached its seasonal minimum and was the eighth lowest on record.[6] The ten lowest summer sea ice minima have all occurred since 2007.[7]

The peak of Arctic winter ice growth typically occurs around mid-March and in 2018, the season's extent of ice growth was another record low (5.59 million square miles), about as low as that measured in 2017 (5.57 million square miles), according to the government's 39-year satellite record and a recently released National Snow and Ice Data Center (NSIDC) report.[8] In fact, the last 4 years have witnessed the four lowest seasonal Arctic ice cover.[9]

The fragility of Arctic ice is measured not only in diminished total surface area, but also in waning thickness and age. Although ice 10–12 ft thick used to cover the North pole and in some areas of the Arctic extended down to 150 ft, NSIDC reports that since 1984 multiyear or thick ice cover has declined from 61% to 34%. The thinning of Arctic ice has implications for its suitability as wildlife habitat. In addition, solar radiation can penetrate thin ice, stimulate early algal blooms and trigger asymmetric trophic effects. Furthermore, first-year ice has a much greater likelihood of disappearing over the summer.

Although growth in ice free areas of the Arctic Ocean promises new commercial opportunities such as faster shipping routes and expanded oil, gas and mineral exploration, these changes are also harbingers of serious disruptions to both local and distant natural cycles such as storm tracks and the regulation of sea level that sustain fundamental aspects of the human enterprise such as food production and flood control. At first glance, accelerated Arctic warming seems to be due to the positive feedback associated with the melting of Arctic snow and ice both at sea and on land, the resulting decrease in surface albedo (diffuse reflection of solar radiation) and the concomitant increase in surface solar radiation absorbance. Open Arctic waters absorb huge amounts of solar radiation causing water temperatures in some places to climb as much as 4 °C above the long-term average.[10]

On land the decrease in surface albedo due to diminished snow cover amplifies the melting of glaciers and permafrost, as well as the increase in temperatures of Arctic rivers, which transport the heat to the coastal areas and warm the relatively

[6] National Snow & Ice Data Center. Arctic Sea Ice news & Analysis; Monthly Archives: September 2017. http://nsidc.org/arcticseaicenews/2017/09/

[7] Harvey C. Keeping the Arctic icy might hinge on half a degree. ClimateWire (04.03.18). https://www.eenews.net/climatewire/2018/04/03/stories/1060077985

[8] National Snow & Ice Data Center. Arctic Sea Ice news & Analysis; Arctic winter warms up to a low summer ice season. http://nsidc.org/arcticseaicenews/2018/05/arctic-winter-warms-up-to-a-low-summer-ice-season/

[9] Hobson M.K. Sea ice hits 2nd-lowest level in 39 years. ClimateWire (03.26.18). https://www.eenews.net/climatewire/2018/03/26/stories/1060077383

[10] Wadhams P. The Global Impacts of Rapidly Disappearing Arctic Sea Ice. Yale Environment 360 (09.26.16). https://e360.yale.edu/features/as_arctic_ocean_ice_disappears_global_climate_impacts_intensify_wadhams

shallow continental shelves. Relative to 15 years ago, Arctic river ice now retreats a month earlier, mid-June rather than mid-July.[11] Rapid glacial retreat is rerouting meltwaters in a phenomenon christened "*river piracy*"[12] that abruptly reorganizes watersheds and alters surface water levels.[13]

As a consequence of these dramatic changes to the Arctic landscape, the potential increase in the release of vast stores of methane from the warming seabed and tundra will have severe impacts (positive feedback) on atmospheric greenhouse gas (GHG) concentrations and hence, warming temperatures locally and globally. Methane is a much more potent GHG than CO_2 having a global warming potential as much as 80 times that of CO_2.[14] The implication of these interactions is not simply that they amplify warming temperatures and the destabilization of the global climate system; rather, it is more that we are rapidly approaching the point when the feedbacks themselves will trigger a cascade of events that may exceed the direct effects of continuing anthropogenic GHG emissions.[15] This, of course, is the tipping point of the global climate system that we appear en route to reach and after which corrective measures such as drastic cuts in fossil fuel use will have little to no effect.

Yet, this relatively simple, localized and positive feedback of increased surface temperatures, melting ice, diminished albedo, and hence, amplified warming does not fully account for Arctic phenomena. If it did, greater warming would be expected in summer months, when the surface albedo is the lowest. Surprisingly, recent wintertime warming exceeds that of summertime. February 2018 marked the fourth winter in a row of alarming Arctic heatwaves.[16] There is deep concern that the temperature extremes occurring during the sunless Arctic winters are eroding the polar vortex, a natural force field of winds circling the pole that centers frigid temperatures and deflects warm air masses in the region. The polar vortex depends on a gradient in temperature between the Arctic and mid-latitudes which appears to be weakening. The influx of warm air not only pushes Arctic winter temperatures far outside normal bounds, it also displaces the polar vortex south altering storm tracks and causing frigid temperatures, in Europe or North America. Winter temperatures in Siberia have reached as much as 35 °C above historical averages and at the northern tip of Greenland, the world's most northerly land weather station, temperatures

[11] Waldman S. Climate change is transforming, rerouting Arctic rivers. ClimateWire (04.19.17). https://www.eenews.net/climatewire/2017/04/19/stories/1060053256

[12] Shugar D.H. et al. (2017) River piracy and drainage basin reorganization led by climate-driven glacier retreat. *Nature Geoscience,* 10:370–375. https://www.nature.com/articles/ngeo2932

[13] Struzik E. How Warming Is Profoundly Changing a Great Northern Wilderness. Yale Environment 360 (04.25.17). https://e360.yale.edu/features/how-warming-is-profoundly-changing-a-great-northern-wilderness

[14] U.S.E.P.A. Greenhouse Gas Emissions. Understanding Global Warming Potentials. https://www.epa.gov/ghgemissions/understanding-global-warming-potentials

[15] Wadhams P. Yale Environment 360. https://e360.yale.edu/features/as_arctic_ocean_ice_disappears_global_climate_impacts_intensify_wadhams

[16] Hobson M.K. Sea ice hits 2nd-lowest level in 39 years. ClimateWire (03.26.18). https://www.eenews.net/climatewire/2018/03/26/stories/1060077383

have been recorded that are warmer than European cities such as London or Zurich located thousands of miles to the south. Although such extreme events happened in the past for a few hours before adjusting back to historically normal ranges, the 2018 winter saw a stretch of 10 consecutive days of temperatures reaching above freezing at the Greenland weather station, less than 500 miles from the north pole.[17]

Arctic winter heatwaves may only be short-term anomalies, but they are so far beyond the established patterns of variation that it is difficult to attribute them to freak weather events rather than the direction the Arctic climate is headed. The seasonality of Arctic warming is likely the result of energy in the atmosphere transported to the poles through large weather systems originating at mid-latitudes,[18] coupled to negative summertime feedbacks associated with increased water vapor from warming Arctic land and water creating greater cloud cover which reflects sunlight and has a cooling effect. Thus, climate perturbations at a distance from the Arctic combined with local phenomena create intensifying and asymmetric impacts near and far.

Another set of astounding changes to the Arctic region with far reaching global consequences is occurring in Greenland. Recent discoveries about Greenland's past and present climate reveal surprising insights into how fast the earth can change as it warms. After the Antarctic continent, Greenland holds the greatest amount of glacial ice on the planet and the melting of either or both of these ice sheets has enormous consequences for global sea levels. For instance, melting ice on Antarctica and Greenland would increase sea level by more than 60 m (200 ft.) and 6 m (20 ft.), respectively.[19]

Over geological time, sea levels widely fluctuate. At the peak of the last ice age, approximately 20,000 years ago, sea level was roughly 120 m (400 ft.) less than present day levels because so much of the earth's surface water was tied up in massive ice sheets. In contrast, during the last interglacial (LIG) period, about 125,000 years ago, sea level was 4–9 m (15–30 ft.) greater than today. Researchers are discovering that during interglacial periods ice melted more quickly than originally thought and as a result sea level rise probably occurred in rapid pulses, with intervals of rapid sea level rise (RSLR) jumping 10–30 m at a time (<500 years).[20]

Among climate scientists some of the most persistent and pressing questions whirl around the intensity and rates of modern ice sheet melting and the resulting effects on sea level rise. Although there is greater uncertainty with respect to Antarctic phenomena, answers to these questions are becoming clearer on Greenland. A recently published study reports that the Greenland ice sheet is

[17] Watts J. The Guardian (02.27.18). https://www.theguardian.com/environment/2018/feb/27/arctic-warming-scientists-alarmed-by-crazy-temperature-rises

[18] Ecochard K. What's causing the poles to warm faster than the rest of the Earth? NASA. https://www.nasa.gov/topics/earth/features/warmingpoles.html

[19] National Snow & Ice Data Center. Quick Facts on Ice Sheets. https://nsidc.org/cryosphere/quickfacts/icesheets.html

[20] Gornitz V. Sea Level Rise, After the Ice Melted and Today. NASA, Goddard Institute for Space Studies. https://www.giss.nasa.gov/research/briefs/gornitz_09/

melting at its fastest rate in at least 400 years and is accelerating.[21] To date the largest contributor to SLR is ocean expansion due to warming, but this has held to a relatively constant rate over the last 2 decades. The acceleration of SLR is attributed in large part to the melting of the Greenland ice sheet, which today contributes more than 25% to SLR.

3 Lessons Learned: The Earth's Climate Is a Complex System

The interconnected cascade of change taking place in the Arctic and triggering change elsewhere is a vivid illustration that the earth's climate is a complex system. Complex systems are not simply systems that are complex; rather, they are systems that display four fundamental properties[22]:

1. A complex system is made up of a *hierarchy of subsystems* that create an internal structure or network of interacting components that can span multiple scales of time and/or space;
2. A complex system exhibits *emergent behavior* that is non-linear and potentially chaotic, arising from subsystem interactions, but is not predicted by the study of a single, simple subsystem;
3. A complex system is *adaptive*, can self-organize and has the ability to evolve in surprising ways;
4. A complex system shows *high uncertainty* and its behavior is difficult to control or predict.

In fact, the earth's climate is an interconnected network of complex subsystems, the Arctic being one such complex subsystem. As such, simple cause and effect relationships do not exist and efforts to adjust the global climate system are fraught with uncertainty and risk.

The Arctic system reveals that the rate and extent of warming are unprecedented over recorded human history and in recent years the rates of change are becoming even faster, which provides little time for biology, including humans, to adapt. Moreover, the effects of warming are far-reaching and highly non-uniform in time and space. The disruption of long established cycles such as freeze/thaw patterns will likely promote changes that will then become the drivers of further change. A chain of small events can propagate to reach a tipping point beyond which it is difficult, if not impossible, to return the system to its original state, meaning it will not

---

[21] Graeter K.A. et al. (2018) Ice Core Records of West Greenland Melt and Climate Forcing. *Geophysical Research Letters,* 45:7:3164–3172. https://doi.org/10.1002/2017GL076641

[22] Guckenheimer J. & Ottino J.M. Foundations for Complex Systems Research in the Physical Sciences and Engineering. NSF Workshop Report, September 2008. http://mixing.chem-biol-eng.northwestern.edu/docs/nsf_complex_systems_FINAL.pdf

be possible to reestablish the climatic conditions to which human societies are presently adapted.

While there is little doubt that human activity is altering the control mechanisms of the global climate system, it is very difficult to understand the full range of interactions that would need to be corrected to restore the historic controls. Rapid sea level rise is a perfect example. Since the effects of climate change are emergent, they may not occur immediately and instead continue to unfold despite the cessation of GHG emissions. In other words, there are lags in the climate system. Once GHG emissions are halted, global temperatures will continue to rise for decades before they plateau and SLR is expected to continue for centuries.[23] Even if global temperatures are stabilized at or below the 2 °C warming threshold of the Paris Climate Agreement, the ice sheets of Greenland and Antarctica may already be destabilized to the point where it is no longer possible to stop or reverse their melting. Thus, under current conditions, we are already committed to significant levels of SLR for future generations and recent modeling predicts that the rate and extent of SLR depend on when peak emissions are reached and the subsequent rates at which emissions fall to zero.[24] Thus, there is a penalty for delayed action (the longer we take to reach peak emissions, the higher seas will rise), even if we eventually arrive at the same targets.

There is a great deal of uncertainty around SLR, however, due to our incomplete understanding of how the polar ice sheets are affected by amplified warming. It has taken decades of observations to determine that recent Greenland ice loss drove the acceleration of SLR over the last 10 years. But newly published findings are altering the picture once again. Since 1992 annual ice loss from the Antarctic Peninsula has more than doubled and it has more than tripled in West Antarctica, although most of the increase in ice loss has happened in the last 5 years.[25] New discoveries of rapid ice loss on Greenland and Antarctic indicate that model estimates of future SLR are likely too low and must be revised to include the fact that Antarctic phenomena are becoming more significant.[26]

Long term monitoring and data acquisition are required in order to identify these changes and understand what they mean for the future. The challenge is to distinguish between natural variation in the data and actual changes, and then, not only determine what factors are driving the change, but also formulate models that can quantify interactions and map out various future scenarios with some precision. Modeling complex systems is a daunting task and a research field in its own right.

---

[23] Harvey C. Waters on track to rise for centuries, even if emissions stop. ClimateWire (02.21.18). https://www.eenews.net/climatewire/2018/02/21/stories/1060074341

[24] Mengel M. et al. (2018). Committed sea-level rise under the Paris Agreement and the legacy of delayed mitigation action. *Nature Communications*, 9:601. https://www.nature.com/articles/s41467-018-02985-8

[25] The IMBIE team (2018). Mass balance of the Antarctic Ice Sheet from 1992 to 2017. *Nature*, 558:219–222. https://www.nature.com/articles/s41586-018-0179-y

[26] Harvey C. Ice is melting 3 times as fast as it did 25 years ago. ClimateWire (06.14.18). https://www.eenews.net/climatewire/2018/06/14/stories/1060084477

Mathematical models are the way we test our understanding of how systems worked historically and if they are still working in similar ways today. They are a lens through which we view the past, present and future. Modeling is a continuous process that requires data and with data from both the deep and recent past, models are refined, reformulated and validated. The accuracy with which models can predict the future is a function of how much data we have, how good the data are, if the correct data have been gathered, and if we have included all the critical relationships that control the processes we are trying to model. Finally, the major difference between past and modern climates is that in the past natural processes controlled events whereas present day climate is driven by anthropogenic phenomena.

While climate scientists want to understand the upper and lower bounds of trends over the next hundreds of years, city managers, public policy makers, and regional infrastructure designers want to know what is likely to happen over the next 10–50 years with high confidence in order to prepare for growth and protect human life and investments. The polar regions influence the weather, the hydrologic cycle and extreme events of mid-latitude regions (and vice versa). As discussed, warming and melting at the poles are driving rapid sea level rise on our coasts. But overlaid on these global phenomena are local factors that make places such as New York City and Miami hot spots for rapid SLR and particularly vulnerable to extreme storm events such as hurricanes.[27] Negotiators, planners and engineers want to know what are the tools and strategies available to slow or possibly even reverse global climate change, make regions resilient and adaptive to extreme events, and protect communities in the face of a future that is rapidly evolving in ways greatly different from the past.

4 Climate Intervention Toolbox: The Basics

There are virtually an infinite number of strategies that can be employed to combat climate change either directly by reducing drivers or indirectly by buffering effects. The tools of climate intervention can be divided into three broad categories: **mitigation, geoengineering, and adaptation**. As detailed above, given the rate and extent at which climate change is unfolding and the resistance to take decisive and targeted actions at any organizational level, future efforts to slow, stabilize or reverse climate change must involve deployment of all possible and practical methods from each category of intervention. The technical and political feasibility of climate action at a global scale rests on deep knowledge of how the climate system can be adjusted by these interdependent and synergistic tools.

Mitigation Anthropogenic climate change is driven by the patterns of modern human resource and energy use. Almost every aspect of the human enterprise adds more carbon to the earth system, in the form of CO_2 or other GHG, than it removes

[27] Goodell J. **The Water Will Come** (Little, Brown & Co. 2017 NY, NY) p. 149 & 234.

or can be absorbed by natural processes. This is due primarily to the fact that fossil fuels constitute 85% of the global primary energy consumed as shown in Fig. 2 and since the industrial revolution fossil fuel combustion and cement manufacturing have released over 400 billion metric tonnes or gigatonnes (Gt) of CO_2 to the atmosphere, half of which has been emitted since the late 1980s.[28] Although the proportions of fossil fuel sources have shifted, the general, global picture depicted in Fig. 2 has remained unchanged for almost 50 years. The total amount of energy consumed, however, has more than doubled over the last half century and in 2017 stood at over 14,000 million tons of oil equivalent or approximately 555 quadrillion BTU, which translates to a historic high 32.5 Gt of energy-related CO_2 emitted.[29] In 2017, after a three-year respite in growth, total global CO_2 emissions from fossil fuels, industry and changes in land use such as deforestation rose by 2% to approximately 41 Gt.[30]

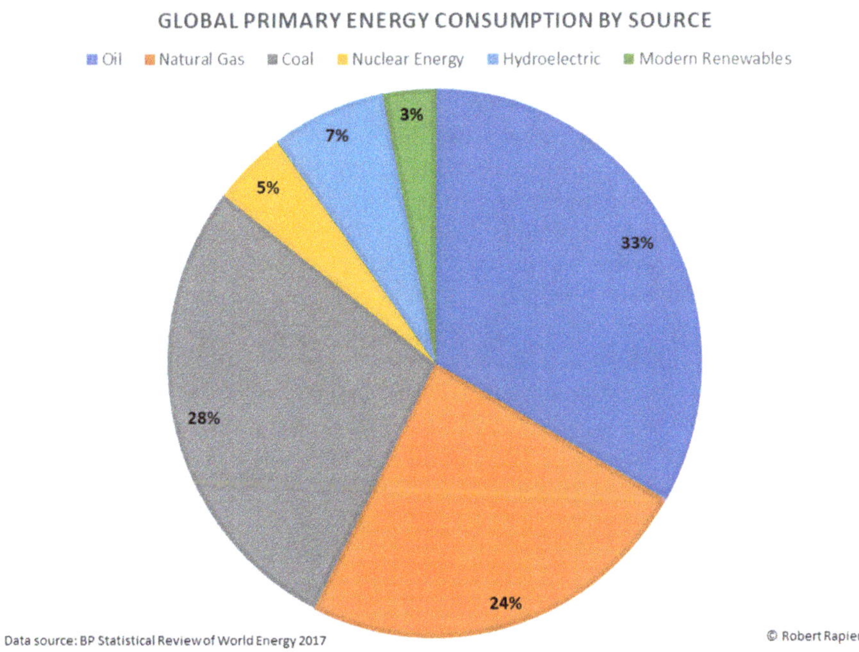

Fig. 2 Fossil fuels comprise 85% of global primary energy consumption. (http://www.rrapier.com/2017/07/renewable-gains-offset-coals-decline-in-2016/)

[28] Carbon Dioxide Information Analysis Center. Global Fossil-Fuel CO_2 Emissions. http://cdiac.ess-dive.lbl.gov/trends/emis/tre_glob_2014.html

[29] International Energy Agency. Global Energy & CO_2 Status Report 2017 (OECD/IEA, March, 2018). https://www.iea.org/publications/freepublications/publication/GECO2017.pdf

[30] Welch C. Carbon Emissions had Leveled Off. Now They're Rising Again. *National Geographic* (11.13.17). https://news.nationalgeographic.com/2017/11/climate-change-carbon-emissions-rising-environment/

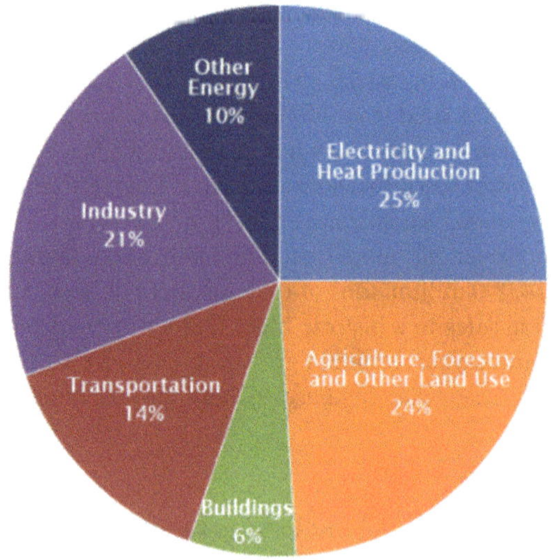

Fig. 3 Global greenhouse gas emissions by economic sector. (https://www.epa.gov/ghgemissions/global-greenhouse-gas-emissions-data)

This energy and emissions picture underpins every sector of the global and national economy: food production, materials for products and structures, transportation of people and goods, communication, heating and cooling etc. Figure 3 shows global GHG emissions by economic sector based on 2010 data to underscore the fact that climate mitigation requires sweeping GHG reductions in all areas of the global economy, as well as a transformative shift from fossil fuel dependency.

The basic strategies of climate mitigation are *resource conservation, energy efficiency, carbon capture and storage (CCS),* and *deep decarbonization.* The low hanging fruit of mitigation efforts involves reducing resource use and improving the efficiency of resource utilization, which in theory should result in reduced resource demand. The production and consumption of goods generate massive waste streams.[31] Final products typically contain only a small fraction, about 5–7%, of the raw materials consumed in their manufacture; 99% of the original materials of most products are waste within 6 weeks of sale; 80% of most products are discarded after a single use. The efficacy of resource recovery and reuse, however, continues to face major economic hurdles because of the low costs of virgin materials, limited investment in recovery supply chains and technologies, costs of acquisition and processing, closing of foreign markets, and lack of regulatory push. Despite the high potential for some materials such as paper, glass, and metals, in the U.S. less than 35% of used materials are captured and recycled. This statistic varies widely, however, depending on location; in New York City, an early pioneer in resource recovery, recycling has fallen to 15% due to high costs, but in San Francisco, guided by a goal of zero waste by 2020, material recycling exceeds 77% .[32] The widespread

---

[31] McDonough W., Braungart M. **Cradle to Cradle**. (Farrar, Straus & Giroux, NY, NY, 2002)

[32] Scheer R. & Moss D. After 40 Years, Has Recycling Lived Up to Its Billing? EarthTalk, Scientific American. *https://www.scientificamerican.com/article/has-recycling-lived-up-to-its-promises/*

adoption of resource recovery and hence, resource conservation requires major changes in material and product design criteria to include, for instance, non-hazardous material selection and the ease of disassembly.[33] Furthermore, clear targets, incentives and/or penalties must be established to create circular economies underpinned by resource recovery, recycle, and reuse activities. Innovative EU policies are leading the way on this and rules governing electronic waste recycling are a good example of such strategies.

A defining characteristic of human resource and energy cycles is their massive inefficiencies. For instance, in transportation which depends on oil and combustion technologies, 80% of the energy is lost to heat and over 60% energy loss is typical for electricity generation. Simple improvements in energy efficiency in the domestic and industrial sectors could reduce emissions 40–50% by 2030.[34] Over the last decade repeated attempts have been made to improve fuel efficiency standards for automobiles from under 25 mpg to over 50 mpg, which has long been technologically feasible and would save approximately 2 billion barrels of oil and about 600 million metric tonnes of CO_2 emissions annually. Since such efficiency requirements necessitate a shift to smaller vehicles, at current fuel prices there is resistance from both consumers and auto makers. Energy efficiency improvements, however, have been easier to achieve with respect to appliances, lighting, and building operations where improved performance and savings are more compelling to consumers and not met with unacceptable trade-offs.

In general, as economies mature they become more efficient which is reflected in a decline of the total energy consumed per unit of GDP or the energy intensity of the economy. Since 1990 the energy intensity of the global economy has dropped an average of 1.6% per year for a total improvement of 32% and the EU, with clear commitments to energy-efficiency targets, exhibits the lowest energy intensity in the world.[35] Due to recent policy reversals, however, the U.S. regressed with respect to energy efficiency and fell from 8th to 10th place in the 2018 efficiency ranking among the 25 largest energy-consuming countries in the world representing nearly 80% of global energy consumed and global GDP.[36] Increasing efficiency, however, has not reduced total resource use due to a myriad of factors such as increased population and expanding consumption habits associated with increased prosperity worldwide. In fact, in 2018 the annual planetary budget of resources and services was spent by August 1st, whereas in 1970 it wasn't expended until the very end of

---

[33] Anastas P.T. & Zimmerman J.B. (2003). Design through the Twelve Principles of Green Engineering. *Environmental Science & Technology*, 37:5:94A-101A.

[34] http://science.sciencemag.org/content/355/6331/1269.full

[35] Global Energy Statistical Yearbook 2018. Energy intensity – slowdown in energy intensity improvement in 2017. https://yearbook.enerdata.net/total-energy/world-energy-intensity-gdp-data.html

[36] Griffin R. The U.S. is Losing Ground in the Race for Energy Efficiency. Bloomberg (06.26.18). https://www.bloomberg.com/news/articles/2018-06-26/the-u-s-is-losing-ground-in-the-race-for-energy-efficiency

December.[37] In the U.S, this "overshoot" day was March 15, 2018 indicating that it would take 5 planets to sustain U.S consumption patterns at a global level.

In the 2017 International Energy Outlook, the Energy Information Agency (EIA) of the U.S. Department of Energy (DOE) projects that global energy consumption will increase by 28% from 2015 to 2040, driven mostly by strong growth in China, India and other non-OECD countries.[38] As illustrated in Fig. 4, this long term projection shows that 75% of global energy consumption will be met by fossil fuels, although there is dampened demand for coal and robust growth in renewables. Concomitant to the projected increase in energy consumption is a 16% rise in energy-related CO_2 emissions to approximately 40 Gt/y, mostly accounted for by a 25% increase in non-OECD emissions to 27 GT and relatively unchanged OECD emissions stabilized at 12 Gt.[39] Yet, in order to keep the planet from warming beyond a global average of 2 °C relative to pre-industrial levels, signatories of the Paris Agreement have pledged very large reductions in CO_2 emissions, which are inconsistent with these long term projections. It is important to note that although continued dominance and growth in fossil fuel consumption is projected to 2040, the

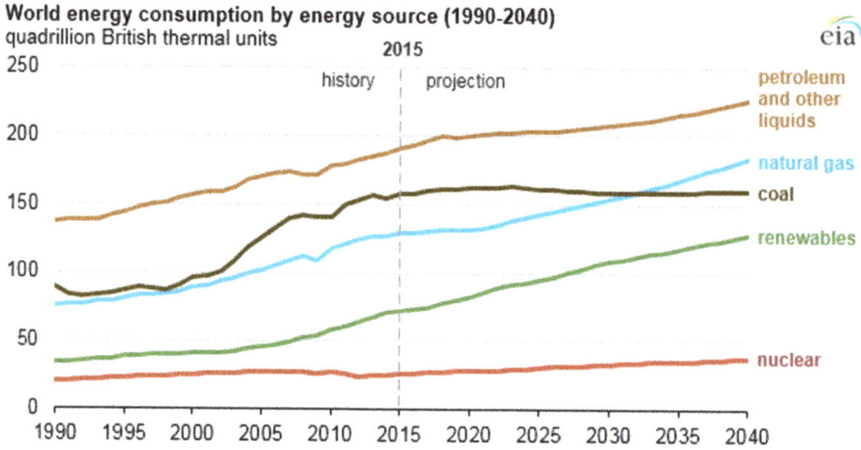

Fig. 4 Historic and projected growth in global energy consumption, 1990–2040.[47] (https://www.eia.gov/todayinenergy/detail.php?id=32912)

[37] McCamy L. On August 1, we'll have consumed more resources than the Earth can regenerate n a year – here's how you can reduce your ecological footprint. Business Insider (07.31.18). https://www.businessinsider.com/earth-overshoot-day-is-august-1-2018-7 Rockström J. et al. (2017). A roadmap for rapid decarbonization. *Science*, 355:6331:1269–1271.

[38] U.S. Energy Information Administration. Today in Energy. EIA projects 28% increase in world energy use by 2040. (09.14.17). https://www.eia.gov/todayinenergy/detail.php?id=32912

[39] U.S. Energy Information Administration. International Energy Outlook 2017. Executive Summary. https://www.eia.gov/outlooks/ieo/exec_summ.php

estimated percent increase has declined in recent years. For instance, in 2012 the EIA forecasted energy consumption to grow by 40% by 2040.

Carbon capture and storage (CCS) has emerged as the technology to rectify these two realities of continued global growth in fossil fuel use and the reciprocal urgency to reduce emissions to zero within this century in order to limit warming. CCS technology captures CO_2 emissions at large stationary sources, such as coal fired electric power generation plants, compresses them to a supercritical fluid, and then injects them into subsurface geological formations as a long term repository.

There are two operating CCS power plants in the world: Petra Nova located near Houston, Texas and the Boundary Dam plant in Saskatchewan, Canada. The Petra Nova plant began operation in January 2017 and captures about 33% of the total emissions from *one* of the four coal-fired units at the WA Parish Generating Station (there are also 4 natural gas units), which is the second largest conventional power station (3700 MW total capacity, 2700 MW from coal) in the U.S. Although the Petra Nova CCS can capture about 90% of the CO_2 delivered to it, it only separates CO_2 from a small side stream of flue gas at a significant energy cost. A 75 MW gas fired plant was constructed to power the CCS system, which exerts a 45 MW parasitic energy load.[40] Furthermore, since the CO_2 captured in this system is then employed for enhanced oil recovery in nearby oil fields, it is not entirely clear how much net CO_2 is actually sequestered. The Boundary Dam plant, which became operational in 2014, captures and compresses only about 18% of the total plant emissions, consuming approximately 25% of the plants power to do so.[41] Here too, most of the compressed CO_2 is used for enhanced oil recovery, although any residual CO_2 is sequestered in dedicated geological storage.[42]

Completed on time and budget, the total cost of the Petra Nova CCS retrofit was about $1B. In contrast, another plant in Kemper, Mississippi was designed to capture about 65% of the plant's total emissions. Due to overruns associated with construction costs in excess of $7.5B, this project was cancelled.[43] A number of "clean coal" projects that use Integrated Gasification Combined Cycle (IGCC) technology to produce electricity and H_2 and considered "CCS ready" have been proposed, but few are brought on-line. For instance, DOE's FutureGen demonstration project, announced in 2003, would have been the most comprehensive demonstration project focused on CCS feasibility, but was canceled in 2015 due to time and budget overruns.[44] At the present time, the capital costs of CCS are anywhere from 3–8 times the capital costs of conventional fossil fuel power plants.

[40] Holmes à Court S. Could Petra Nova, The Leading CCS Power Station, Provide A Model for Australia. Clean Technica (06.12.17). https://cleantechnica.com/2017/06/12/petra-nova-leading-ccs-power-station-provide-model-australia/

[41] Ibid.

[42] Global CCS Institute. Projects Database. Boundary Dam Carbon Capture and Storage. https://www.globalccsinstitute.com/projects/boundary-dam-carbon-capture-and-storage-project

[43] U.S. EIA. Today in Energy (10.31.17). https://www.eia.gov/todayinenergy/detail.php?id=33552

[44] Daniels S. FutureGen 'clean-coal' plant is dead. Crain's Chicago Business (02.03.15). http://www.chicagobusiness.com/article/20150203/NEWS11/150209921/futuregen-clean-coal-plant-in-illinois-is-killed-by-obama-administration

In summary, then, CCS technology is technically feasible, but under current market conditions of low natural gas and petroleum prices the economics are not. When CCS is coupled to enhanced oil recovery, carbon sequestration is not achieved to a sufficient degree. In general, switching from coal to natural gas reduces CO_2 emissions by 50% and is the direction of current market trends.

Numerous roadmaps have been devised that chart feasible pathways to slash GHG emissions and decarbonize the global economy. In 2006 Socolow and Pacala introduced the "wedge concept" to halt growth in carbon emissions without constraining economic growth and to stabilize emissions at 2006 levels using existing technologies and deliberate carbon policy.[45] It is instructive to reflect on the success of this eminently practical proposal which identified 15 technologies or wedges that, when phased in over 50 years, would each prevent the release of 25 billion tons of carbon. The measures largely relied on increased efficiency, resource conservation, halting deforestation, displacing coal with renewable or nuclear energy, and employing other carbon-neutral fuel switching methods. Three of their wedges depend on the installation of CCS at large coal-fired plants, which has yet to show much traction as a carbon strategy. Totaling an area greater than that of Poland, 2016 and 2017 witnessed the worst years in recorded history for loss of forest cover.[46] Persistent deforestation and forest damage contribute about 10% to global GHG emissions and have converted tropical forests to a net carbon source to the atmosphere. Intrinsic to Socolow and Pacala's plan was the pricing of carbon(C) at \$100–200/ton (\$27–54/ton CO_2) and other enforceable policies. Over the last decade and in some countries, diminished C-intensity has been realized largely as a result of natural market forces, not due to deliberate policy and targeted intent as Socolow and Pacala envisioned.

Our understanding of the earth system has evolved a great deal over the last decade and it is now apparent that the pace of change espoused by Socolow and Pacala is far too slow and incremental. In the absence of marked improvements in the business-as-usual (BAU) trajectory, *deep* decarbonization has emerged as the only mitigation strategy that can achieve the drastic cuts in the 41 Gt of annual anthropogenic GHG emissions to set the course to net-zero emissions by 2050. The Deep Decarbonization Pathways Project (DDPP) is an internationally coordinated, goal-oriented, long-term approach to *operationalizing* the 2 °C warming limit and involves 16 countries which are responsible for 74% of global energy-related CO_2 emissions.[47]

---

[45] Socolow R.H. & Pacala S.W. (2006) A Plan to Keep Carbon in Check. *Scientific American,* 305:968–972.

[46] Harvey C. Forests had a really bad year. ClimateWire (06.28.18). https://www.eenews.net/climatewire/2018/06/28/stories/1060087181

[47] Deep Decarbonization Pathways Project, http://deepdecarbonization.org

As detailed by the DDPP, steep declines in C-intensity across all sectors require more than anything else, a profound transformation of energy systems.[48] Yet, DDPP relies on the same three pillars to transform them that we have already discussed: energy conservation and efficiency, decarbonizing electricity and fuels, and switching end-uses to low-C supplies. Aviation, freight and industrial sectors such as steel and cement face the most difficult challenges in reducing their C-intensity by 2050.[49] For example, decarbonization of the cement industry, which produces the most widely used construction material in the world and accounts for about 8% of all global carbon emission, is a formidable task not only due to its high fuel demands, but also due to the chemical processing of limestone, which releases large amounts of CO_2 as a byproduct. Diminishing the C-intensity of cement involves developing alternative fuels, incorporating CCS technology, and as importantly, discovering new formulations of cement and new ways to manufacture it – in other words, virtually a complete transformation of the industry.[50] Yet, over its life cycle concrete (cement + water + aggregate) naturally absorbs CO_2, potentially more than 40% of it original emissions, but on a slow time scale.[51] If the rate and extent of CO_2 absorption or cement carbonation could be enhanced and emissions captured and sequestered, cement manufacturing could one day be carbon negative.

Despite the intrinsic difficulties, the DDPP delineated various technically feasible pathways to reduce carbon intensity by almost 90% and energy-related CO_2 by 50% allowing for projected population growth and a 3% annual increase in GDP. To meet these ambitious goals there are a number of key technologies that are commercialized but not yet mature or operating at large scales. These are advanced energy storage, flexible load management, very high performing appliances, controls and low-C materials for buildings, zero-emission vehicles and sustainable liquid biofuels for air or marine transportation.[52] Nevertheless, the DDPP projects that the transition to a low-C economy by 2050 will require C-free electricity produced by a mix of renewable sources and CCS installed with any remaining fossil fuel plant, the replacement of petroleum by biofuels or H_2, and the complete phase out of coal. Significant investments are necessary, but the DDPP predicts that current investment can be displaced and redirected from traditional high-C to low-C portfolios and projects.

---

[48] Deep Decarbonization Pathways Project (2015). *Pathways to deep decarbonization 2015 report – executive summary,* SDSN – IDDRI. http://deepdecarbonization.org/wp-content/uploads/2015/12/DDPP_EXESUM-1.pdf

[49] Davis S. et al. (2018). Net-zero emissions energy systems. *Science,* 360:6396; DOI: https://doi.org/10.1126/science.aas9793
http://science.sciencemag.org/content/360/6396/eaas9793.full

[50] Harvey C. Cement's CO_2 is everywhere. Will it sink climate goals? ClimateWire (07.09.18). https://www.eenews.net/climatewire/stories/1060088153

[51] Xi F. et al. (2016). Substantial global carbon uptake by cement carbonation. *Nature Geoscience,* 9:880–883 https://www.nature.com/articles/ngeo2840

[52] Sachs J. et al. (2014) *Pathways to deep decarbonization, 2014 report,* Deep Decarbonization Pathways Project, SDSN – IDDRI. http://deepdecarbonization.org/wp-content/uploads/2015/06/DDPP_Digit.pdf

Finally, the aggressive decarbonization scenarios mapped out by the DDPP are believed to reduce energy poverty, to enhance energy access and to be consistent with individual countries' socio-economic and environmental goals. Neither individual countries nor market forces alone, however, can achieve the necessary decarbonized targets. Directed technological change is needed that requires organized, sustained and funded collaboration between governments, academics and businesses at a global scale. This coordination does not aim to pick winners, but rather to have a sufficient pool of contestants from which winners will emerge.

In principle, while the deep decarbonization strategies outlined above are technically and economically feasible, they are still not proving to be deep or rapid enough to prevent catastrophic warming in this century. As illustrated in Fig. 5 and at this point in time (2018), the decarbonization pathway that can achieve the Paris Agreement goals of limiting warming to less than 2 °C dictates that human carbon emissions from fossil fuels, industry and land use must peak at 2020 and reach net-zero by 2050, meaning carbon emissions have to be balanced by carbon uptake. In an analogy to Moore's Law, a general axiom about the pace of innovation, *rapid* decarbonization could be accomplished by invoking a **"carbon law"** that requires halving gross anthropogenic CO_2 emissions every decade.[53] In part, this could be achieved by implementing non-linear renewable energy growth trajectories based on global 2005–2015 trends and maintaining historical doubling times of 5.5 years constant over the next 30 years, resulting in the end of coal by 2035, the end of oil by 2045 and full decarbonization by 2050, if not before. Intrinsic to these projections is the cessation of fossil fuel subsidies, currently $500–600 billion annually, by 2020 rather than 2025 as agreed upon by the G7 nations.

Also integral to the success of this proposed carbon law is enhancing C-storage through natural C-uptake on land and in the oceans, as well as through human induced carbon sinks facilitated by halting land use emissions and engineering carbon removal from the atmosphere. Figure 5 illustrates that beyond the 2050 net-zero point, anthropogenic CO_2 removal can bring cumulative CO_2 emissions down from 700 Gt to less than 200 Gt by 2100, returning atmospheric CO_2 levels to

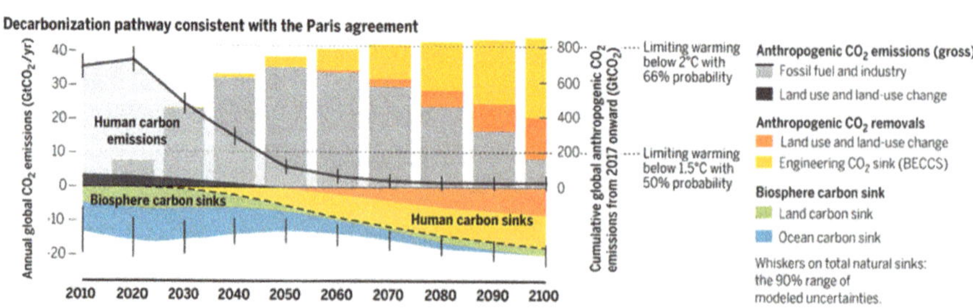

Fig. 5 A global carbon law and roadmap to make Paris goals a reality.[53] (http://science.sciencemag.org/content/355/6331/1269/tab-figures-data)

[53] Rockström J. et al. (2017). *Science*, 355:6331:1269–1271. http://science.sciencemag.org/content/355/6331/1269.full

approximately 380 ppm. A carbon law mandates transformation motivated by innovation rather than resisted by inertia. It directs non-linear, disruptive technological advances not simply toward a zero emissions world but also to a point of synergy between natural and anthropogenic C-uptake bringing GHG levels gradually back to levels to which present day biota on earth are adapted.

Geoengineering In 2015 as part of the Paris Agreement under the United Nations Framework Convention on Climate Change (UNFCCC), 197 countries pledged to take action to limit warming of the earth to less than 2 °C relative to pre-industrial levels and to make every effort to cap warming to 1.5 °C. The 1.5–2 °C limit of increase in average global temperature reflects a belief that it marks a range of manageable risk, within which humans will be able to adapt to climate changes. Yet, in 2018 there are alarming gaps between the targets set by climate scientists and the commitments set by policy makers in each country.[54] While some of the numbers are relatively straightforward – we know the current atmospheric level of CO_2 relative to pre-industrial levels, have a good handle on global emissions, and have agreed internationally on a "safe limit" for resultant warming, uncertainty persists about the sensitivity of the earth system to CO_2 levels. For instance, models estimate that warming can be kept to less than 2 ° C if atmospheric concentrations do not exceed 430–480 ppm CO_2 equivalents.[55] Current emissions, however, put the world on course to reach global mean temperatures 3.7 °C to 4.8 °C higher compared to pre-industrial levels by 2100; yet, when taking full climate uncertainty into account, this range extends from 2.5 °C to 7.8 °C.[56] Furthermore, we lack a specific time table as to how and when the planet will respond to specific CO_2 levels. As detailed in the previous section, although the challenges to reversing established carbon emissions patterns are massive, there is no dearth of potential mitigation techniques or roadmaps. There is, however, a distressing lack of political will.

The window for action is rapidly closing, particularly for the more aggressive 1.5 °C goal. Current trends predict that the planet will cross the 1.5 °C warming threshold in the 2040s and no model under any scenario gives more than a 66% probability of holding warming to 1.5 °C.[57] The 2018 IPCC special report, Global Warming of 1.5 °C, stresses that while meeting this goal is possible and safer for human societies, it is unlikely to be achieved unless strategies to rapidly reduce anthropogenic GHG emissions by 2030 and reach net-zero by 2050 accompany ramped up efforts to *remove* carbon dioxide from the atmosphere.[58] Thus,

[54] Irfan U. World must pull CO_2 from the sky to meet Paris goals. ClimateWire (03.24.17). https://www.eenews.net/climatewire/stories/1060052028/

[55] Smith P. et al. (2016). Biophysical and economic limits to negative CO_2 emissions. *Nature Climate Change,* 6:42–50. https://www.nature.com/articles/nclimate2870

[56] Sachs J. et al. (2014). DDPP 2014 report. http://deepdecarbonization.org/wp-content/uploads/2015/06/DDPP_Digit.pdf

[57] Hood M. 1.5 C climate goal 'very unlikely' but doable: draft UN report. Phys.org. (https://phys.org/news/2018-01-climate-goal-doable.html

[58] IPCC Special Report (2018). Global Warming of 1.5 °C. (https://www.ipcc.ch/sr15/)

geoengineering the global system is emerging center stage as a requisite component of climate action. Geoengineering, however, entails two broad classes of endeavors with entirely different aims: *carbon dioxide removal* and *albedo modification*. Carbon dioxide removal, also referred to as "negative emission technologies," involves scrubbing CO_2 from the atmosphere and storing it in a permanent repository to bring levels back quickly to a safer range well below 400 ppm (current levels are greater than 410 ppm and ascending quickly). In contrast, albedo modification or solar radiation management seeks to limit the amount of sunlight reaching the earth's surface, and hence warming, through increased reflectivity or albedo of the atmosphere by introducing aerosols or whitening clouds. In this way, albedo modification treats the symptoms of climate change by hindering the warming effects of elevated GHG, rather than directly addressing and modifying causative factors.

Carbon Dioxide Removal At only 0.04% of the total gases making up dry air, CO_2 is a small but critical component of the earth's atmosphere. After water vapor, CO_2 is the second most abundant GHG and thus, this tiny fraction plays a critical role in creating a habitable temperature for life on earth. CO_2 levels in the atmosphere are equilibrated by biogeochemical cycles that control and balance CO_2 exchange through air, water, soils and rocks. CO_2 chemistry mediates critical processes in all phases (gas, liquid, solid and biota) of the environment and entails dissolution into aquatic systems (about 30% of CO_2 emissions dissolve in the oceans) to influence mineral precipitation and pH, weathering of soils and rocks, uptake by plants to create organic carbon, and metabolization/decomposition of the organic matter back to CO_2.

Since the Industrial Revolution this intricate carbon cycle has been disrupted by the massive release of CO_2 and other GHG into the atmosphere due to fossil fuel combustion, which is then exacerbated by diminished CO_2 uptake by forests due to deforestation. Quantitative understanding of the global carbon budget and cycle indicates that 500–600 Gt will need to be removed from atmosphere (Fig. 5) and offers a number of carbon dioxide removal (CDR) strategies to reestablish CO_2 homeostasis supportive of life as we know it on the planet. There are three interrelated and general categories of CDR: *Geochemistry of soils and rocks; Photosynthesis; Direct Air Capture*.

Geochemistry of Soils and Rocks CDR methods seeking to manipulate and accelerate geochemical processes in rock (lithosphere) and soils (pedosphere) fall into two classes: *enhanced chemical weathering* and *altered agricultural practices*. When exposed to the atmosphere (O_2, CO_2, etc.) and water, rocks naturally undergo various chemical transformations that break them down and release minerals, but also form new materials. A major chemical weathering pathway is carbonation in which CO_2 and water react to dissolve exposed rocks and reform carbonate materials (the same reaction occurs in cement/concrete). For instance, the carbonation of calcium-, magnesium- or iron-rich silicates yields insoluble carbonate minerals such as magnesite ($MgCO_3$) or calcite ($CaCO_3$), a principle component of limestone. This process occurs spontaneously on geological time scales of hundreds to thousands of years.

Recently, however, progress has been made on *enhanced weathering* (EW) that could capture and store billions of tons of CO_2 per year.[59] EW is based on the carbonation of olivine, a magnesium iron silicate mineral ($(Mg^{2+}Fe^{2+})_2SiO_4$) that can be deployed either *in situ* or *ex situ* to store captured CO_2 in the subsurface or to scrub CO_2 from water or air to form carbonate rocks. Olivine makes up 60–80% of the Earth's upper mantle, which in turn comprises about a quarter of the Earth, and is present is a wide variety of rocks that undergo carbonation at varying rates depending on conditions. Basalt is a volcanic or igneous rock containing olivine and comprising about 10% of the earth's surface, mostly on the ocean floor. In Iceland the CarbFix pilot project has demonstrated that gaseous mixtures of CO_2 and H_2S associated with geothermal power production can be injected into wells passing through basaltic formations at depths 400–800 m to sequester the 95% of the CO_2 as stable carbonate minerals in under 2 years, a rate far faster than originally proposed.[60] Strictly speaking, this particular application of EW is not a CDR strategy, but rather mitigation of CO_2 emissions by CCS. Yet, there are two major advantages to this improvement on conventional CCS. First, since the costs of CCS are dominated by CO_2 capture and separation, the injection of a CO_2 mixture represents substantial cost reductions. Secondly, the CO_2 is sequestered as rock obviating the risks associated with CO_2 gas leakage, a serious concern for conventional CCS where CO_2 is contained in the void fraction of geological formations. A serious disadvantage of this EW method, however, is that large amounts of water, 25 tons of water per ton of CO_2, are required.

Other sources of olivine can improve on the basalt application and truly achieve CDR. For instance, peridotite formations in Oman which contain large proportions of olivine display high rates of carbonation, a million times greater than natural rates, and at much lower ratios of water (1.5 tonne H_20: 1 tonne C equivalent (Ceq) removed).[61] Accelerated carbonation of the peridotite was achieved at depth under increased pressures and temperature (e.g, 185 °C).[62] Once the reaction was initiated, the elevated temperature could be self-sustained due to its exothermic properties. *In situ* carbonation experiments in Oman alone demonstrated the potential to sequester more than one billion tons of CO_2 per year. Based on these results, the researchers propose a continuous CDR process that would involve drilling and fracturing rock formations in shallow offshore waters and then pumping sea water saturated with CO_2 to a depth where the water reaches 185 ° C and the CO_2 precipitates out of

[59] Kelemen P.B. Lamont-Doherty Earth Observatory. Carbon Sequestration. Mineral carbonation in peridotite for CO_2 capture and storage (CCS). Earth Institute, Columbia University. https://www.ldeo.columbia.edu/gpg/projects/carbon-sequestration

[60] Rockström J. et al. (2017). *Science*, 355:6331:1269–1271. http://science.sciencemag.org/content/355/6331/1269.full

[61] Carrington D. CO_2 turned into stone in Iceland in climate change breakthrough. The Guardian (06.09.16). https://www.theguardian.com/environment/2016/jun/09/co2-turned-into-stone-in-iceland-in-climate-change-breakthrough

[62] Kelemen P.B. & Matter J. (2008). In situ carbonation of peridotite for CO_2 storage. *Science,* 105:45:17295–17300. http://www.pnas.org/content/105/45/17295

solution. The water would return to the surface via fractures to a second drilled hole.[63] The key factor, then, to attaining practical EW with these materials is elevated temperature and/or pressure.

Dunite, a member of the peridotite group of mantle derived rocks, is about 90% olivine and is considered a good candidate for engineering *ex situ* mineral carbonation which involves mining and grinding the materials to fine powder. There are proposals to simply disperse the powders on croplands where potential CDR could be as great as 95 Gt CO_2 per acre with a co-benefit of fertilizing the fields by mineral and nutrient release.[64] There is skepticism concerning the overall efficacy of *ex situ* application due to slower kinetics at ambient temperature and pressure conditions and it is likely that the technology is limited to use in warm and humid regions of India, Brazil, Southeast Asia and China. Alternatively, fine dunite powders may be activated by heat-treatment, held at elevated pressure and temperature or deployed in engineered carbonation reactors under high pressure and temperature. In general, due to transportation, energy, and high capital costs, techno-economic assessments (TEAs) show that *ex situ* EW is 2–10 times more expensive per ton of CO_2 removed than CDR strategies based on enhanced photosynthetic CO_2 uptake discussed below.[65]

Agricultural practices since the start of the Industrial Revolution have grossly eroded and degraded soils releasing an estimated 135 Gt of CO_2 into the atmosphere, which at current rates is equivalent to a decade of all anthropogenic emissions.[66] *Carbon farming* is a CDR strategy that reverses the carbon depleting effects of intensive agriculture to employ land management practices that convert the soil into a carbon sponge. C-farming techniques include agroforestry, which involves growing trees and crops together, and no-till agriculture, which, by eliminating plowing, retards soil erosion and preserves beneficial soil ecosystems including worms, fungi, roots, and their associated microbial communities. In addition, other techniques include maintaining plant coverage in fields and using certain soil amendments to enhance net primary productivity. Internationally, the largest carbon farming program, 4 per 1000 Initiative, was conceived by the French government and aims to increase carbon storage in crop and rangelands soils by 0.4% annually, albeit voluntarily. Overall, it is estimated that soils could store about 5–15% of

---

[63] Bullis K. Carbon-Capturing Rock. Geologists discover that certain rock formations could sequester large amounts of carbon dioxide. MIT Technology Review (11.04.2008) https://www.technologyreview.com/s/411129/carbon-capturing-rock/

[64] Strefler J. et al. (2018). Potential and costs of carbon dioxide removal by enhanced weathering of rocks. *Environmental Research Letters,* 13:034010 http://iopscience.iop.org/article/10.1088/1748-9326/aaa9c4/meta

[65] Ibid.

[66] Velasquez-Manoff M. Can Dirt Save the Earth. The New York Times Magazine (04.18.18). https://www.nytimes.com/2018/04/18/magazine/dirt-save-earth-carbon-farming-climate-change.html

current annual CO_2 emissions.[67] Field experiments on California rangelands demonstrated that the use of compost combined with manure enhanced grass growth and carbon storage in soils.[68] Extrapolating these results to 5% of California's 56 million acres of rangeland could offset 80% of all of the states' agriculturally associated GHG emissions.

Biochar is another soil amendment that creates a C-sink while simultaneously enhancing soil fertility and producing energy.[69] The pyrolysis (treatment at 400–500 °C in the absence of oxygen) of biomass (plant debris and animal wastes) produces biochar, a form of charcoal that is rich in carbon and extremely stable, as well as bio-oil and syngas ($CO + H_2$), both of which are fuels that can replace fossil fuels. Overall, this biomass conversion process is both energy positive, producing 3–9 times the amount of energy required for production, and C-negative.[70] Beyond its CDR potential, as an amendment biochar improves soil quality by retaining and making nutrients and water available to plants. It can adsorb pollutants and prevent their release into surface waters. There have also been reports that it reduces the emissions of N_2O and methane from soils through unknown interactions with soil microbiota.[71] The CDR properties of biochar, however, vary depending on the residues and pyrolysis conditions used to produce it, as well as the field conditions of application. Due to its dark color, however, biochar may diminish the albedo and increase the absorbance of solar radiation of soils.

Integrated assessment modeling estimates that C-farming and biochar combined could remove approximately 1.4 Gt Ceq per year by 2100 with potentially much lower biophysical impacts on land, water use, nutrients, albedo, energy requirement and cost than other CDR techniques.[72] There are questions, however, about the permanence of the carbon sequestered in soils and also if there is a saturation limit beyond which soils cannot show net C-uptake.

Photosynthesis As illustrated in Fig. 3, agriculture, forestry and other land uses contribute about one quarter of current CO_2 emissions and deep decarbonization

---

[67] Erickson B.E. (2016) Regenerating degraded dirt. Efforts to boost soil carbon aim to improve crop yields and combat climate change. *Chemical & Engineering News,* 94:10:40–44. https://cen.acs.org/articles/94/i10/Regenerating-degraded-dirt.html

[68] DeLonge M.S. et al. (2013). A Lifecycle Model to Evaluate Carbon Sequestration Potential and Greenhouse Gas Dynamics of Managed Grasslands. *Ecosystems,* 16:6:962–979. https://link.springer.com/article/10.1007/s10021-013-9660-5

[69] Erickson B.E. (2016). Interest in biochar surges. *Chemical & Engineering News,* 94:10:40–44. https://cen.acs.org/articles/94/i10/Interest-biochar-surges.html

[70] Gaunt J.L. & J. Lehmann (2008). Energy Balance and Emissions Associated with Biochar Sequestration and pyrolysis Bioenergy Production. *Environmental Science & Technology,* 42:4152–4158. https://pubs.acs.org/doi/pdf/10.1021/es071361i

[71] Lehmann J. (2007). Bio-energy in the black. *Frontiers in Ecology and the Environmental,* 5:7:381–387. https://esajournals.onlinelibrary.wiley.com/doi/full/10.1890/1540-9295%282007%295%5B381%3ABITB%5D2.0.CO%3B2

[72] Smith P. (2016). Soil carbon sequestration and biochar as negative emission technologies. *Global Change Biology,* 22:1315–1324; doi: https://doi.org/10.1111/gcb.13178. https://onlinelibrary.wiley.com/doi/pdf/10.1111/gcb.13178

mitigation efforts require radical changes to these sectors. Taking another cue from nature, the simple act of planting more trees is an obvious CDR strategy, but there are significant biophysical constraints for large-scale deployment. Afforestation is the establishment of forests on new land and reforestation is the restoration of damaged or destroyed forests. Afforestation/reforestation (AR) activities are estimated to have the potential to remove between 1.1–3.3 Gt Ceq/y in 2100 with a land requirement of 320–970 Mha,[73] which represents a large fraction of arable and cultivated land, 2–4 times greater than current stores of abandoned or marginal land. Thus, at a scale necessary for CDR, AR would likely compete with food and bioenergy production, as well as other ecosystem services. Although AR has a very low energy demand and is relatively inexpensive, it would exert a high water demand and may have negative nutrient and albedo impacts.

Two major factors affect forests' C-storing potential: their age and the influence of environmental factors like temperature and precipitation.[74] Hence, as forests age, their growth slows and eventually plateaus reaching growth saturation. On-the-ground validation of the model estimates described above has recently revealed that the forests in North America, restored after eighteenth and nineteenth century clear cutting, have already reached 78% of their C-sequestration capacity.[75] For North American forests, then, future growth over the next 60 years will likely display growth saturation with only 22% of their C-storage capacity remaining in the best case. Increased temperatures and changed precipitation patterns may also diminish growth and C-storage capacity in the future. In general, it will be difficult to expand large scale AR in North America or Europe and AR efforts should shift to the tropics where deforestation is still widespread.

Tropical forests, mangroves and peatlands offer a number of valuable opportunities to counter climate change that are distinct from the general, global averaged discussion of AR above. From the perspective of carbon effects, the conservation, restoration and improved management of these tropical systems could achieve 23% of cost-effective mitigation action by 2030 to limit warming to 2 °C, which is two thirds of C-uptake potential of forests and other land sector solutions.[76] A more significant non-carbon climate impact of tropical AR, however, is its influence on the water cycle. The evapotranspiration of large forest cover pumps water into the atmosphere creating clouds and rainfall hundreds to thousands of miles away. Large scale deforestation in the Congo Basin, Southeast Asia and the Amazon disrupts the

---

[73] Smith P. et al. (2015). *Nature Climate Change,* 6:42–50. https://www.nature.com/articles/nclimate2870.pdf

[74] Harvey C. Trees are losing their ability to soak up CO_2. ClimateWire (07.13.18). https://www.eenews.net/climatewire/2018/07/13/stories/1060088955

[75] Zhu K. et al. (2018). Limits to growth of forest biomass carbon sink under climate change. *Nature Communications,* 9:2709. https://www.nature.com/articles/s41467-018-05132-5

[76] Wolosin M. & Harris N. (2018) Tropical Forests and Climate Change: The Latest Science. Working Paper June 2018. World Resources Institute. https://wriorg.s3.amazonaws.com/s3fs-public/ending-tropical-deforestation-tropical-forests-climate-change.pdf

water cycle of critical agricultural areas in the U.S., China and India.[77] AR and protection of tropical forests produces cascading effects on climate by reducing local temperatures due to shading and evapotranspiration, enhancing water availability locally, regionally, and also at distance, and hence, stimulating photosynthesis and agriculture regionally and globally.[78]

Future emissions scenarios that include CDR, such as those relied on by the IPCC, typically emphasize the large scale CO_2 removal potentially achieved by bioenergy with carbon capture and storage (BECCS). As the label implies, BECCS involves growing bioenergy crops such as perennial grasses (e.g., switchgrass or miscanthus) and trees (e.g., poplar and willow), burning them in electric generation power stations, and then capturing and storing the resultant CO_2 by CCS. Currently about 10% of global primary energy is met with bioenergy derived at small scales primarily in the form of wood and agricultural residues. A very small portion, about 3%, is produced by dedicated energy crops. There is potential for BECCS to deliver 3.3 Gt Ceq per year in CDR while generating about 160 quadrillion (10^{15}) BTU by 2100,[79] but there are an enormous number of caveats that accompany this estimate. It is likely that crops would have to be planted solely for CDR, totaling approximately 21% of the current net primary productivity utilized in agriculture or 4% of the total global potential. The land intensity demanded by this level of BECCS is immense, although it varies widely depending on the type and conditions of biomass cultivation. Some estimates calculate that 430–580 M hectares would be required, which is about one third of the total arable land globally or half the land area of the U.S.[80] Remarkable improvements in productivity are needed to enhance the genetic machinery of plants to capture CO_2, otherwise food demands may clash with the BECCS push. With one third of the earth's ice free surface already devoted to agriculture, the implementation of BECCS will potentially intensify the competition for land availability with the unintended consequences of forest and grassland conversion which would ultimately release more CO_2 than uptake. Freshwater requirements for both crop production and CCS will also be huge, about 3% of current human water use. BECCS will likely benefit from improvements in CCS offered by EW. Despite the serious biophysical issues associated with large-scale application, BECCS is typically selected as a cost-effective part of both a future energy and CDR mix.[81]

---

[77] Ibid.

[78] Pearce F. Rivers in the Sky: How Deforestation Is Affecting Global Water Cycles. Yale Environment 360 (07.24.18). https://e360.yale.edu/features/how-deforestation-affecting-global-water-cycles-climate-change

[79] Smith P. et al. (2015). *Nature Climate Change,* 6:42–50. https://www.nature.com/articles/nclimate2870.pdf

[80] Williamson P. (2016) Emissions reduction: Scrutinize CO_2 removal methods. *Nature,* 530:153–155; doi:https://doi.org/10.1038/530153a. https://www.nature.com/news/emissions-reduction-scrutinize-co2-removal-methods-1.19318

[81] Smith P. et al. (2015). *Nature Climate Change,* 6:42–50. https://www.nature.com/articles/nclimate2870.pdf

The oceans have long attracted attention as a potential photosynthetic pump of CO_2 into long term storage and various strategies have been tossed about to enhance phytoplankton growth in the coastal or open ocean by overcoming nutrient limitations. Ocean fertilization, however, has largely been dismissed as a feasible strategy to promote algal blooms and net CO_2 removal because most of the CO_2 is simply released by the decomposition of the algae rather than stored in marine sediments.[82] Other issues such as fisheries impacts and oxygen depletion have prompted organizations such as the Convention on Biological Diversity to reject enhanced ocean productivity as an acceptable CDR method.

Direct Air Capture While there is no dearth of ideas for ways to adapt natural processes to accelerated CDR, there are serious limitations to the schemes discussed thus far when the theoretical picture is considered alongside biophysical, ecological, and social constraints. Thus, the engineering of direct air capture (DAC) technologies is shifting from a fantastic to a realistic realm. In order to circumvent the limitations of land, water and growth kinetics of AR, the global deployment of 100 million "artificial" trees, such as the synthetic urban tree or Treepod concept,[83] has been proposed.[84] DAC technology is based on the high CO_2 binding affinity of certain materials (e.g. amine-modified mesoporous silicas, zeolites, metal-organic frameworks)[85] or chemical solutions (e.g., aqueous solutions of monoethanolamine or alkali/hydroxides)[86] allowing the extraction of CO_2 from the ambient air. Once saturated, the sorbents would be regenerated by desorbing and concentrating the CO_2, which would then be permanently stored. In most schemes, DAC entails a large energy and materials handling demand. Separation technologies for CO_2 in industrial or natural gas streams are well established, but they operate under much higher concentrations of CO_2 than the ultra-dilute, ambient air and their expense is recovered in the value of the product stream from which CO_2 is removed. Early analysis of DAC predicted that at best it could achieve 10–15% reduction in atmospheric CO_2 levels in 100 years and thus was not matched to CDR targets.[87] More recent reviews provide an exhaustive list of possible solvents and solid sorbents and processing schemes for sorption and desorption cycles, but as yet practi-

---

[82] Williamson P. (2016) *Nature*, 530:153–155. https://www.nature.com/news/emissions-reduction-scrutinize-co2-removal-methods-1.19318

[83] Boston Treepods 2011. Shift Boston. http://www.shiftboston.org/competitions/2011_treepods.php

[84] Biello D. How Far Can Technology Go to Stave Off Climate Change? Yale Environment 360 (01.18.17). https://e360.yale.edu/features/how_far_can_technology_go_to_stave_off_climate_change

[85] Kumar A. et al. (2015). Direct Air Capture of CO_2 by Physisorbent Materials. *Angewandte Chemie*, 54:14372–14,377. https://onlinelibrary.wiley.com/doi/full/10.1002/anie.201506952

[86] Socolow R. et al. (2011) Direct Air Capture of CO_2 with Chemicals. A Technology Assessment for the APS Panel on Public Affairs. American Physical Society (06.01.2011). https://infoscience.epfl.ch/record/200555/files/dac2011.pdf

[87] Ibid.

cal sorbent-air contacting technology are not sufficiently developed for large scale use.[88]

There are a few demonstration-scaled projects testing DAC technologies around that world. Climeworks near Zurich, Switzerland began operation in May 2017 and aims to capture 1% of global CO_2 emissions annually by 2025 by drawing ambient air through a filter made from porous granules modified with amines that bind CO_2 and moisture.[89] CO_2 free air is released and once the filter reaches saturation, it is heated to 100 °C to release the CO_2, which is then concentrated and either sold to customers or routed for storage. Although today the Climeworks DAC process costs about $600/tonne of CO_2, the goal is to reduce the costs to $100/t$CO_2$ by 2030. Carbon Engineering is a Canadian-based DAC startup in operation since 2015 to capture atmospheric CO_2 for synthesizing affordable transportation fuels with ultralow C-intensity, producing other materials, enhancing oil recovery or sequestering CO_2 permanently to achieve CDR.[90] The technology produces a concentrated CO_2 gas stream by capturing atmospheric CO_2 in a concentrated hydroxide (e.g., KOH) solution which is processed to produce $CaCO_3$ and then thermally treated to liberate the CO_2 and regenerate lime (CaO) for recycling.[91] The cost of processing a tonne of CO_2 from the atmosphere with this technology reportedly ranges from $94–$232 depending on a various design options and economic assumptions[92] and the company hopes to realize a scale that captures 1 Mt CO_2 annually.

In comparison to other CDR, DAC energy and costs requirements are very high, although technology breakthroughs and scale-up may reduce them. In contrast, the environmental footprint of DAC in terms of land, water, nutrients, or albedo is negligible in comparison to the other geochemical or photosynthetic CDR. The large-scale implementation of any CDR will be determined by highly variable biophysical, biochemical, energy, and economic factors.[93] At current technology readiness and under current economic conditions, however, none of the CDR alone or in combination could be deployed to meet the 2 °C warming target without prohibitive biophysical or economic impacts. Thus, as illustrated in Fig. 5, aggressive mitigation of GHG emissions remains the centerpiece of a Climate Action Plan A.

[88] Sanz-Perez E. et al. (2016). Direct Capture of CO_2 from Ambient Air. *Chemical Reviews,* 116:19:11840–11,876; **DOI:** https://doi.org/10.1021/acs.chemrev. https://pubs.acs.org/doi/full/10.1021/acs.chemrev.6b00173

[89] Climeworks. Capturing CO_2 from air. http://www.climeworks.com

[90] Carbon Engineering. Direct Air Capture. http://carbonengineering.com/about-dac/

[91] Keith D.W. et al. (2018) Process for Capturing CO_2 from the Atmosphere. *Joule,* 2:8:1573–1594. https://www.cell.com/joule/fulltext/S2542-4351(18)30225-3

[92] Tollefson J. (2018). Sucking carbon dioxide from air is cheaper than scientists thought. *Nature,* 558:173; doi: https://doi.org/10.1038/d41586-018-05357-w. https://www.nature.com/articles/d41586-018-05357-w

[93] Smith P. et al. (2015). *Nature Climate Change,* 6:42–50. https://www.nature.com/articles/nclimate2870.pdf

Albedo Modification In contrast to the elimination of CO_2 and other GHG emissions (mitigation) or reduction of atmospheric CO_2 levels (carbon dioxide removal, CDR), *solar radiation management* (SRM) techniques are aimed at augmenting the reflectivity and light scattering properties of the atmosphere to cool the earth. SRM is not a new idea at all. Decades ago climate physicists mused about the possibility of harnessing the properties of particulate haze to bounce a portion of sunlight back into space preventing it from penetrating the atmosphere and reaching the earth's surface.[94] Perhaps the most compelling case is made by nature and the evidence provided by volcanic eruptions. In 1991 the cataclysmic explosion of Mount Pinatubo on the island of Luzon in the Philippines launched a plume, more than 5 km^3 in volume, of molten rock, ash and gases 40 km into the sky. In total the 1991 volcanic activity of Mount Pinatubo injected nearly 20 million tons of sulfur dioxide (SO_2) into the stratosphere, which was then dispersed around the planet decreasing average global temperatures about 0.5 °C for the next 2 year.[95] More than any modeled simulations, this event suggests that perhaps humans can artificially and temporarily modify the upper atmosphere in much the same way by dispersing a thin layer of sulfate aerosols to shade and cool the planet.

Approximately 100 M tonnes of SO_2 are released into the atmosphere annually, mostly from anthropogenic sources; 76% is emitted from the combustion of fossil fuels and biomass.[96] The atmospheric chemistry of SO_2 is complicated and varies depending on interactions with other atmospheric constituents and where in the atmosphere the reaction is taking place (e.g., height). In general, SO_2 undergoes oxidation via many different mechanisms to produce sulfate (SO_4^{2-}) which then forms condensation nuclei for aerosols and clouds influencing the optical properties of clouds, the acidity of rain, and regional precipitation patterns.[97] In the troposphere, which extends from the earth's surface to a height of 6–10 km, the SO_2 emissions and resultant sulfate aerosols are very short-lived having a lifetime of a few days. In the stratosphere, which extends from the tropospheric boundary to about 50 km above the earth's surface, sulfate aerosols are much longer lived, with lifetimes ranging from weeks to years. In both regions, however, sulfate aerosols whiten the atmosphere, increase atmospheric albedo, and exert a cooling effect on surface temperatures by preventing a portion of sunlight from hitting the earth's surface.

---

[94] Morton O. (2007). Is This What It Takes To Save The World? *Nature*, 447:132–136. https://www.nature.com/articles/447132a.pdf

[95] U.S. Geological Survey. The Cataclysmic 1991 Eruption of Mount Pinatubo, Philippines. Fact Sheet 113–97. https://pubs.usgs.gov/fs/1997/fs113-97/

[96] Kilmont Z. et al. (2013). The last decade of global anthropogenic sulfur dioxide: 2000–2011 emissions. *Environmental Research Letters,* 8:014003. http://iopscience.iop.org/article/10.1088/1748-9326/8/1/014003

[97] Crutzen P.J. (2006). Albedo Enhancement by Stratospheric Sulfur Injections: A Contribution to Resolve a Policy Dilemma. *Climate Change,* 77:211–219; DOI: https://doi.org/10.1007/s10584-006-9101-y. https://link.springer.com/content/pdf/10.1007/s10584-006-9101-y.pdf

Detailed understanding of how much of the anthropogenic SO_2 released at ground level reaches the tropospheric/stratospheric boundary and then exerts various chemical and physical effects is lacking. With the advent of air pollution controls, global SO_2 emissions declined about 17% from 1990 to 2011.[98] In contrast, over the same time interval, global CO_2 emissions increased 44%.[99] The radiative forcing resulting from SO_2 removal could result in a future increase in average global temperatures of about 1 °C, which prompted the Nobel Laureate Paul Crutzen to recommend that in the face of no other climate action, it may be necessary to initiate a rapid response and disperse stratospheric SO_2 to cool a planet that is warming too quickly beyond a tolerable window.[100]

Since Crutzen's 2006 paper in *Climate Change*, climate scientists have hotly debated the feasibility of SRM brought about by injecting reflective particles derived from SO_2 into the upper atmosphere to create a "sunshade" for the earth below. Initially there was skepticism about the climate compensation of SO_2 for CO_2 since there is a mismatch in time and space between their relative cooling and warming effects. CO_2 warms all day, every day and its effects are greater at the poles. The cooling effects of SO_2 aerosols, on the other hand, only occur with daylight, are larger in the summer and greater in the tropics. Climate models, however, have demonstrated that SRM due to stratospheric aerosol dispersal can cancel out increases in global average temperature caused by climbing levels of CO_2. Uncertainty arises, though, when trying to discern regional and local climatic effects, as well as system feedbacks, associated with cooling.

The SRM strategy of stratospheric aerosol injection is deceptively simple, cheap, immediate and certain or as some assert, "the underlying science is sound".[101] Solar dimming could be achieved by dispersing submicron SO_2 aerosols from an airplane flying at 60,000 ft. ($\approx$20 km) where the tiny particles would remain for 1–2 years. It is possible to offset the radiative forcing of millions of tons of CO_2 with just tons of SO_2. For instance, calculations demonstrate that in order to counter half the radiative forcing caused by 240 Gt CO_2 (1.7 W/m^2), only 1 Mt of sulfur (−0.85 W/m^2) would be needed.[102] There are many scenarios in which SRM using SO_2 aerosols might be deployed, but all recommend an approach that proceeds slowly, carefully, iteratively, and reversibly. For example, one might begin at the Arctic, deploy only a year's worth of SO_2 low in the stratosphere, and observe the effects.[103] Cooling Arctic temperatures could increase ice formation and hence, surface albedo, in turn reflecting more incoming solar radiation. Others have proposed to inject 25,000

[98] Kilmont Z. et al. (2013). *Environmental Research Letters,* 8:014003. http://iopscience.iop.org/article/10.1088/1748-9326/8/1/014003

[99] Global Historical GHG Emissions. Climate Watch. https://www.climatewatchdata.org/ghg-emissions?source=31&version=1

[100] Crutzen P.J. (2006). *Climate Change,* 77:211–219. https://link.springer.com/content/pdf/10.1007/s10584-006-9101-y.pdf

[101] Keith D. **A Case for Climate Engineering** (MIT Press, Boston, MA, 2013) 224 pp.

[102] Ibid.

[103] Morton O. (2007). *Nature,* 447:132–136. https://www.nature.com/articles/447132a.pdf

tonnes of sulfur aerosols today and then ramp up to 250,000 tonnes by 2030 in order to begin a slow reversal of the effects of CO_2 emissions and allow time for GHG reductions to take place.[104] At these levels the costs of SRM is estimated to be less than $1B per year and might not reach $1B until dispersing 1 Mt. by 2070. Although it may be possible to reverse global average temperatures to pre-industrial levels, it will not be possible to restore a pre-industrial global climate.

Other chemicals besides SO_2 such as alumina or calcium carbonate, could be used since the role is mostly to hinder the evaporation of water from the aerosol, but there is a preference for SO_2 since its behavior in volcanic emissions is reasonably well understood. A variety of deployment mechanisms could be employed; a tethered balloon delivery system has been proposed, but this lacks the flexibility of injecting aerosols by jet.[105] Another SRM proposal involves trillions of nearly transparent, very thin "fliers" the size of dustbin lids launched into orbit at 1.85 M km from the earth's surface to reduce sunlight by about 1.8%. The cost and scale of the operation, however, is prohibitive – about $5 trillion for 16 trillion fliers that would take about 100 years to produce.[106]

SRM can also be achieved through tropospheric phenomena involving *marine cloud whitening* or *brightening* to exert local cooling effects. In this scenario very fine sea salt sprays produced near the subtropical ocean surface would be introduced into stratocumulus or low-lying marine clouds making them whiter and perhaps longer lived by enhancing the number of cloud condensation nuclei they contain.[107] There are two major limitations to this strategy, however; marine clouds are short-lived so that sea salt sprays would have to be pumped continuously, and our understanding of marine cloud processes is limited and there are no models that can accurately predict the behavior of the clouds in response to the introduction of sea salt particles.

The application of SRM is fraught with concerns about the unintended consequences that cooling at a global scale could exert at local and regional scales to adversely affect precipitation patterns, extreme weather events, and agricultural productivity. There are also serious moral and ethical questions surrounding the fact that SRM is only masking the effects of climate change. While for some this extends the time for action, for others it eliminates the urgency to reduce GHG emissions and remove CO_2 from the atmosphere. In the worse case, SRM could allow us to continue our patterns of fossil fuel use to the point of no return. In order to offset CO_2 warming, SRM needs to be deployed at a global scale, which requires a level of international cooperation that has yet to be achieved in any other situation. SRM could be weaponized or adopted to favor technologically advanced societies over poor or enemy states. Even in the absence of nefarious intent, there are serious

---

[104] Keith D. **A Case for Climate Engineering** (MIT Press, Boston, MA, 2013).

[105] Temple J. Harvard Scientists Moving Ahead on Plans for Atmospheric Geoengineering Experiments. *MIT Technology Review* (03.24.17). https://www.technologyreview.com/s/603974/harvard-scientists-moving-ahead-on-plans-for-atmospheric-geoengineering-experiments/

[106] Morton O. (2007). *Nature*, 447:132–136. https://www.nature.com/articles/447132a.pdf

[107] Keith D. **A Case for Climate Engineering** (MIT Press, Boston, MA, 2013).

issues around equity. If for some reason stratospheric aerosol dispersal were halted, global temperatures would quickly rebound to elevated levels which could have catastrophic consequences. If SO_2 were used to create aerosols, ozone recovery in the stratosphere would be slowed, acid rain would be created, and premature deaths from fine particle air pollution would occur, although it is unlikely that these effects would exceed those under current conditions of anthropogenic SO_2 emissions. Despite all these risks and uncertainty, the overwhelming justification for SRM is that too little progress has been made on combatting the causes of a rapidly changing climate and the world is pulling up to a crisis point at which an emergency response will be required. And that response will be to inject SO_2 aerosols into the stratosphere in perpetuity.

Adaptation The changing global climate threatens to rapidly unravel the conditions to which human societies are adapted. Human societies have flourished over many millennia under relatively stable climatic conditions facilitating the development of cities and infrastructure that promote innovation, exchange, wealth, culture, and quality of life for their inhabitants and food supplies on land or water or by trade that sustain their populations. Great cities are resilient and persist despite the shocks of disastrous storms, economic calamity, devastating wars, and social unrest. But the unrelenting and global pressures of climate change require a completely different set of technical, political, economic, and social responses. In this context "*adaptation*" connotes technological, behavioral, and policy approaches to climate change that seek to lessen future risks and vulnerabilities. Essentially, since we are highly unlikely to halt global climate changes and since climate change is already producing shifts that cannot be reversed, we should figure out how to minimize the damages, buffer the most deleterious effects and compensate for losses. The special challenge of adaptation is the uncertainty circling climate change outcomes and time tables. Consider the following case in point: Adaptation to rapid sea level rise (RSLR).

Approximately 40% of global population, about 3 billion people, live within 100 km of a coast[108] and 8 of the 10 largest cities in the world are situated in coastal regions. About 1B people living in low lying coastal areas are seriously threatened by sea level rise and storm surges.[109] In the U.S. there are already 90 coastal communities battling chronic flooding, defined as 10% or more of usable land flooding 26 times per year. Most of these communities are in Louisiana and Maryland where

---

[108] Percentage of total population living in coastal areas. http://www.un.org/esa/sustdev/natlinfo/indicators/methodology_sheets/oceans_seas_coasts/pop_coastal_areas.pdf

[109] Usery E.L. et al. (2010) Modeling Sea-level Rise and Surge in Low-lying Urban Areas using Spatial Data, Geographic Information Systems, and Animation Methods, in **Geospatial Techniques in Urban Hazard and Disaster Analysis**, P. Showalter & Y. Lu, eds. (Springer Netherlands 2019) Chapter 2, p. 11–30; DOI: https://doi.org/10.1007/978-90-481-2238-7_2. https://cegis.usgs.gov/pdf/sea_level_rise_text.pdf

SLR is exacerbated by land subsidence.[110] With time, chronic coastal flooding will extend from Maine to Texas and include California, with projections of 170 communities in 20 years and 670 by the end of the century. California is particularly vulnerable to the melting of the West Antarctic ice sheets and in addition to more extensive coastal flooding during storms, SLR will also cause periodic tidal flooding and increased coastal erosion.[111] In general, with a 10–20 cm SLR the frequency of serious flooding doubles.[112] In 2016 the global average increase in sea level was 8 cm (3.2 inches) above the 1993 average.[113] Yet, in Miami, ground zero for RSLR, sea level has risen more than 12 cm (5 inches) over the last 20 years. Much of this increase has occurred since 2011 due to a remarkable upswing in the rate of SLR with many areas regularly experiencing dry weather flooding due to tides. Faced with biweekly flooding inhabitants of low-lying areas are beginning to choose to relocate, a trend that is expected to swell dramatically over the next two to three decades as tens of millions of people globally are displaced by RSLR.[114]

There are basically 3 approaches to RSLR adaptation: *hardening* shorelines, *softening* coastal areas, or *retreating* altogether. The traditional engineering approach to protecting shorelines from the encroachment of the sea is to **armor** or **harden** them by building levees, dykes, walls and flood gates. These are only effective if they are high enough and strong enough to withstand rising water levels, wave action and storm surges. Furthermore, walls only protect property directly behind them diverting water to those areas where the wall ends to cause greater damage. Over half the New Orleans levee system was damaged or breached in Hurricane Katrina which resulted in the flooding of 85% of greater New Orleans, 1600 deaths, and the homelessness of half million people.[115] Model simulations revealed that the system of hurricane protection levees and raised roads of the Mississippi River Gulf Outlet amplified the effects of hurricane storm surges into

[110] Parker L. Sea Level Rise Will Flood Hundreds of Cities in the Near Future. National Geographic (07.12.17). https://news.nationalgeographic.com/2017/07/sea-level-rise-flood-global-warming-science/

[111] Mulkern A. Rising Sea Levels Will Hit California Harder Than Other Places. Scientific American (04.27.17). https://www.scientificamerican.com/article/rising-sea-levels-will-hit-california-harder-than-other-places/

[112] Vitousek S. et al. (2017). Doubling of coastal flooding frequency within decades due to sea-level rise. Scientific Reports, 7:1399. https://www.nature.com/articles/s41598-017-01362-7#citeas

[113] Lindsay R. Climate Change: Global Sea Level. Climate.gov, NOAA (08.01.18) https://www.climate.gov/news-features/understanding-climate/climate-change-global-sea-level

[114] McLeman R. Migration and displacement risks due to mean sea-level rise. Bulletin of the Atomic Scientists (05.04.18). https://thebulletin.org/2018/05/migration-and-displacement-risks-due-to-mean-sea-level-rise/

[115] van Heerden I.L. (2018). Setting the Stage for the Katrina Catastrophe: Environmental Degradation, Engineering Miscalculation, Ignoring Science and Human Mismanagement, in **Creating Katrina, Rebuilding Resilience**, Lessons from New Orleans on Vulnerability and Resiliency, M.J. Zakour, N.B. Mock, P. Kadetz, eds. (Butterworth-Heinemann, 2018) Chapter 6, p133–158. https://www.sciencedirect.com/science/article/pii/B9780128095577000065

New Orleans and vicinity.[116] Strategies such as lowering levee height to allow some flooding and replacing long linear stretches of levees with "citadel" levees built directly around communities work more synergistically with nature to create surge protection.[117]

The city of Venice has long been plagued by flooding and the combined effects of subsidence, tides, storms and rising sea levels have worsened inundation to a crisis level. In order to protect the city from both recurring tidal flooding and extreme events defined by a historic 1966 storm, the engineering megaproject called *Modulo Sperimentale Elettromeccanico* (MOSE), or "Experimental Electromechanical Module" was envisioned as the ultimate flood barrier. MOSE involves a series of high-tech, retractable barriers filling with water to sink and with air to deploy. Stretching across each of the three inlets to the Venetian Lagoon, individual barriers consist of 20 gates. The MOSE project, one of the world's largest and highest-profile civil engineering undertakings, took 50 years to approve, design, fund and partially build. It is billions of dollars over budget and several years overdue – in 2017 only the gates for one lagoon inlet had been installed.[118] When the design was finalized in 2002, the intent of MOSE was to protect Venice from tidal surges of 9 ft (2.7 m) and to be closed temporarily for about 5 hours approximately 10 times per year for the next 50 years of its design life. With less than 1 foot (8 inches or 20.3 cm) added to compensate for SLR, the reality is that MOSE is woefully under-designed and growing obsolete due to RSLR. With 50 cm (about 20 inches) of SLR the barriers would need to be closed daily and beyond that the barriers will likely be closed more often than opened, which essentially defeats the purpose of the system.[119] At less than 2 ft of SLR (60 cm) the system would no longer be protective as floodwaters would find other paths to Venice. The MOSE project illustrates an unfortunate truth about adaptation based on extreme engineering – it is often not very adaptable because designs are set to a specific set of historic conditions and are not easily adjusted to evolving or worsening situations.

Yet, plans for large scale coastal flood barriers are sprouting all over the East and Gulf coasts of the U.S. For instance, in the aftermath of Hurricane Harvey, a $15B project, the Ike Dike, would protect Galveston Bay with 55 miles of sand dunes and sea walls and an 800-foot wide retractable barrier at the mouth of the Galveston shipping channel. The U.S. Army Corps of Engineers is planning a similar barrier

---

[116] Westerink J. et al. (2006). Note on the Influence of the Mississippi River Gulf Outlet on Hurricane Induced Storm Surge in New Orleans and Vicinity. http://www.columbia.edu/itc/journalism/cases/katrina/Army/Army%20Corps%20of%20Engineers/Influence%20of%20the%20MRGO%20on%20Storm%20Surge.pdf

[117] Gilroy W.G. Changes proposed to New Orleans area levee systems. Science Daily (07.24.13). https://www.sciencedaily.com/releases/2013/07/130724200557.htm

[118] Goodell J. Rising Waters: Can a Massive Barrier Save Venice from Drowning. Yale Environment 360 (12.05.17). https://e360.yale.edu/features/rising-waters-can-a-massive-sea-barrier-save-venice-from-drowning

[119] Rossi M. Will a Huge New Flood Barrier Save Venice? CityLab (04.03.18) https://www.citylab.com/environment/2018/04/will-a-huge-new-flood-barrier-save-venice/556226/

around Jamaica Bay in New York City to protect nearby vulnerable neighborhoods.[120] In response to the destruction of Hurricane Sandy and $4B worth of damage to the subway system alone, major efforts to flood-proof critical infrastructure such as subways and power stations entail sealing stations and tunnels and installing flexible fabric (e.g., Kevlar) flood closures. In Miami Beach, city managers and developers want to raise the level of roads, buildings and homes by 2 ft, invoking the justification that in the 1860s Chicago raised its streets and buildings 8 ft.[121]

An alternative adaptation approach to armoring shorelines is to **soften** them by restoring ecological buffers such as wetlands, berms, beaches, dunes, etc. which absorb the rains and dampen the energy of winds, waves and surges from storms. Coastal ecosystems are designed evolutionarily to flood and withstand high wind and wave action. Mature systems typically self-organize to repair themselves after severe storms. In other words, they are highly resilient. Among the most effective and inexpensive protections against hurricanes and storm surges are native coastal ecosystems such as coastal marshes. The steady loss of this "protective ecological apron," however, exacerbates the devastating impacts of hurricanes. This is vividly evident along the Gulf coast, particularly coastal Louisiana home to one quarter of all wetlands in the lower 48 states, 40% of salt marshes, and where more than 500,000 hectares of coastal wetlands have been lost since 1930.[122] Damage to the Louisiana coast is caused by massive alteration to the deltaic landscape associated with the oil and gas industry and the thousands of miles of pipelines through which oil moves. In addition, urban development and over-engineering the hydrology of the Mississippi delta disrupts the natural processes of sediment deposition. Coast 2050, proposed and approved in 1998, was a $14 B plan to restore Louisiana's coastal wetlands and barrier islands over 30 years and essentially replumb the hydrology of the region to enhance storm defenses. Despite laying the foundation for further study and a series of specific restoration projects, which Congress approved, no funding for the restoration was ever appropriated.[123]

In general, while the effectiveness of coastal ecosystem restoration is widely accepted for both storm surge and SLR protection, until recently it had not been rigorously studied, especially relative to reduction of economic damages associated with storms. Focusing on the impact of coastal wetlands along the northeastern U.S. coast, researchers recently determined that the ecological goods and services of coastal wetlands prevented $625 M in direct flood damages during Hurricane

---

[120] Goodell J. Yale Environment 360 (12.05.17). https://e360.yale.edu/features/rising-waters-can-a-massive-sea-barrier-save-venice-from-drowning

[121] Goodell J. **The Water Will Come** (Little, Brown & Co. 2017 NY, NY) Chapter 11, p238.

[122] van Heerden I.L. (2018). in **Creating Katrina, Rebuilding Resilience**, Lessons from New Orleans on Vulnerability and Resiliency, M.J. Zakour, N.B. Mock, P. Kadetz, eds. (Butterworth-Heinemann, 2018) Chapter 6, p133–158. https://www.sciencedirect.com/science/article/pii/B9780128095577000065

[123] Robichaux E. Coast 2050's Lasting Impacts on Coastal Restoration. Delta Dispatches. Restore the Mississippi River Delta (11.05.15). http://mississippiriverdelta.org/coast-2050s-lasting-impacts-on-coastal-restoration/

Sandy.[124] An important design constraint of coastal softening through ecosystem restoration is that while it is much less expensive than extreme engineering endeavors such as movable storm barriers, it can still take many years to restore and allow ecosystems to mature.

Looking to a low-lying country that has long battled the sea for its very existence, the Dutch are the world experts in flood control technology as evidenced by an elaborate network of dikes, canals, and sea walls. The masterpiece of Dutch engineering is the Delta Works, commissioned as the result of a devastating North Sea storm in 1953 and consisting of three locks, six dams and four storm surge barriers to create their largest flood defense system. One of the storm surge gates near Rotterdam, the Maeslant Barrier, is an engineering marvel designed to remain open to allow ship traffic and to close to withstand a storm tide of 5 m, which is associated with a historic storm probability of 0.01% or 1 in 10,000 years. Yet, upon completion of this 40-year project, Dutch planners realized that climate change and RSLR require the dikes and sea walls to be made even higher and wider and that they need to be designed for a 1 in 100,000-year storm, which has a 0.001% probability of occurring based on the historic record. Thus, realizing that ultimately barriers will be inadequate defenses against rising seas, Dutch engineers now design for letting the water in.[125] They have revised their approach by returning flood-defense to the basics of utilizing natural materials, mimicking natural systems, and harnessing nature's power to protect. For instance, existing dikes are being transformed into ecologically enhanced "rich levees" that mimic rocky coasts, but are also monitored by embedded sensors that relay real-time status of conditions to decision makers. Dikes support parks that couple storm water protection with social welfare and neighborhood improvement. Structures, in general, do double duty; garages, parks, plazas, meet the demands of daily life, but in times of floods, they serve as storage reservoirs. As an example, the Sand Engine project harnesses ocean currents to rearrange 28 million cubic yards of dredged sand to fortify a 12-mile buffer protecting the Dutch coast.[126]

Repeated flooding, the sense of futility in rebuilding and repairing again and again, and rising flood insurance premiums all point to what used to be considered unimaginable, that the time has come for residents of coastal communities to *retreat*. Over 20 years ago, as discussed above, the Dutch came to the realization that they cannot continue to build higher dikes and barriers and that they need to "make room" for flood waters by moving dikes back from the edge of rivers to allow for

---

[124] Narayan S. et al. (2017). The Value of Coastal Wetlands for Flood Damage Reduction in the Northeastern USA. *Scientific Reports,* 7:9463. https://www.nature.com/articles/s41598-017-09269-z

[125] Kimmelman M. & Haner J. The Dutch Have Solutions to Rising Seas. The World Is Watching, in Changing Climate, Changing Cities, The New York Times (06.15.17). https://www.nytimes.com/interactive/2017/06/15/world/europe/climate-change-rotterdam.html

[126] Katz C. To Control Floods, The Dutch Turn to Nature for Inspiration. Yale Environmental 360 (02.21.13). https://e360.yale.edu/features/to_control_floods_the_dutch_turn_to_nature_for_inspiration

rivers to swell, thereby, designing for inundation in some areas of cities to protect more highly populated districts.[127] The "Room for the River" project in Nijmegen, a 2000 year old Dutch city on the Rhine River, necessitated demolishing 50 houses and compensating the property owners. The *depoldering* projects here and elsewhere in the Netherlands abandoned traditional flood protection, bought property owners out, moved neighborhoods, and in their place constructed ecologically based public spaces where water could spread in times of flood.[128] In the U.S. retreat is slowly unfolding, mostly as a result of Hurricane Sandy. Historically, it had been easier to obtain funding to put homes on stilts rather than to relocate, but post-Sandy programs have emerged for large buyouts of clustered properties in coastal communities such as New Jersey and Staten Island. Retreat, then, has made way for restoration of new coastal ecologies of biodiverse salt marshes.[129] In post-Hurricane Harvey Houston, applications for buyouts increased to 3000, a number exceeding the total of the last 25 years.

In the U.S. buyouts of chronically flooding properties are not coordinated with post-retreat planning. Given our penchant for building to the waters edge, retreat is intrinsic to efforts to ecologically softening coastlines. Adaptation strategies seek to future-proof communities and vital resources to the vagaries of climate change over the long term, and few adaptation projects are implemented quickly and inexpensively. Without additional adaptation measures and under a high emissions scenario, global annual flood costs could climb to $14 trillion per year, reaching 2.8% of GDP by 2100.[130] Yet, at warming restricted to 1.5 °C and without additional adaptation to SLR, models project that global coastal flooding could be $10.2 trillion per year (1.8% global GDP) by 2100. In contrast, adaptation could decrease these costs substantially, by an order of magnitude, to about $1 trillion.

Perhaps out of frustration, the focus of climate action seems to be shifting from mitigation to adaptation. Indirectly, however, adaptation can be antagonistic or synergistic with the mitigation of CO_2 emissions. Ideally, coastal adaptation measures should embrace ecological and human designed elements integrated with population retreat from low lying areas and sustainable development of human settlements.[131] Such an approach is at work in Manhattan with the BIG U project designed by Bjarke Ingels Group (BIG). Here, a series of levees, a floodwall, and a park with

---

[127] ClimateWire. How the Dutch Make "Room for the River" by Redesigning Cities. Scientific American (01.20.12). https://www.scientificamerican.com/article/how-the-dutch-make-room-for-the-river/

[128] Bentley C. Holland is relocating homes to make more room for high water. Public Radio International (PRI) (06.22.16). https://www.pri.org/stories/2016-06-22/holland-relocating-homes-make-more-room-high-water

[129] Schwartz J. Surrendering to Rising Seas. Scientific American (08.01.18). https://www.scientificamerican.com/article/surrendering-to-rising-seas/

[130] Jevrejeva A. et al. (2018). Flood damage costs under the sea level rise with warming of 1.5 °C and 2 °C. *Environmental Research Letters,* 13:074014. http://iopscience.iop.org/article/10.1088/1748-9326/aacc76/pdf

[131] van Heerden I.L. (2018). in **Creating Katrina, Rebuilding Resilience**, Lessons from New Orleans on Vulnerability and Resiliency, M.J. Zakour, N.B. Mock, P. Kadetz, eds. (Butterworth-

vegetated berms will wrap around the southern half of the island to protect the area from the next Sandy-like storm.[132] The estimated price tag for the BIG U project is about $1 B, substantially less than extreme engineering projects such as the MOSE. Six years after Hurricane Sandy with its $19 B damage to NYC alone, most of the regions infrastructure has been repaired and some improved. Yet, the more transformative plans to stormproof New York City, such as the BIG U project, remain just that, plans.[133] And development has continued to encroach on the water's edge.

5 Cautionary Tale: 2. Engineered Solutions to Engineered Problems

Poyang Lake, the largest freshwater lake in China, is located in the Yangtze River Basin and has a complex hydrology characterized by seasonally fluctuating water levels that can vary as much as 30 ft between the dry winters and rainy summers. Five tributaries feed the lake, which then flows into the Yangtze River via a free and natural connection. In the wet, summer season, however, the flow reverses and the Yangtze River feeds Poyang Lake.[134] The lake supports invaluable ecological resources that are tuned to this widely oscillating system. It provides habitat to the rare Yangtze finless porpoise and in the winter its mud flats are a migratory feeding ground for thousands of birds, including the endangered Siberian crane and more than a dozen other threatened species.

Massive hydraulic and urban development projects, however, threaten the ecological integrity of the lake and its habitat by disrupting the intricate hydrological cycles causing drastic shrinkage of the lake's area and depth.[135] Three Gorges Dam, the world's largest dam and power station, is located upstream of Poyang on the Yangtze River and exerts an enormous impact on the water budget of downstream systems, some of which is in the service of flood control. Water storage by the dam for electric power generation reduces the water levels of the Yangtze River inducing

Heinemann, 2018) Chapter 6, p133158. https://www.sciencedirect.com/science/article/pii/B9780128095577000065

[132] Garfield L. Manhattan plans to build a massive $1 billion wall and park to guard against the next inevitable superstorm. Business Insider (04.27.18). https://www.businessinsider.com/new-york-city-flooding-manhattan-coastal-barriers-2018-4

[133] McGeehan P. & Hu W. Five Years After Sandy, Are We Better Prepared? The New York Times (10.29.17). https://www.nytimes.com/2017/10/29/nyregion/five-years-after-sandy-are-we-better-prepared.html

[134] Harris J. (2016) Poyang Lake, Yangtze River Basin, China, in **The Wetland Book**, C.M. Finlayson, G.R. Milton, R.C. Prentice, N.C. Davidson, eds. (Springer Nature Switzerland 2018) https://link.springer.com/referenceworkentry/10.1007%2F978-94-007-6173-5_34-2

[135] Ives M. As China's Largest Freshwater Lake Shrinks, Solution Faces Criticism. New York Times (12.28.16) https://www.nytimes.com/2016/12/28/world/asia/china-lake-poyang-finless-porpoise.html

greater discharge from the lake into the river and disrupting the summer flow reversal.[136] In addition, extensive dredging in the lake to mine sand for construction projects along the Yangtze and as far away as Shanghai has lowered the lake's bed further disrupting its hydrological flows. Finally, the region has experienced recent droughts causing even more perturbations to the lake's hydrology.

Human uses of the lake are made difficult by the lake's natural fluctuations compounded by its recent shrinkage. Thus, the local government has proposed a large scale engineering solution to what is essentially a problem caused by local and regional large scale engineering projects. Engineers believe that they can keep more water in the lake in the winter by building a 10,000 ft. sluice gate at the lake's connection to the Yangtze River, which will essentially function as a dam and also serve to stabilize drinking water supply and promote shipping. At a cost of $1.9B, the project will eliminate the mud flats, and the nudge the Siberian crane and finless porpoise closer to extinction.[137] This situation illustrates the dilemma created by an "engineering mentality" in which the unintended consequences of engineering projects are tackled with more engineering, when less intensive interventions may be far more beneficial and effective.

A similar quandary exists on the Mississippi River, where engineered hydraulic controls have been designed to minimize flooding and maintain navigation. Ironically, floods on both the Upper and Lower Mississippi River have become more frequent and severe due largely to the infrastructure designed to prevent them. It has long been known that when river water is constrained by levees from spreading over the flood plain at times of high flows, it flows faster and higher and increases flood risks for those downstream or in areas not protected by levees. Yet, this knowledge has not deterred the "engineering mentality" that farmland and towns along the Mississippi need to be protected by levees. Although the inevitable floods are more severe than they would have been in the absence of levees, flood plain development drives a "hydrologic spiral" of flooding, levees, more flooding, and higher levees.[138] The alternative, as recommended by the Dutch, is to "make room for the river" and retreat from operating in flood plains. A sign posted near the Mississippi River during a recent flood said it well, "It's called flood plain because it is plain that it floods."[139]

Lessons Learned: The Unintended Consequences of the "Engineering Mentality" The rapidly changing climate with its increased frequency of extreme

[136] Zhang Z. et al. (2016). Analysis of Poyang Lake water balance and its indication of river-lake interaction. *SpringerPlus*, 5:1:1555: doi: https://doi.org/10.1186/s40064-016-3239-5. https://www.ncbi.nlm.nih.gov/pmc/articles/PMC5021641/

[137] Ives M. The New York Times (12.28.16) https://www.nytimes.com/2016/12/28/world/asia/china-lake-poyang-finless-porpoise.html

[138] Hersher R. Levees make Mississippi River Floods Worse, But We Keep Building Them. Environment & Energy Collaborative, National Public Radio (NPR), (05.21.18). https://www.npr.org/2018/05/21/610945127/levees-make-mississippi-river-floods-worse-but-we-keep-building-them

[139] Ibid.

events only escalates the vulnerabilities we have unintentionally woven into our engineered landscape. Human battles to control nature have been continuously fought throughout the history of civilization and massive engineering projects that transform the landscape, make resources available, produce energy, and ensure safety and comfort are the latest attempts at bending the environment to human will and convenience. These two examples of engineering vast waterways for economic benefit and ostensibly for the enhanced safety of coastal communities reveal an uncomfortable fact: The unintended consequences of large scale engineering feats are themselves large and difficult to correct. It is not surprising that the engineering mentality of industrial societies leads us to devise technological fixes for the problems technology inadvertently introduces. Every engineer knows that there is no unique solution to a design problem; instead, there is a menu of options with varying costs and benefits. The valuation of these costs and benefits, however, is more subjective than absolute, which brings us back to the bias in favor of the techno-fix and the illusion of control and least cost in the short term. It should come as no surprise, then, that decision makers are seriously contemplating climate intervention strategies that entail endeavors deployed at massive scales in both time and space in order to buy time, despite the many options that could be employed to actually solve the climate problem.

6 Global Scale Climate Engineering: Feasibility, Risks and Benefits

There was a time not too long ago when a changing climate was regarded as only a threat to generations far in the future. Climate action was reasoned to be the job of those future generations, since they would be the ones experiencing the effects and would be in a better economic position to address them. Although who knew for sure if any action would even be needed, since the consequences of climate change are shrouded in so much uncertainty. While some may cling to these beliefs, the influence of climate change on present day life is becoming ever clearer.

In the summer of 2018 heat waves, droughts and wildfires ravaged large swaths of the Northern Hemisphere from India to Siberia and California. Peak temperatures exceeding 50 °C (120 °F) were recorded in Algeria and Southern California. In parts of Japan temperatures rose above 38 ° C (100 °F) and in Siberia and Scandinavia they soared above 32 °C (90 °F). In France, nuclear power plants had to be shut down because rivers that provide cooling waters to the plants had become too warm and elsewhere in the world power grids teetered on the brink of collapse due to surging demand. What distinguishes these extreme temperatures is that unlike past events, these elevated temperatures lasted for extended periods of time due to warm ocean surface temperatures and a weakened jet stream influenced by Arctic warming and reduced temperature gradients between the poles and mid-latitudes. In Sweden, Greece, and California the combination of heat and drought caused some of the worst wildfires on record. In mid July 2018, more than 80 fires raged in Sweden destroying 25,000 hectares of forest, ten times the average wildfire loss.[140] In California, seven of the twelve most deadly wildfires occurred in the last 3 years[141] and the largest in the state's history, the Mendocino Complex fire, engulfed over 450 square miles (117,000 hectares) in August 2018.[142] As these events illustrate, climate change is increasing the severity and frequency of extreme events for the present generation.

In U.S. cities, extreme heat kills more people than all other weather events combined.[143] By mid-century the number of cities around the world suffering average summertime temperature highs of 35 °C (95 °F) will increase nearly three-fold, from 354 to 970 and a large proportion of these cities will be in India.[144] Recent reports indicate that extreme heat and humidity, the combination of which is measured by the wet-bulb temperature index, are particularly dangerous for South Asia's largest cities. If warming trends continue unabated, by the end of the century heat and humidity conditions may reach levels where people would not be able to survive six or more hours of direct exposure.[145] By 2050 extreme temperatures greater than 46 °C will be five times more likely in the Middle East than in 2000 and by 2100 wet-bulb temperatures in the Persian Gulf may render the region uninhabitable and even more unstable politically.[146] The World Bank recently warned that the living conditions of 800 million people in six South Asian countries, already home to some of the worlds' most impoverished populations, will severely deteriorate due to the climbing temperatures and changing rainfall patterns associated with unchecked climate change.[147]

Today the centerpiece of international climate action, as well as our best hope that the worst of these events can be avoided, is the 2015 Paris Agreement to limit voluntarily warming to a range of 1.5–2 °C above average pre-industrial levels.

---

[140] Total fire ban enforced in several counties across Sweden. The Local (07.24.18). https://www.thelocal.se/20180724/sweden-wildfires-25000-hectares-of-forest-still-burning

[141] Top 20 Most Destructive California Wildfires. http://www.fire.ca.gov/communications/downloads/fact_sheets/Top20_Destruction.pdf

[142] Arango T. & Medina J. California Fire Now the Largest in State History: 'People are on Edge.' The New York Times (08.07.18) https://www.nytimes.com/2018/08/07/us/california-fires-mendocino.html

[143] Extreme Heat. https://www.ready.gov/heat

[144] For Cities, The Heat Is On. The Future We Don't Want. Heat Extremes. C40 Citieshttps://www.c40.org/other/the-future-we-don-t-want-for-cities-the-heat-is-on

[145] Chandler D.L. Deadly heat waves could hit South Asia this century. Without action, climate change could devastate a region home to one-fifth of humanity study finds. MIT News (08.02.17). http://news.mit.edu/2017/deadly-heat-waves-could-hit-south-asia-century-0802

[146] Climate change is making the Arab world more miserable. The Economist (05.31.18)https://www.economist.com/middle-east-and-africa/2018/05/31/climate-change-is-making-the-arab-world-more-miserable

[147] Sengupta S. & Popovich N. Global Warming in South Asia: 800 Million at Risk. The New York Times (06.28.18). https://www.nytimes.com/interactive/2018/06/28/climate/india-pakistan-warming-hotspots.html?action=click&module=RelatedLinks&pgtype=Article

There is a substantial difference between the probabilities of extreme events when comparing the 1.5 °C and 2 °C warming limits, with the 2 °C limit being much less protective. Realistically, though, both warming limits appear elusive. In order to keep average global temperatures to only a 1.5 °C increase, an extremely rapid and far-reaching systems (energy, agricultural, urban, industrial) transition would have to occur during the next 10–20 years, which, although possible, is highly unlikely.[148] The 2° warming limit requires that we reach the same endpoint, but more gradually. Current national level commitments to the Paris Agreement, however, are not even sufficient to keep the planet at the 2 °C threshold.[149] In fact, despite all the grave warnings, current emission reduction policies will likely allow warming up to 3 °C above preindustrial averages in this century. The nature and magnitude of the global decarbonization challenge are such that there is no quick and easy fix. Given the unexpected rates at which climate changes are unfolding and the growing fears of passing various earth systems thresholds, there are profound political risks associated with failing to meet yet another international agreement to limit emissions. It is doubtful that there is time for the international community to develop a new set of quantitative targets.[150]

As this chapter has outlined, there are a vast number of mitigation tools available to lessen energy and resource demands and shift energy sources away from fossil fuels to alternatives. There are many synergies among mitigation, geoengineering, and adaptation strategies, but they are only realized with long term planning. The foundation of a Plan A of climate action is the decarbonization of the human enterprise, which has been ramped up to be *deep* decarbonization and then ramped up again to be *rapid* and *deep* decarbonization of the global economy. With a detailed roadmap to a low-C economy,[151] serious action is finally taking place throughout the EU with, for instance, phase out schedules for the internal combustion engine,[152] substantial adoption of renewable energy (17% in 2016, on target to be 20% in 2020),[153] and widespread promotion of passive homes and buildings (e.g. Passivhaus). Although recent federal policies around climate action in the U.S. have been weakened (e.g. Clean Power Plan, fuel economy standards), strong commitments persist at various local (e.g., NYC, San Francisco, Seattle, Chicago) and state

[148] Mathiesen K. et al. Climate Home News (06.27.18). http://www.climatechangenews.com/2018/06/27/new-leaked-draft-of-un-1-5c-climate-report-in-full-and-annotated/

[149] Harvey C. Even 2 °C of warming could turn Earth into 'hothouse.' ClimateWire (08.07.18). https://www.eenews.net/climatewire/2018/08/07/stories/1060092901

[150] Sachs J. et al. DDPP 2014 report. http://deepdecarbonization.org/wp-content/uploads/2015/06/DDPP_Digit.pdf

[151] 2050 low-carbon economy. Climate Action. European Commission. https://ec.europa.eu/clima/policies/strategies/2050_en

[152] Gray A. Countries are announcing plans to phase out petrol and diesel cars. Is yours on the list? World Economic Forum (09.26.18). https://www.weforum.org/agenda/2017/09/countries-are-announcing-plans-to-phase-out-petrol-and-diesel-cars-is-yours-on-the-list/

[153] Renewable energy statistics. Eurostat (06.2018) http://ec.europa.eu/eurostat/statistics-explained/index.php/Renewable_energy_statistics

(e.g., California, Massachusetts, Oregon, New York) levels. Despite anemic commitments to renewable energy nationally in the U.S., market forces are paving the way to a renewable future; by 2020 renewable sources of electricity are projected to displace fossil fuel sources based on cost alone.[154] In fact, in 2018 with the cost of power from utility-scale solar farms equivalent to that of natural gas and the unsubsidized cost of wind energy even lower, plans for some natural gas plants, currently the major U.S. power source, are being replaced by wind and solar, particularly in the Western U.S. where state renewable energy mandates prevail and help to nudge the transition.[155] Fundamentally, though, as evidenced so vividly in the U.S., it is the lack of political will on the part of global leaders and national governments, not technological know-how, that undermines long-term efforts and extensive adjustments to the prevailing socio-economic frameworks that are required to address the root causes of global climate change. And it is the dearth of political will that has brought us to this point in the 40-year history of climate inaction where the deployment of geoengineering tools is now essential in order to stave off the worst consequences of climate change. Thus, without having fully exercised Plan A, we are advancing serious and controversial discussions about Plan B, which invokes geoengineering strategies, many of which are archetypes of extreme engineering.

Geoengineering entails two very different approaches to climate intervention, but intrinsic to neither is the requisite dramatic reduction in anthropogenic GHG emissions. Carbon dioxide removal, CDR, is essentially environmental remediation, except the contaminated site is the atmosphere of the entire planet. Under the best of circumstances, site clean-up involves contaminant destruction, but usually contaminated materials are excavated or pumped to transfer them elsewhere, or sites are sealed and stabilized to reduce human exposure to the contaminants. In cases of diffuse pollution of large scale systems (e.g. Green Bay sediments) contaminants are left in place for the system to "naturally attenuate" or bury or dilute, because it just is not feasible to treat or remove them or to cap the entire site. In other words, when contaminants released into the environment result in low level, but nonetheless dangerous, pollution over large areas or vital ecosystems the only economically and technologically feasible method of clean up is to nudge natural processes to reduce, over long periods of time, exposure risks to humans. In any event, though, the source of pollution to the system is turned off. The same principles should hold for global scale geoengineering.

CDR strategies to create anthropogenic carbon sinks are now essential tools in the climate action game plan. It is puzzling that there is any hesitation or reluctance to employ ecologically based approaches to CDR: afforestation/reforestation, carbon farming/biochar, BECCS. The direct benefits to these actions are two fold: they

---

[154] Dudley D. Renewable Energy Will Be Consistently Cheaper Than Fossil Fuels by 2020, Report Claims. Forbes (01.13.18). https://www.forbes.com/sites/dominicdudley/2018/01/13/renewable-energy-cost-effective-fossil-fuels-2020/#66c3f0e14ff2

[155] Penn I. It's the No. 1 Power Source, but Natural Gas Faces Headwinds. The New York Times (03.28.18). https://www.nytimes.com/2018/03/28/business/energy-environment/natural-gas-power.html

halt or dampen the carbon emissions associated with deforestation, agricultural practice, and fossil fuel use, and they each stimulate carbon uptake at roughly estimated rates of 1–3 Gt/y. As importantly, though, the indirect benefits of ecologically based CDR are immense: local and long range positive cascading effects on hydrological cycles, enhanced soil fertility, boosted agricultural productivity. There are costs, of course, associated with changing established practices in the agricultural and energy sectors, and there are real biogeophysical constraints (e.g., water, temperature, nutrients, etc.) of employing photosynthetically-based technologies. Moreover, estimated rates of carbon uptake need to be verified in the field and at expanded scales, and will vary widely as a function of environmental conditions (e.g. water, temperature, nutrients, etc.). BECCS is potentially the most controversial, albeit the most frequently invoked CDR, in that at large scale its deployment will likely collide with food production and AR efforts. Moreover, bioenergy production must be integrated with CCS, which has yet to achieve economic feasibility when the carbon storage or sequestration step is actually implemented. Once a business case can be made, *in situ* EW, which is essentially reverse mineral mining, seems a very promising path to subsurface carbon storage. Thus, the co-location of BECCS facilities with olivine rich peridotite rock formations will allow its integration with *in situ* EW reducing one of the hurdles to the CDR technology.

Figure 5 delineates a general path to achieving both net-zero C-emissions by 2050 and absolute reduction of atmospheric CO_2 concentrations. The latter challenge is daunting, however, in that even in the face of decades of C-emissions drawdown and the cessation of fossil fuel use, post 2050 it will be necessary to remove over 500 Gt of CO_2 from the atmosphere. Over 50 years, then, approximately 10 Gt of CO_2 annually would need to be removed – this is in addition to whatever human carbon sinks were developed to get to the net-zero emissions mark at 2050. Borrowing from the carbon wedge approach of Socolow and Pacala, ten 1 Gt wedges of action would be needed annually. Integrated assessment modeling has estimated that AR, C-farming, and BECCS could each achieve roughly 1–3 Gt CO_2 uptake per year.[156] A back of the envelop calculation suggests that it might be possible, then, to achieve a CDR rate of 5–6 Gt/y or 5 to 6 wedges. Thus, in the best case with current technology, it is likely that ecologically based CDR will only get us half-way to our target of 10 Gt of CO_2 uptake per year.

Although there are potentially many cascading benefits associated with nature-based CDR, the strategies have clear, if unknown, biogeophysical limits and will be highly non-uniform in their performance depending on the location and type of ecosystem. Furthermore, while they represent a distributed approach, nature-based CDR strategies have to be deployed collectively at a very large scale to realize their CDR target potential. The remaining wedges of CDR, then, must be achieved in other ways, which is why direct air capture of CO_2, DAC technology, is emerging as a realistic contender. There are a number of serious challenges, though, that must be

[156] Smith P. et al. (2015). *Nature Climate Change*, 6:42–50. https://www.nature.com/articles/nclimate2870.pdf

solved if DAC technology is to be viable. One of the more basic issues is "the flow capture problem," determining where to locate DAC units in order to capture the CO_2 transported in heterogeneous flow environments.[157] The placement, design and operation of DAC must take into account local heterogeneity and chaos in flow conditions, even under conditions of high flow volumes and high CO_2 removals. For instance, over short times, DAC units perform best when located in regions of high flow velocity, but over longer times care must be taken to make sure that the units are not just recycling "clean" air. There are many air flow and operational conditions that carry high risks of failure and there can be marked differences between short term and long term results. DAC is significantly more costly than BECCS or other CDR strategies and it, too, is a technology that relies on long term carbon storage technology. While DAC is an unproven technology, there are a number of demonstration projects investigating its efficacy, and integrated modeling assessments estimate that DAC could achieve approximately 3 Gt of CO_2 removal per year by 2100 putting it on par with assessments of BECCS.[158] It is possible, then, that under a best case scenario the combined use of all these CDR methods may have the potential to achieve nearly 8–10 Gt of CO_2 removal annually, but it is unclear at what point prior to 2100 these high rates could be reliably achieved.

There are a number of technology leaps, though, that would ease some of the limitations and improve the likelihood that these CDR strategies could be in full force at 10 Gt/y of CO_2 removal by 2050. Advances in synthetic biology and genetic engineering may create plants that are more heat tolerant and photosynthesize at higher rates on marginal soils reducing competition with food production. Engineering mobile, smaller scale, highly efficient DAC units may minimize the flow capture problem. The captured and concentrated CO_2 may be directly sequestered using scaled up EW strategies or DAC CO_2 could be converted to fuels and then recaptured at the end of its energy cycle for storage or sequestration in a manner analogous to BECCS. All of these strategies, however, require a market which would be created by pricing (or taxing) CO_2 emissions. Market incentives are needed to plant forests rather than clear them, to nudge fuel switching from fossil to bio-fuels, to sequester captured CO_2 safely rather than use it for enhanced oil recovery. It is extremely important to stress, that CDR alone can not achieve the warming targets of 1.5–2 °C, but must be deployed in combination with aggressive mitigation efforts if there is any chance to limit warming in this century to a range to which human societies can adapt.

Realistically, given the pace of mitigation and the readiness of CDR technologies, we are unlikely to hit net-zero CO_2 emissions targets by 2050 or the 500+ Gt reduction in cumulative atmospheric CO_2 levels by 2100. Uncertainty about the consequences of continuing down the business-as-usual path or any path that does not include aggressive decarbonization augmented by a building CDR strategy, is

---

[157] Smith L. et al. (2018). Chaos and the Flow Capture Problem: Polluting is Easy, Cleaning is Hard. *Physical Review Applied*, in press

[158] Smith P. et al. (2015). *Nature Climate Change,* 6:42–50. https://www.nature.com/articles/nclimate2870.pdf

receding and the certainty that the planet is on a trajectory of increasingly frequent, extreme climatic events, the evidence of which is clearly visible in the present, is gathering force. There is growing concern that planetary thresholds may be crossed even at 2 °C warming, beyond which a cascading series of climatic phenomena will unfold to propagate intensified warming, escalating sea level rise, and other serious climate disruptions.[159] A planetary threshold indicates a tipping point that triggers processes that cannot be reversed or controlled.

New research suggests that positive feedbacks such as permafrost melting may overtake anthropogenic CO_2 emissions to become the dominant driving force of the earth system putting it on a "Hothouse Earth" pathway at some point in the near future.[160] While there is no debate that the regulation of the earth climate system has thresholds and tipping points, there is uncertainty about the exact point at which thresholds will be crossed to cause runaway feedbacks and irreversible climate changes. The significance of the current research is that certain thresholds may be exceeded despite a 2 °C limit on warming causing an uncontrollable unraveling of the current earth climate system that can never be returned to its prior state. Thus, there is a growing risk that Plan A of climate action may not be protective enough to steer the planet away from climatic cataclysm. Therefore, emergency measures, as outlined in a Plan B for climate action, are necessary to hold the earth steady while human societies finally and forcefully implement the mitigation and CDR strategies of Plan A bringing the earth system back to the homeostatic point to which human societies are acclimated.

Hence, we have arrived at the place where global scale geoengineering involving albedo modification to halt the effects, not the primary drivers, of climate change is materializing as an inexorable strategy in our defensive arsenal for climate protection. The term "geoengineering" connotes a designed outcome based on mathematically determined prediction and implies a more precisely tailored and controllable response than may be the case for global scale climate intervention. Despite the fact that global scale SRM achieved by stratospheric aerosol injection is considered to be easy, cheap, and feasible, its engineering design faces three fundamental challenges. First, the design of SRM strategies is based on climate circulation models that at this point are more learning tools with which we test and interrogate our understanding of the planetary climate system than they are predictive tools that tell us with a high probability what climatic events will happen, when they will happen and where they will happen—essential information if engineering solutions are to be designed to prevent them. As explained, the earth's climate is a complex system and we are still learning about the couplings, feedbacks, and nonlinearity among its network of subsystems.

Secondly, the aim of SRM is to limit the increase in global *average* annual temperature, but this is a very imprecise design goal as a vast number of spatial and

---

[159] Harvey C. ClimateWire (08.07.18). https://www.eenews.net/climatewire/2018/08/07/stories/1060092901

[160] Steffen W. et al. (2018). Trajectories of the Earth System in the Anthropocene. *PNAS*, 115:33:8252–2859. http://www.pnas.org/content/115/33/8252

temporal patterns can yield the same global mean value annually. While we have a general understanding that increases in atmospheric CO_2 cause increases in global temperature, the temperature response is highly non-uniform across the earth's surface. Recent research findings indicate that despite stabilized average global temperatures, rising CO_2 levels may directly lead to more extreme weather and climate disasters.[161] Although a range of atmospheric CO_2 concentrations are consistent with 1.5 °C warming, as determined using an ensemble of model simulations, in certain parts of the world (Northern Hemisphere) higher CO_2 levels resulted in greater risks of extreme heat and tropical precipitation.[162] These findings are explained by the fact that CO_2 levels may alter precipitation and atmospheric circulation independent of average warming and suggest that there may be CO_2 thresholds as well as temperature thresholds that need to be addressed with climate policy. In addition, these results reveal that we do not have a firm understanding of climate sensitivity to CO_2. Therefore, it may be prudent not only to seek lower caps on global average temperature, but also on the planet's carbon budget.

Finally, since climate modeling and global average temperatures do not provide local resolution of climate responses, we are unable to predict with certainty what areas will benefit and what areas will suffer from global scale SRM deployment. There are general claims that lowered surface temperatures will enhance agricultural productivity and plant photosynthesis by lessening heat stress and promoting CO_2 fertilization.[163] Yet, detailed analyses reveal highly variable results. For instance, in a modeling study of SRM impacts on Chinese agriculture, rice production was not found to be affected, even with a sudden termination of SRM, but corn production showed potential increases with SRM and decreases with SRM termination.[164] A new study asserts that SRM will do little to protect the world's major crops due to the fact that positive cooling effects are offset by negative light scattering effects which reduce the total amount of light hitting the earth's surface.[165] By analyzing crop yields as a function of the quality and quantity of sunlight after recent volcanic eruptions, this research reveals a number of intriguing insights. Unmanaged ecosystems respond differently than crops to the effects of altered sunlight (scattered & diffuse versus total insolation), as do different types of crops (C3 – soy, rice,

---

[161] Harvey C. CO_2 can sharpen extreme weather without higher temps. ClimateWire (06.21.18). https://www.eenews.net/climatewire/2018/06/21/stories/1060085723

[162] Baker H.S. et al. (2018). Higher CO_2 concentrations increase extreme event risk in a 1.5 °C world. *Nature Climate Change,* 8:604–608. https://www.nature.com/articles/s41558-018-0190-1

[163] Pongratz J. et al. (2012). Crop yields in a geoengineered climate. *Nature Climate Change,* 2:101–105. https://www.nature.com/articles/nclimate1373

[164] Xia L. et al. (2014). Solar radiation management impacts on agriculture in China: A case study in the Geoengineering Model Intercomparison Project (GeoMI). *Journal of Geophysical Research: Atmospheres,* 119:8695–8711; DOI: https://doi.org/10.1002/2013JD020630. https://agupubs.onlinelibrary.wiley.com/doi/epdf/10.1002/2013JD020630

[165] Harvey C. Manipulating sun rays won't help crops grown. ClimateWire (08.09.18). https://www.eenews.net/climatewire/2018/08/09/stories/1060093717

wheat & C4 – corn), and failure to account for the dimming of total sunlight overestimates SRM benefits to agriculture.[166]

There are also concerns that SRM could adversely affect precipitation to threaten regional food security. Many general climate models show that SRM effects on precipitation are small and are compensated by the positive effects of cooling and CO_2 fertilization on photosynthesis. But actual data, rather than modeled results, from the 1991 Mount Pinatubo eruption, the natural experiment that forms the basis of the SRM proposal, show that there was a substantial decrease in precipitation over land and a record decline in runoff and river discharge to the ocean.[167] The dimming of solar radiation by SRM does not simply reduce incoming sunlight or solar energy and hence outgoing radiation; SRM disrupts the flow of energy associated with the evaporation and condensation of water that drives the hydrologic cycle. The absorbance of solar energy, in the form of sunlight, at the earth's surface drives the evaporation (requiring energy) of water to water vapor which can then be transported long distances, hundreds of kilometers, before it condenses to form rain or snow releasing energy back into the atmosphere as latent heat. Basically, with less sunlight hitting the earth's surface, there is less evaporation. The integration of atmospheric and surface energy budgets and the hydrologic cycle are not captured well by climate models and as a result the details and long range dynamics of precipitation patterns are poorly simulated, particularly in the tropics.[168] For instance, the same amount of precipitation can occur in an area, but may be delivered in a series of small storms or one large event. In the latter case, flooding is likely to occur and less water will be stored locally. The physics underlying the hydrologic cycle indicate that SRM may have deleterious impacts on regional precipitation patterns, although the interactions are not fully incorporated into climate models.[169] A more recent effort to improve the modeling of aerosol effects on regional precipitation illustrate that increased aerosol levels cool land surfaces and reduce precipitation over land, with the most significant regional changes happening in the tropics.[170]

Despite these challenges, there is a strong attraction to SRM methods, particularly stratospheric aerosol injection, because it could be deployed quickly and cheaply with immediate, albeit temporary, cooling effects, none of which is achievable by mitigation and/or CDR. In fact, much of the justification of SRM is rooted

---

[166] Proctor J. et al. (2018). Estimating global agricultural effects of geoengineering using volcanic eruptions. *Nature,* 560:480–483. https://www.nature.com/articles/s41586-018-0417-3

[167] Trenberth K.E. & Dai A. (2007). Effects of Mount Pinatubo volcanic eruption on the hydrological cycle as an analog of geoengineering. *Geophysical Research Letters,* 34:L15702; doi:https://doi.org/10.1029/2007GL030524
https://agupubs.onlinelibrary.wiley.com/doi/abs/10.1029/2007GL030524

[168] Trenberth K.E. (2011). Changes in precipitation with climate change. *Climate Research,* 47:123–138; doi: https://doi.org/10.3354/cr00953. https://www.int-res.com/articles/cr_oa/c047p123.pdf

[169] Ibid.

[170] Richardson T.B. et al. (2016). Understanding the Rapid Precipitation Response to CO_2 and Aerosol Forcing on a Regional Scale. *Journal of Climate,* 29:583–594; DOI: https://doi.org/10.1175/JCLI-D-15-0174.1. https://journals.ametsoc.org/doi/pdf/10.1175/JCLI-D-15-0174.1

in the hopelessness of the alternatives.[171] But, additionally, our faith in the power of extreme engineering projects to redirect natural processes lulls us into believing that it will be relatively simple to install a sunshade in the earth's upper atmosphere allowing us to tune sunlight's penetration to only favorable outcomes (and thus, eliminating the negative effects of the spike loadings of volcanos) and providing a respite to human societies as they struggle to get their acts together on the far more difficult tasks of decarbonizing economies and reversing biosphere degradation. Proposals to improve on stratospheric aerosol injection are appearing. For instance, cirrus cloud thinning may be a technique to convert cirrus cloud warming to a net cooling potential without adverse precipitation effects since a part of the cooling mechanism is due to water vapor removal. Furthermore, this approach may also lend itself more to smaller scale, geographical targeting and more temporary effects than stratospheric aerosols.[172]

Overall, the efficacy of SRM strategies has been based primarily on modeled simulations and the field results that natural experiments in the form of volcanic eruptions have afforded scientists. There are a number of discrepancies, though, between what the models tell us and what we have observed with aerosols formed by volcanoes, particularly with respect to altered precipitation patterns. Theoretically, there are major, if unknown, differences between aerosol injection by a massive volcanic eruption and the methodic, systematic dispersal of a thin layer of aerosols in the stratosphere. Question remain, however, about the precise dimming response per quantity of a particular type of aerosols and whether it is possible to modulate the degree to which sunlight is dimmed relative to specific locations. Since many climate phenomena are not simply a function of local temperatures, but rather are caused by complex interactions at a distance and by gradients in temperatures, the notion that stratospheric aerosol injection can be deployed successfully with ease is misleading. Moreover, developing the ability to tune the cooling response of aerosol injection to subsequently reverse climate changes such as Arctic thawing is complicated by the lengthy time lags that characterize the complex earth system. Extensive experimentation to gather data with which to formulate and then validate models is imperative. The experimental program to iron out all these questions requires long-term funding commitments that prompt additional questions as to which agencies would fund the research or would a new agency be formed, who profits, and how are intellectual property issues resolved.

Large scale experiments rarely take place without surprising and unexpected consequences. Let's consider the process of urbanization alone, which, in the absence of adaptive, sustainable design and apart from GHG effects, is expected to

[171] Kolbert E. Can Carbon-Dioxide Removal Save the World? Annals of Science, The New Yorker (11.20.17). https://www.newyorker.com/magazine/2017/11/20/can-carbon-dioxide-removal-save-the-world

[172] Lohmann U. & Gasparini B. (2017). A cirrus cloud climate dial? Cirrus cloud seeding may help to reduce climate warming, but large uncertainties remain. *Science*, 357:6348:248–249; DOI: https://doi.org/10.1126/science.aan3325. http://science.sciencemag.org/content/sci/357/6348/248.full.pdf

increase regional temperatures 1–2 °C by 2100,[173] compounding global scale climate changes. Using a suite of regional climate models, researchers interrogated the effects of a variety of albedo enhancing and cooling strategies common to urban adaptation efforts across a range of geographies and climates in the U.S. for minimum and maximum urban expansion conditions. In all cases, the maximum extent of urbanization relative to a base condition showed at least 1 °C increase in summertime temperatures and in some cases 2–3 °C. Yet, the maximum extent of warming due to urban development varied geographically with the Great Lakes region warming more than southern Florida. Similarly, the response to modifying the urban surface either with green roofs, cool (highly reflective) roofs or a combination shows variable effects on cooling (negligible to about 2 °C) as a function of region, although in all cases the surfaces modified to have high albedo (cool roofs) showed the greatest increment of cooling and those that were highly transpiring (green) and high albedo (cool and reflective) resulted in the greatest overall decrease in temperature.

Urban adaptation measures, however, also exerted surprising effects on precipitation. Where little to no effects were observed in arid regions such as Southern California or Arizona, the implementation of cool, reflective surfaces diminished precipitation between 2–4 mm/day along the eastern seaboard of the U.S. from Florida to Maine, whereas precipitation increased around 1 mm/d from Texas to the Great Lakes region. Enhancing the albedo of urban surfaces in regions of burgeoning urbanization is an effective way to cool both the local and regional temperatures, but there are unintended consequences and tradeoffs to this "easy, cheap, and feasible" tactic. Cool surfaces may reduce energy demands in the summer, but will increase them in the winter. At the earth's surface, albedo enhancement has a profound and variable effect on precipitation patterns over large distances. Imagine the challenge of tuning the albedo of the upper atmosphere to achieve cooling effects with minimum adverse effect on precipitation at the earth's surface.

The sole aim of SRM is to reduce the risks of climbing temperatures beyond an unknown threshold and to dampen the threats of extreme events associated with elevated global temperatures. Yet, without a deeper understanding of the responses of the earth system to stratospheric aerosol injection there is a long list of potential, unintended, and negative consequences that could be experienced.[174] Stratospheric aerosols only remain in the upper atmosphere for 1–2 years and if not replaced, their effect is temporary. Once initiated, then, aerosol injection must be continued indefinitely until GHG levels are brought back down. Since the sudden suspension of SRM will cause an abrupt jump in global temperatures and climatic disasters, there is potential for the creation of another large-scale technological lock-in, analogous

[173] Georgescu M. et al. (2014). Urban Adaptation can roll back warming of emerging megapolitan regions. *PNAS,* 111:8:2909–2914; https://doi.org/10.1073/pnas.1322280111 . http://www.pnas.org/content/111/8/2909

[174] Robock A. (2008). 20 reasons why geoengineering may be a bad idea. *Bulletin of the Atomic Scientists,* 64:2:14–18 (59); DOI: https://doi.org/10.2968/064002006. http://climate.envsci.rutgers.edu/pdf/20Reasons.pdf

to the very fossil fuel lock that currently grips us. Aerosol injection will slow the recovery of the ozone depletion in the stratosphere. The use of sulfate to form aerosols will contribute over time to both ocean acidification and acid deposition at the earth's surface. There are unknown and unquantified effects on cirrus cloud formation at the boundary of the tropo- and stratospheres and fears that the aerosols could seed ice crystals in these thin, wispy clouds which have a strong warming, rather than cooling effect. The dimming of sunlight by aerosols will whiten the skies eliminating its blue color, but creating spectacular sunsets. Such dimming reduces insolation intensity and thus, will have a negative effect on solar power generation, at the very time we need to be shifting to renewable sources of power. And then there are all the potential shortcomings inherent to human endeavors on any scale: errors, cooperation versus dissent, commercial versus public control, international versus national governance, peaceful versus weaponized uses.

But perhaps the greatest hazard of SRM is a moral one, that any success in cooling global temperatures will undermine international resolve to reduce CO_2 emissions, end fossil fuel use, and remake the human enterprise in ecologically inspired ways. The grand scale of SRM galvanizes the adoption of other extreme undertakings such as geoengineering polar glaciers to halt or delay their flow into the oceans and thereby, retard sea level rise.[175] The steady and rapid rise in sea level is well underway and as stated above, there is serious doubt that cooling global temperatures by SRM will stem the tide of ice loss in Greenland and Antarctica over the next decades to ward off severe coastal flooding and devastating storm surges. In order to defend against current and near term SLR, nations will spend collectively trillions of dollars to construct and maintain sea walls and flood barriers in order to protect coastal populations and assets. Why not redirect some of this investment to buttress ice shelves and disrupt glacial flow to the oceans? Although very little attention has been given to this task, there are a variety of possible ways to dampen the fast movement of Greenland and Antarctic ice sheets. Embankments or berms could be constructed across fjords or bays to stop glacial movement, allowing thinning glaciers to thicken and refreeze. There would be ecological penalties due to diminished inputs of nutrients to marine ecosystems and temporary disturbances associated with construction. But such smaller scale efforts could serve as test cases to determine the efficacy of glacier geoengineering.

A second approach would be to reinforce ice shelves from calving ice bergs, as was observed in 2017 with the Larsen C ice shelf, by pinning them in place with artificial islands or outcroppings built up from the sea floor. Massive quantities of material would be required and construction in the cold, rough Antarctic seas could be perilous. Another technique might be to remove the thin layer of melt water that lubricates and heats with friction the base of glaciers, speeding glacier slippage. In Greenland, the melt water at the base of glaciers is much greater than with Antarctic ice sheets; thus, in the latter case, it may be possible to establish pumping stations

[175] Moore J.C. et al. (2018). Geoengineering polar glaciers to slow sea-level rise. *Nature*, 555:303–305. https://www.nature.com/magazine-assets/d41586-018-03036-4/d41586-018-03036-4.pdf

to extract or freeze the water at the glacier base and thus hinder slippage. Overall, the costs of these endeavors are estimated to be comparable to typical large scale energy or civil engineering projects. There are innumerable unknowns accompanying these possible actions, particularly since so much of glacial dynamics occurs below the surface, beyond our ability to observe and measure. We do not know exactly what is happening 1000s of meters into glaciers' depths, how oceans circulate beneath ice shelves, how glaciers slide and erode at their base, and how much stress a pinned ice shelf can sustain before it fractures. Only experimentation with these ideas will yield answers to all the questions and determine if glacier geoengineering is a wild distraction from the resolute work of decarbonizing the world economy or like SRM, urgently vital to buying time.

7 Cautionary Tales: 3. Our Ability to Recognize a Crisis

Located in a region of the world that is beginning to show the strains of climate change, Cape Town, the second largest city in South Africa, had been heralded both for its progressive environmental policies and its management of water. Since the early 2000s the city's population grew by 30%, but its water demand remained relatively constant. Adapted to decadal cycles of rain, drought, water restrictions and recovery, the most recent of which occurred in 2004–2005, Cape Town relied on 6 dammed surface waters to supply approximately 97% of its water and on a warning system triggered when water levels dropped below a certain elevation. In 2007, the national Department of Water and Sanitation warned the city and the Western Cape province that it needed to diversify and develop new sources of water by 2015, but the city was so efficient in its response to these warnings, eliminating leaks, conserving, repairing the distribution and plumbing systems, that it pushed the projections of shortages until 2019 and did not bring any new sources on line. Reservoirs were full in the wet years of 2013–2014 and the following year the Cape Town received an adaptation implementation prize from the C40 Cities Climate Leadership Group.[176] But then, for the next 3 years, the most severe drought ever experienced pressed down on Cape Town and the region.

As 2017 closed and with the largest water supply reservoir falling below 15% capacity, Cape Town made global headlines by declaring that *Day Zero*, the day taps would run dry, loomed just months away. How does a cosmopolitan city recognized for its environmental stewardship suddenly become branded by the distinction of watching its water sources evaporate? The explanation is only partly that city officials did not immediately recognize that current conditions differed substantially from those of the past. If rains did not replenish the very shallow, large surface area reservoirs that were subject to high evaporative losses, the region's water supply

[176] Onishi N. & Sengupta S. Dangerously Low on Water, Cape Town Now Faces 'Day Zero.' The New York Times (01.30.18). https://www.nytimes.com/2018/01/30/world/africa/cape-town-day-zero.html

was not reliable. Water managers began to install small water desalination units, which are energy intensive, expensive and slow to put in place. Eventually, Cape Town officials realized that they needed to develop groundwater and recycled supplies, but the national government denied them funding and continued to make full allocations of water to the agricultural sector, which in the Western Cape province were powerful vineyard, orchard and livestock owners The national government refused to provide emergency funding for the crisis as it loomed claiming that it had yet to happen.[177] Ultimately, Cape Town avoided Day Zero in 2018, but only because of the extraordinary conservation efforts of its citizens who brought down their water use to less than 50 liters per day.

8 Lessons Learned: The New Normal

As one expert wrote, "Cape Town teaches us that water crises are rarely a matter of rainfall".[178] And as is often the case, the political wrangling between opposing political parties played a role in Cape Town's skirmish with disaster. The Western Cape province and Cape Town are governed by the Democratic Alliance which is the opposing political party to the African National Congress which governs the rest of the country. While the two political forces were not united in their efforts, each in their own way failed to recognize early indicators that the drought was far outside normal cycles.

There are few places that execute water conservation to maximize a limited resource supply as well as water managers in Cape Town. Yet, they advanced water conservation, a practice that served them well in the past, rather than diversify and develop new water sources that were less vulnerable to extreme drought. The national government refused to acknowledge the growing disaster and exacerbated it by refusing to revise the agricultural allocations. They chose not to provide relief in prevention of the disaster, only once the full crisis unfolded. In other words, they favored reaction over action. While extensive planning characterizes water management in Cape Town, it was planning for past events, not the new normal of a changing climate.

The drama of Day Zero is a vivid illustration of the escalating risks of extreme, recurring droughts that are a particular threat to the African continent with climate change. It is difficult to attribute this recent drought, no matter how extreme, exclusively to climate change, but climate models do predict that the Western Cape region is becoming warmer and drier and rainfall patterns are changing. Once Cape Town officials realized the magnitude of the problem, they ignored low-hanging fruit in the form of groundwater resources and water recycling in favor of a high tech

[177] Olivier D.W. Cape Town's water crisis: driven by politics more than drought. The Conversation (12.12.17). https://theconversation.com/cape-towns-water-crisis-driven-by-politics-more-than-drought-88191

[178] Ibid.

solution, water desalination, that takes years to reach operation, even in the case of the small scale units being explored in Cape Town. In the panic of the crisis even such extreme ideas as towing an ice berg from Antarctica were discussed.

About half the world's population faces water scarcity issues for some portion of the year and about a billion lack outright access to water. By 2030, global demand for fresh water is expected to outstrip supply by 40% due to climate change, anthropogenic deterioration of the biosphere, and population growth. Cape Town is not alone in facing down the threat of running out of drinking water in a modern era; worldwide, there are at least 11 other cities facing similar crises, including post-industrial cities such as London, Tokyo, and Miami, cities plagued by pollution such as Beijing, Bangalore, Cairo, and Moscow, and cities with inadequate infrastructure and growing water scarcity such as São Paulo, Jakarta, Istanbul, and Mexico City.[179] In all cases, climate change exacerbates other drivers such as contamination, subsidence, and salt water intrusion due to sea level rise.

9 Conclusions

The geoengineering of the planet has been underway at an ever proliferating rate since before the industrial revolution of the mid-nineteenth century. We can observe its mark in altered geochemical cycles, in astounding extinction rates and biodiversity losses, in desertification, in ozone depletion, in aquifer drawdown, in contaminant bioaccumulation, and in climate change. In the view of many earth scientists, the culmination of these anthropogenic changes is the transition to a new epoch, the Anthropocene, in which atmospheric, geologic, hydrologic, biospheric, and other earth system processes are now driven by human actions.[180] Since the earth is a complex system of highly interconnected networks, anthropogenic biosphere deterioration not only contributes significantly to climate change (e.g., deforestation, urbanization, industrial agriculture, etc.), but also severely undercuts the ability of human and natural systems to adapt to a changing climate. When considered in this larger context, climate change is a "threat multiplier" exacerbating the tangle of scientific, technological, social, political, and economic challenges facing societies and threatening national security. In 2014 the U.S. Secretary of Defense, Chuck Hagel, declared in a speech at the Conference of Defense Ministers of the Americas in Arequipa, Peru, "Rising global temperatures, changing precipitation patterns, climbing sea levels and more extreme weather events will intensify the challenges of global instability, hunger, poverty and conflict. They will likely lead to food and

[179] The 11 cities most likely to run out of drinking water- like Cape Town. BBC News (02.11.18). https://www.bbc.com/news/world-42982959

[180] Welcome to the Antrhopocene. http://www.anthropocene.info

water shortages, pandemic disease, disputes over refugees and resources and destruction by natural disasters in regions across the globe."[181]

Our understanding about the relationship between CO_2 and climate has been been evolving since the the 1850s when John Tyndall, an Irish physicist, discovered that CO_2 absorbed infrared radiation, a phenomenon forming the basis of the greenhouse effect. In the late 1890s Svante Arrhenius, a Swedish chemist and Nobel prize winner, hypothesized that the CO_2 released in the burning of fossil fuels could increase global temperatures, although he thought this would take centuries. In the 1950s oil companies studied the very climate effects that Arrhenius predicted, understood their seriousness and were puzzled that something so far reaching was being ignored by public, who instead, were more obsessed by discrete pollution issues, such as benzene emissions.[182] Provocatively, Nathaniel Rich wrote in the New York Times Magazine that nearly everything about climate change has been known since 1979, although we know these facts with greater certainty today. "All the facts were known, and nothing stood in our way. Nothing, that is, except ourselves."[183]

As this chapter outlines, climate change is not a particularly difficult technical problem to solve, given enough time. There are an enormous number of mitigation actions and a plethora of synergies and cascading benefits to be exploited among mitigation, ecologically based CDR and adaptation endeavors. Rather, climate change is an insurmountable political problem. It has been dubbed the quintessential super wicked problem, one that not only defies resolution because of its complex nature (e.g., enormous interdependencies, uncertainties, circularities) and clashing stakeholders, but also, because it possesses three additional complicating features: (1) the longer the problem persists, the more difficult it becomes to address; (2) those in the best position to solve it have the least incentive to do so and will likely suffer the least; and (3) institutional governmental frameworks to coordinate efforts to tackle a global problem stretching over such far reaching spatial and temporal scopes do not exist.[184]

In not managing the imposition of the many solutions we have long recognized as imperative, we are faced with managing a crisis. And as far as crisis management goes, the geoengineering of direct air capture, solar radiation management or glacier containment would be the "mother of all engineering projects" for three basic

---

[181] Banusiewicz J.D. Hagel to Address 'Threat Multiplier' of Climate Change. DoD News (10,13,14), Defense Media Activity. U.S. Department of Defense. https://dod.defense.gov/News/Article/Article/603440/

[182] Rich N. Losing Earth: The Decade We Almost Stopped Climate Change. The New York Times (08.01.18). https://www.nytimes.com/interactive/2018/08/01/magazine/climate-change-losing-earth.html

[183] Ibid.

[184] Lazarus R. J. (2009). Super Wicked Problems and Climate Change: Restraining the Present to Liberate the Future. *Cornell Law Review*, 94:1153–1233. https://scholarship.law.georgetown.edu/cgi/viewcontent.cgi?referer=http://scholar.google.com/&httpsredir=1&article=1152&context=facpub

reasons – the massive scale at which the technologies must be deployed, the need to integrate these actions with mitigation, adaptation and other CDR measures and the necessity of maintaining flexible designs since they will have to be adjusted as we learn how the climate system responds. Geoengineering is the ultimate exercise of the engineering mentality. SRM offers a seductively simple pause button to the super wicked problem of climate change and the promise of lessening injury while we search for other technical fixes. In a sense, geoengineering is emergency, life-support engineering at a technological, economic and political scale that defies both logic and perhaps, feasibility.

There are many ironies that surround our response to climate action. For instance, the most significant public works engineering projects of the twenty-first century are likely to be aimed at correcting the environmental damage of the major public works engineering projects of the twentieth century. The ultimate irony of the climate action debate, however, is that inaction is justified by the costs to the current economic order, despite the fact that, according to the DDPP, "The economic, social, and environmental risks of unabated climate change are immense. They threaten to roll back the fruits of decades of growth and development, undermine prosperity, and jeopardize countries' ability to achieve even the most basic socio-economic development goals in the future, including the eradication of poverty and continued economic growth. These risks affect all developed and developing countries alike."[185] Thus, international governance must place climate stabilization on par with economic development, human rights, democracy and peace.[186]

For exactly these reasons there is an enormous procrastination penalty associated with the slow pace of real climate action, such that it has brought us to the point where geoengineering in its many forms is now "vital without necessarily being viable.[187] As Elizabeth Kolbert wrote in a 2017 New Yorker piece, geoengineering encompasses technologies of last resort and by virtue of its nature and scale, it is also paradoxical. It may be impossible to manage and govern, but it is beginning to seem that it may also be impossible to manage and govern without.

References

1. Anastas, P.T., Zimmerman, J.B.: Design through the twelve principles of green engineering. Environ. Sci. Technol. **37**(5), 94A–101A (2003)
2. Arango, T., Medina, J.: California fire now the largest in state history: 'people are on edge.' The New York Times (08.07.2018). https://www.nytimes.com/2018/08/07/us/california-fires-mendocino.html

[185] Sachs J. et al. (2014) DDPP 2014 report, p VII. http://deepdecarbonization.org/wp-content/uploads/2015/06/DDPP_Digit.pdf

[186] Rockström J. et al. (2017). *Science*, 355:6331:1269–1271. http://science.sciencemag.org/content/355/6331/1269.full

[187] Kolbert E. The New Yorker (11.20.17). https://www.newyorker.com/magazine/2017/11/20/can-carbon-dioxide-removal-save-the-world

3. Baker, H.S., et al.: Higher CO_2 concentrations increase extreme event risk in a 1.5 °C world. Nat. Clim. Change **8**, 604–608 (2018) https://www.nature.com/articles/s41558-018-0190-1
4. Bentley, C.: Holland is relocating homes to make more room for high water. Public Radio International (PRI) (06.22.2016). https://www.pri.org/stories/2016-06-22/holland-relocating-homes-make-more-room-high-water
5. Biello, D.: How far can technology go to stave off climate change? Yale Environment 360 (01.18.2017). https://e360.yale.edu/features/how_far_can_technology_go_to_stave_off_climate_change
6. Bullis, K.: Carbon-Capturing Rock. Geologists discover that certain rock formations could sequester large amounts of carbon dioxide. MIT Technology Review (11.04.2008). https://www.technologyreview.com/s/411129/carbon-capturing-rock/
7. Carrington, D.: CO_2 turned into stone in Iceland in climate change breakthrough. The Guardian (06.09.2016). https://www.theguardian.com/environment/2016/jun/09/co2-turned-into-stone-in-iceland-in-climate-change-breakthrough
8. Chandler, D.L.: Deadly heat waves could hit South Asia this century. Without action, climate change could devastate a region home to one-fifth of humanity study finds. MIT News (08.02.2017). http://news.mit.edu/2017/deadly-heat-waves-could-hit-south-asia-century-0802
9. Climate change is making the Arab world more miserable. The Economist (05.31.2018). https://www.economist.com/middle-east-and-africa/2018/05/31/climate-change-is-making-the-arab-world-more-miserable
10. ClimateWire.: How the Dutch Make "Room for the River" by Redesigning Cities. Scientific American (01.20.2012). https://www.scientificamerican.com/article/how-the-dutch-make-room-for-the-river/
11. Crutzen, P.J.: Clim. Change. **77**, 211–219 (2006) https://link.springer.com/content/pdf/10.1007/s10584-006-9101-y.pdf
12. Daniels, S.: FutureGen 'clean-coal' plant is dead. Crain's Chicago Business (02.03.2015). http://www.chicagobusiness.com/article/20150203/NEWS11/150209921/futuregen-clean-coal-plant-in-illinois-is-killed-by-obama-administration
13. Davis, S., et al.: Net-zero emissions energy systems. Science. **360**, 6396 (2018). https://doi.org/10.1126/science.aas9793. http://science.sciencemag.org/content/360/6396/eaas9793.full
14. DeLonge, M.S., et al.: A lifecycle model to evaluate carbon sequestration potential and greenhouse gas dynamics of managed grasslands. Ecosystems. **16**(6), 962–979 (2013) https://link.springer.com/article/10.1007/s10021-013-9660-5
15. Erickson, B.E.: Interest in biochar surges. Chem. Eng. News. **94**(10), 40–44 (2016) https://cen.acs.org/articles/94/i10/Interest-biochar-surges.html
16. Garfield, L.: Manhattan plans to build a massive $1 billion wall and park to guard against the next inevitable superstorm. Business Insider (04.27.2018). https://www.businessinsider.com/new-york-city-flooding-manhattan-coastal-barriers-2018-4
17. Gaunt, J.L., Lehmann, J.: Energy balance and emissions associated with biochar sequestration and pyrolysis bioenergy production. Environ. Sci. Technol. **42**, 4152–4158 (2008) https://pubs.acs.org/doi/pdf/10.1021/es071361i
18. Georgescu, M., et al.: Urban Adaptation can roll back warming of emerging megapolitan regions. PNAS. **111**(8), 2909–2914 (2014). https://doi.org/10.1073/pnas.1322280111. http://www.pnas.org/content/111/8/2909
19. Gilroy, W.G.: Changes proposed to New Orleans area levee systems. Science Daily (07.24.2013). https://www.sciencedaily.com/releases/2013/07/130724200557.htm
20. Goodell, J.: The water will come, pp. 149–234. Little, Brown & Co, New York (2017)
21. Goodell, J.: The water will come, p. 238. Little, Brown & Co, New York). Chapter 11 (2017)
22. Goodell, J.: Rising waters: can a massive barrier save Venice from drowning. Yale Environment 360 (12.05.2017). https://e360.yale.edu/features/rising-waters-can-a-massive-sea-barrier-save-venice-from-drowning
23. Graeter, K.A., et al.: Ice Core Records of West Greenland Melt and Climate Forcing. Geophys. Res. Lett. **45**(7), 3164–3172 (2018) https://doi.org/10.1002/2017GL076641

24. Gray, A.: Countries are announcing plans to phase out petrol and diesel cars. Is yours on the list? World Economic Forum (09.26.2018). https://www.weforum.org/agenda/2017/09/countries-are-announcing-plans-to-phase-out-petrol-and-diesel-cars-is-yours-on-the-list/
25. Griffin, R.: The U.S. is losing ground in the race for energy efficiency. Bloomberg (06.26.2018). https://www.bloomberg.com/news/articles/2018-06-26/the-u-s-is-losing-ground-in-the-race-for-energy-efficiency
26. Guckenheimer, J., Ottino, J.M.: Foundations for complex systems research in the physical sciences and engineering. NSF workshop report (September 2008). http://mixing.chem-biol-eng.northwestern.edu/docs/nsf_complex_systems_FINAL.pdf
27. Harris, J.: Poyang Lake, Yangtze River Basin, China. In: Finlayson, C.M., Milton, G.R., Prentice, R.C., Davidson, N.C. (eds.) The Wetland Book, vol. 2018. Springer Nature Switzerland (2016) https://link.springer.com/referenceworkentry/10.1007%2F978-94-007-6173-5_34-2
28. Harvey, C.: Cement's CO_2 is everywhere. Will it sink climate goals? ClimateWire (07.09.2018). https://www.eenews.net/climatewire/stories/1060088153
29. Harvey, C.: CO_2 can sharpen extreme weather without higher temps. ClimateWire (06.21.2018). https://www.eenews.net/climatewire/2018/06/21/stories/1060085723
30. Harvey, C.: Even 2 °C of warming could turn Earth into 'hothouse.' ClimateWire (08.07.2018). https://www.eenews.net/climatewire/2018/08/07/stories/1060092901
31. Harvey, C.: Forests had a really bad year. ClimateWire (06.28.2018). https://www.eenews.net/climatewire/2018/06/28/stories/1060087181
32. Harvey, C.: Ice is melting 3 times as fast as it did 25 years ago. ClimateWire (06.14.2018). https://www.eenews.net/climatewire/2018/06/14/stories/1060084477
33. Harvey, C.: Manipulating sun rays won't help crops grown. ClimateWire (08.09.2018). https://www.eenews.net/climatewire/2018/08/09/stories/1060093717
34. Harvey, C.: Trees are losing their ability to soak up CO_2. ClimateWire (07.13.2018). https://www.eenews.net/climatewire/2018/07/13/stories/1060088955
35. Harvey, C.: Waters on track to rise for centuries, even if emissions stop. ClimateWire (02.21.2018). https://www.eenews.net/climatewire/2018/02/21/stories/1060074341
36. Hobson, M.K.: Sea ice hits 2nd-lowest level in 39 years. ClimateWire (03.26.2018). https://www.eenews.net/climatewire/2018/03/26/stories/1060077383
37. International Energy Agency: Global Energy & CO_2 Status Report 2017 (OECD/IEA, March, 2018). https://www.iea.org/publications/freepublications/publication/GECO2017.pdf
38. Irfan, U.: World must pull CO_2 from the sky to meet Paris goals. ClimateWire (03.24.2017). https://www.eenews.net/climatewire/stories/1060052028/
39. Ives, M.: As China's largest freshwater Lake Shrinks, solution faces criticism. New York Times (12.28.2016). https://www.nytimes.com/2016/12/28/world/asia/china-lake-poyang-finless-porpoise.html
40. Jevrejeva, A., et al.: Flood damage costs under the sea level rise with warming of 1.5 °C and 2 °C. Environ. Res. Lett. **13**, 074014 (2018) http://iopscience.iop.org/article/10.1088/1748-9326/aacc76/pdf
41. Katz, C.: To control floods, the Dutch turn to nature for inspiration. Yale Environmental 360 (02.21.2013). https://e360.yale.edu/features/to_control_floods_the_dutch_turn_to_nature_for_inspiration
42. Keith, D.: A Case for Climate Engineering, 224 pp. MIT Press, Boston, MA (2013)
43. Keith, D.W., et al.: Process for Capturing CO_2 from the Atmosphere. Joule. **2**(8), 1573–1594 (2018) https://www.cell.com/joule/fulltext/S2542-4351(18)30225-3
44. Kelemen, P.B., Matter, J.: In situ carbonation of peridotite for CO_2 storage. Science. **105**(45), 17295–17300 (2008) http://www.pnas.org/content/105/45/17295
45. Kilmont, Z., et al.: The last decade of global anthropogenic sulfur dioxide: 2000–2011 emissions. Environ. Res. Lett. **8**, 014003 (2013) http://iopscience.iop.org/article/10.1088/1748-9326/8/1/014003

46. Kimmelman, M., Haner, J.: The Dutch have solutions to rising seas. The world is watching, in changing climate, changing politics. The New York Times (06.15.2017). https://www.nytimes.com/interactive/2017/06/15/world/europe/climate-change-rotterdam.html
47. Kolbert, E.: Can carbon-dioxide removal save the world? Annals of Science, The New Yorker (11.20.2017). https://www.newyorker.com/magazine/2017/11/20/can-carbon-dioxide-removal-save-the-world
48. Kumar, A., et al.: Direct Air Capture of CO_2 by Physisorbent Materials. Angew. Chem. **54**, 14372–14,377 (2015) https://onlinelibrary.wiley.com/doi/full/10.1002/anie.201506952
49. Lazarus, R.J.: Super Wicked Problems and Climate Change: Restraining the Present to Liberate the Future. Cornell Law Rev. **94**, 1153–1233 (2009) https://scholarship.law.georgetown.edu/cgi/viewcontent.cgi?referer=http://scholar.google.com/&httpsredir=1&article=1152&context=facpub
50. Lehmann, J.: Bio-energy in the black. Front. Ecol. Environ. **5**(7), 381–387 (2007) https://esajournals.onlinelibrary.wiley.com/doi/full/10.1890/1540-9295%282007%295%5B381%3ABITB%5D2.0.CO%3B2
51. Lindsay, R.: Climate change: global sea level. Climate.gov, NOAA (08.01.2018) https://www.climate.gov/news-features/understanding-climate/climate-change-global-sea-level
52. Lohmann, U., Gasparini, B.: A cirrus cloud climate dial? Cirrus cloud seeding may help to reduce climate warming, but large uncertainties remain. Science. **357**(6348), 248–249 (2017). https://doi.org/10.1126/science.aan3325. http://science.sciencemag.org/content/sci/357/6348/248.full.pdf
53. McDonough, W., Braungart, M.: Cradle to Cradle. Farrar, Straus & Giroux, NY, NY (2002)
54. McGeehan, P., Hu, W.: Five years after sandy, are we better prepared? The New York Times (10.29.2017). https://www.nytimes.com/2017/10/29/nyregion/five-years-after-sandy-are-we-better-prepared.html
55. McLeman, R.: Migration and displacement risks due to mean sea-level rise. Bulletin of the Atomic Scientists (05.04.2018). https://thebulletin.org/2018/05/migration-and-displacement-risks-due-to-mean-sea-level-rise/
56. Mengel, M., et al.: Committed sea-level rise under the Paris Agreement and the legacy of delayed mitigation action. Nat. Commun. **9**, 601 (2018) https://www.nature.com/articles/s41467-018-02985-8
57. Moore, J.C., et al.: Geoengineering polar glaciers to slow sea-level rise. Nature. **555**, 303–305 (2018) https://www.nature.com/magazine-assets/d41586-018-03036-4/d41586-018-03036-4.pdf
58. Morton, O.: Nature. **447**, 132–136 (2007) https://www.nature.com/articles/447132a.pdf
59. Mulkern, A.: Rising sea levels will hit California harder than other places. Scientific American (04.27.2017). https://www.scientificamerican.com/article/rising-sea-levels-will-hit-california-harder-than-other-places/
60. Narayan, S., et al.: The value of coastal wetlands for flood damage reduction in the Northeastern USA. Sci. Rep. **7**, 9463 (2017) https://www.nature.com/articles/s41598-017-09269-z
61. Olivier, D.W.: Cape Town's water crisis: driven by politics more than drought. The Conversation (12.12.2017). https://theconversation.com/cape-towns-water-crisis-driven-by-politics-more-than-drought-88191
62. Onishi, N., Sengupta, S.: Dangerously low on water, cape town now faces 'day zero.' The New York Times (01.30.2018). https://www.nytimes.com/2018/01/30/world/africa/cape-town-day-zero.html
63. Parker, L.: Sea level rise will flood hundreds of cities in the near future. National Geographic (07.12.2017). https://news.nationalgeographic.com/2017/07/sea-level-rise-flood-global-warming-science/
64. Pearce, F.: Rivers in the sky: how deforestation is affecting global water cycles. Yale Environment 360 (07.24.2018). https://e360.yale.edu/features/how-deforestation-affecting-global-water-cycles-climate-change

65. Penn, I.: It's the no. 1 power source, but natural gas faces headwinds. The New York Times (03.28.2018). https://www.nytimes.com/2018/03/28/business/energy-environment/natural-gas-power.html
66. Pongratz, J., et al.: Crop yields in a geoengineered climate. Nat. Clim. Change. **2**, 101–105 (2012) https://www.nature.com/articles/nclimate1373
67. Proctor, J., et al.: Estimating global agricultural effects of geoengineering using volcanic eruptions. Nature. **560**, 480–483 (2018) https://www.nature.com/articles/s41586-018-0417-3
68. Rich, N.: Losing earth: the decade we almost stopped climate change. The New York Times (08.01.2018). https://www.nytimes.com/interactive/2018/08/01/magazine/climate-change-losing-earth.html
69. Richardson, T.B., et al.: Understanding the Rapid Precipitation Response to CO_2 and Aerosol Forcing on a Regional Scale. J. Clim. **29**, 583–594 (2016). https://doi.org/10.1175/JCLI-D-15-0174.1. https://journals.ametsoc.org/doi/pdf/10.1175/JCLI-D-15-0174.1
70. Robichaux, E.: Coast 2050's lasting impacts on coastal restoration. Delta dispatches. Restore the Mississippi River Delta (11.05.2015). http://mississippiriverdelta.org/coast-2050s-lasting-impacts-on-coastal-restoration/
71. Robock, A.: 20 reasons why geoengineering may be a bad idea. Bull. At. Sci. **64**(2), 14–18 (59) (2008). https://doi.org/10.2968/064002006. http://climate.envsci.rutgers.edu/pdf/20Reasons.pdf
72. Rockström, J., et al.: Science. **355**(6331), 1269–1271 (2017) http://science.sciencemag.org/content/355/6331/1269.full
73. Rossi, M.: Will a huge new flood barrier save Venice? CityLab (04.03.2018). https://www.citylab.com/environment/2018/04/will-a-huge-new-flood-barrier-save-venice/556226/
74. Sachs J., et al.: Pathways to deep decarbonization, 2014 report, Deep Decarbonization Pathways Project, SDSN – IDDRI. (2014). http://deepdecarbonization.org/wp-content/uploads/2015/06/DDPP_Digit.pdf
75. Sachs J., et al.: DDPP 2014 report (2014). http://deepdecarbonization.org/wp-content/uploads/2015/06/DDPP_Digit.pdf
76. Sanz-Perez, E., et al.: Direct Capture of CO_2 from Ambient Air. Chem. Rev. **116**(19), 11840–11,876 (2016). https://doi.org/10.1021/acs.chemrev. https://pubs.acs.org/doi/full/10.1021/acs.chemrev.6b00173
77. Schwartz, J.: Surrendering to rising seas. Scientific American (08.01.2018). https://www.scientificamerican.com/article/surrendering-to-rising-seas/
78. Sengupta, S., Popovich, N.: Global warming in South Asia: 800 million at risk. The New York Times (06.28.2018). https://www.nytimes.com/interactive/2018/06/28/climate/india-pakistan-warming-hotspots.html?action=click&module=RelatedLinks&pgtype=Article
79. Shugar, D.H., et al.: River piracy and drainage basin reorganization led by climate-driven glacier retreat. Nat. Geosci. **10**, 370–375 (2017) https://www.nature.com/articles/ngeo2932
80. Smith, P.: Soil carbon sequestration and biochar as negative emission technologies. Glob. Change Biol. **22**, 1315–1324 (2016). https://doi.org/10.1111/gcb.13178.. https://onlinelibrary.wiley.com/doi/pdf/10.1111/gcb.13178
81. Smith, P., et al.: Nat. Clim. Change. **6**, 42–50 (2015) https://www.nature.com/articles/nclimate2870.pdf
82. Smith, P., et al.: Biophysical and economic limits to negative CO_2 emissions. Nat. Clim. Change. **6**, 42–50 (2016) https://www.nature.com/articles/nclimate2870
83. Smith, L., et al.: Chaos and the flow capture problem: polluting is easy, cleaning is hard. Phys. Rev. Appl. (2018) in press
84. Socolow, R.H., Pacala, S.W.: A Plan to Keep Carbon in Check. Sci. Am. **305**, 968–972 (2006)
85. Socolow, R., et al.: Direct air capture of CO_2 with chemicals. A technology assessment for the APS panel on public affairs. Am. Phys. Soc. (2011) https://infoscience.epfl.ch/record/200555/files/dac2011.pdf
86. Steffen, W., et al.: Trajectories of the Earth System in the Anthropocene. PNAS. **115**(33), 8252–2859 (2018) http://www.pnas.org/content/115/33/8252

87. Strefler, J., et al.: Potential and costs of carbon dioxide removal by enhanced weathering of rocks. Environ. Res. Lett. **13**, 034010 (2018) http://iopscience.iop.org/article/10.1088/1748-9326/aaa9c4/meta
88. Struzik, E.: How warming is profoundly changing a great Northern Wilderness. Yale Environment 360 (04.25.2017). https://e360.yale.edu/features/how-warming-is-profoundly-changing-a-great-northern-wilderness
89. Temple, J.: Harvard Scientists Moving Ahead on Plans for Atmospheric Geoengineering Experiments. MIT Technology Review (03.24.2017). https://www.technologyreview.com/s/603974/harvard-scientists-moving-ahead-on-plans-for-atmospheric-geoengineering-experiments/
90. The IMBIE team: Mass balance of the Antarctic Ice Sheet from 1992 to 2017. Nature. **558**, 219–222 (2018) https://www.nature.com/articles/s41586-018-0179-y
91. Tollefson, J.: Sucking carbon dioxide from air is cheaper than scientists thought. Nature. **558**, 173 (2018). https://doi.org/10.1038/d41586-018-05357-w. https://www.nature.com/articles/d41586-018-05357-w
92. Trenberth, K.E.: Changes in precipitation with climate change. Clim. Res. **47**, 123–138 (2011). https://doi.org/10.3354/cr00953. https://www.int-res.com/articles/cr_oa/c047p123.pdf
93. Trenberth, K.E., Dai, A.: Effects of Mount Pinatubo volcanic eruption on the hydrological cycle as an analog of geoengineering. Geophys. Res. Lett. **34**, L15702 (2007). https://doi.org/10.1029/2007GL030524. https://agupubs.onlinelibrary.wiley.com/doi/abs/10.1029/2007GL030524
94. Usery, E.L., et al.: Chapter 2: Modeling Sea-level Rise and Surge in Low-lying Urban Areas using Spatial Data, Geographic Information Systems, and Animation Methods. In: Showalter, P., Lu, Y. (eds.) Geospatial techniques in urban hazard and disaster analysis, vol. 2019, pp. 11–30. Springer Netherlands (2010). https://doi.org/10.1007/978-90-481-2238-7_2. https://cegis.usgs.gov/pdf/sea_level_rise_text.pdf
95. van Heerden, I.L.: Chapter 6: Setting the Stage for the Katrina Catastrophe: Environmental Degradation, Engineering Miscalculation, Ignoring Science and Human Mismanagement. In: Zakour, M.J., Mock, N.B., Kadetz, P. (eds.) Creating Katrina, Rebuilding Resilience, Lessons from New Orleans on Vulnerability and Resiliency, vol. 2018, pp. 133–158. Butterworth-Heinemann (2018) https://www.sciencedirect.com/science/article/pii/B9780128095577000065
96. Velasquez-Manoff, M.: Can dirt save the earth. The New York Times Magazine (04.18.2018). https://www.nytimes.com/2018/04/18/magazine/dirt-save-earth-carbon-farming-climate-change.html
97. Vitousek, S., et al.: Doubling of coastal flooding frequency within decades due to sea-level rise. Sci. Rep. **7**, 1399 (2017) https://www.nature.com/articles/s41598-017-01362-7#citeas
98. Wadhams, P.: The global impacts of rapidly disappearing arctic sea ice. Yale Environment 360 (09.26.2016). https://e360.yale.edu/features/as_arctic_ocean_ice_disappears_global_climate_impacts_intensify_wadhams
99. Waldman, S.: Climate change is transforming, rerouting Arctic rivers. ClimateWire (04.19.2017). https://www.eenews.net/climatewire/2017/04/19/stories/1060053256
100. Waldman, S.: Atmospheric CO_2 sets record high. ClimateWire (05.03.2018). https://www.eenews.net/climatewire/2018/05/03/stories/1060080715
101. Watts, J.: Arctic warming: scientists alarmed by 'crazy' temperatures. The Guardian (02.27.2018). https://www.theguardian.com/environment/2018/feb/27/arctic-warming-scientists-alarmed-by-crazy-temperature-rises
102. Welch, C.: Carbon emissions had leveled off. Now they're rising again. National Geographic (11.13.2017). https://news.nationalgeographic.com/2017/11/climate-change-carbon-emissions-rising-environment/
103. Westerink J., et al.: Note on the Influence of the Mississippi River Gulf Outlet on Hurricane Induced Storm Surge in New Orleans and Vicinity. (2006). http://www.columbia.edu/itc/

journalism/cases/katrina/Army/Army%20Corps%20of%20Engineers/Influence%20of%20the%20MRGO%20on%20Storm%20Surge.pdf
104. Williamson, P.: Nature. **530**, 153–155 (2016b) https://www.nature.com/news/emissions-reduction-scrutinize-co2-removal-methods-1.19318
105. Wolosin, M., Harris, N.: Tropical forests and climate change: the latest science Working paper June 2018. World Resources Institute (2018) https://wriorg.s3.amazonaws.com/s3fs-public/ending-tropical-deforestation-tropical-forests-climate-change.pdf
106. Xi, F., et al.: Substantial global carbon uptake by cement carbonation. Nat. Geosci. **9**, 880–883 (2016) https://www.nature.com/articles/ngeo2840
107. Xia, L., et al.: Solar radiation management impacts on agriculture in China: A case study in the Geoengineering Model Intercomparison Project (GeoMI). J. Geophys. Res. Atmos. **119**, 8695–8711 (2014). https://doi.org/10.1002/2013JD020630. https://agupubs.onlinelibrary.wiley.com/doi/epdf/10.1002/2013JD020630
108. Zhang, Z., et al.: Analysis of Poyang Lake water balance and its indication of river-lake interaction. SpringerPlus. **5**(1), 1555 (2016). https://doi.org/10.1186/s40064-016-3239-5. https://www.ncbi.nlm.nih.gov/pmc/articles/PMC5021641/
109. Zhu, K., et al.: Limits to growth of forest biomass carbon sink under climate change. Nat. Commun. **9**, 2709 (2018) https://www.nature.com/articles/s41467-018-05132-5

A Moral Framework for Commons-Based Geoengineering

Lisa L. Ferrari and Elizabeth L. Chalecki

1 Introduction

As climate conditions worsen and mitigation efforts continue to fall short, the likelihood increases that States or other international actors will look to geoengineering technologies for a remedy. The most basic of these methods, such as planting trees or sequestering carbon underground, can take place within sovereign territory. However, there are also many types of geoengineering that could be deployed in the global commons. This means that their impact would clearly be felt beyond one specific State or region, and, with certain approaches, such as stratospheric aerosol injection, likely worldwide. These commons-based geoengineering (CBG) technologies are largely untested and, in almost all cases, speculative. This leaves considerable uncertainty about the results of deployment. It also provides an opportunity for States to make deliberate, prospective choices in policies that maximize the global common good.

In this chapter we propose a set of guidelines for States' future decisions about geoengineering. To clarify our argument, we focus on three of the most widely discussed forms of CBG – aerosol injection into the atmosphere, marine-based cloud brightening, and ocean iron fertilization. The moral urgency of the situation rests on two factors. First, the potential changes that could be wrought by large-scale use of geoengineering approaches are of remarkable gravity. Deliberately altering the earth's habitability is about as momentous of a matter as one can imagine in

L. L. Ferrari (✉)
Department of Politics and Government, University of Puget Sound,
Tacoma, Washington, USA
e-mail: lferrari@pugetsound.edu

E. L. Chalecki
Department of Political Science, University of Nebraska, Omaha, Nebraska, USA
e-mail: echalecki@unomaha.edu

© Springer Nature Switzerland AG 2021

W. Burns et al. (eds.), *Climate Geoengineering: Science, Law and Governance*,
AESS Interdisciplinary Environmental Studies and Sciences Series,
https://doi.org/10.1007/978-3-030-72372-9_4

international relations. In this context, it is nearly impossible to avoid some basic moral questions about fundamental effects on life and how humans treat one another. The precautionary principle is on point here: Before taking potentially irreversible action, States should err on the side of caution. But could there be a more robust framework for a choice of such significance?

The second source of moral urgency is the complexity of potential outcomes. Given the intricate webs of life in ecosystems, CBG would have significant political, social, and ecological implications. The global continuity of the environment means that any CBG method deployed by one State may have notable effects on conditions in other States. Such potential widespread effects have proven difficult to model in climate research.[1] However, some current modelers have suggested that experimentation on any scale short of the global level would not yield meaningfully predictive results.[2] This means that any CBG efforts are likely to generate multiple effects, some of which will be unintended consequences.

In other words, decisions about geoengineering place States at an important ethical crossroads. CBG represents a complex interaction of scientific and political concerns. However, the international community has not articulated anything more than general guidelines, such as the Oxford and Asilomar Principles, for policymaking about CBG.[3] International legal agreements do not address the issue. Only two environmental treaty regimes have taken up geoengineering specifically; and each does little more than say that States should refrain from geoengineering until the relevant technologies, and their implications, are understood more clearly.[4] This is neither an outright ban nor a set of guidelines on which to build policies about developing

[1] L.J. Wilcox, E.J. Highwood, and N.J. Dunstone, *The influence of anthropogenic aerosol on multi-decadal variations of historical global climate*, 8(2) Envtl. Research Letters, art. 024033 (2013). Historical evidence from the 1815 eruption of Mt. Tambora indicates that repeated injections of sulfur might 'knock the climate' for longer than anticipated. *See* Clive Oppenheimer, *Climatic, environmental, and human consequences of the largest known historic eruption: Tambora volcano (Indonesia) 1815,* 27(2) Progress in Physical Geography 230, 256 (2003). For examples of the difficulty modeling the wide range of possible outcomes for certain CBG technologies, *see* P. J. Irvine, et al., *Towards a comprehensive climate impacts assessment of solar geoengineering*, 5 Earth's Future, 93 (2017). For an argument about the potential for geoengineering to have a negative impact on biodiversity, *see* C.H. Trisos, et al., *Potentially dangerous consequences for biodiversity of solar geoengineering implementation and termination*, 2 Nature Ecology and Evolution 475–82 (2018).

[2] For example, *see* Alan Robock, et al. *A Test for Geoengineering?* 327 Sci. 530 (2010); Aaron L. Strong, John J. Cullen, and Sallie W. Chisholm, *Ocean Fertilization: Science, Policy, and Commerce* 22 (3) Oceangraphy 236–261 (2009).

[3] Asilomar: Asilomar statement on Climate intervention Technologies, http://www.climateresponsefund.org/index.php?option=com_content&view=article&id=152&itemid=89. Oxford: http://www.geoengineering.ox.ac.uk/www.geoengineering.ox.ac.uk/oxford-principles/principles/

[4] The two relevant efforts are the Decision Adopted by the Conference of the Parties to the Convention on Biological Diversity at its Tenth Meeting, October 29, 2010, UNEP/CBD/COP/DEC/X/33 and an Annex adopted by the Contracting Parties to the 1996 Protocol to the 1972 Convention on the Prevention of Marine Pollution by Dumping of Wastes and Other Matter, October 18, 2013, LC 36/16 Annex I.

technologies. For example, Annex I of the 1996 Protocol to the 1972 London Convention on the Prevention of Marine Pollution by Dumping of Wastes and Other Matter does list among the waste products it covers "carbon dioxide streams from carbon dioxide capture processes for sequestration." However, Annex I primarily addresses the terms under which such streams qualify as dumped waste and therefore fall under the same provisions as other forms of dumping. Annex II then provides detailed information about waste management options, dump site selection, monitoring and assessment of effects, and a permitting process. These terms apply to all forms of dumping and do not specifically mention carbon dioxide streams. For the purposes of this chapter, though, it is most important to note that the Protocol does little to address decisions about whether to attempt carbon dioxide sequestration in the first place. While we are concerned about the possibility of dumping as a consequence of geoengineering, our primary concern is with decisions to engage in geoengineering at all. For this purpose, existing treaty law offers negligible guidance.

2 Similarities and Dissimilarities Between War and CBG

Since international environmental law is largely silent on geoengineering per se, scholars have the choice of developing a new and unique analytical framework or seeking a different area of existing law that is both analogous to geoengineering and the subject of more extensive analysis and agreement. We argue that war – the large-scale use of violent force for political ends, putting another actor's sovereignty at risk – is such an analog, at least from the perspective of a State considering geoengineering as a response to the deleterious effects of climate change on both populations and territory. This is not to suggest that geoengineering is a form of warfare. Rather, we argue only that CBG as a response measure to climate change has important similarities with warfare. Therefore, when considering which principles might guide CBG deployment, we would do well to consider any lessons that can be learned from such similarities.

We note four points of similarity. First, both war and CBG are likely to have widespread and significant physical impact, some of which will cause irreparable harm. Second, the purpose of a State's action is to coerce another agent. In the case of war, these agents are primarily other States. In the case of geoengineering, the agent is the environment itself. In either case, the acting State's goal is to achieve its objectives by compelling change. The effects of both blast and radiation in Hiroshima and Nagasaki, Agent Orange in Vietnam, and oilfield destruction in Kuwait illustrate war's potential for severe environmental harm. With CBG, the potential is for altering global climate effects in States that are distant, in terms of geography, politics, or interests, from the locus of decision making. There is potential for equally widespread impact on biodiversity and ecosystem health.[5]

Third, almost any large-scale coercive act will involve a certain amount of 'collateral damage.' In war, collateral damage includes any harm to non-military targets, including non-combatant casualties resulting from a military action. Such casualties may be the result of deliberate targeting, but collateral damage can also include accidental harms resulting from the fog of war. In the case of CBG, collateral damage can be conceived of as any harmful consequence of geoengineering that is not a stated goal of the deployment. In both war and CBG, the actor's intent is a morally salient variable, but may be challenging to discern.

Finally, war and CBG are similar in that both can have a notable impact at the level of the international system. Regardless of the number of belligerent parties, all wars have implications for the global balance of power by their influence on alliance structures, global trade flows, and the domestic political and economic (in)stability of the combatants. Even a war that is nominally confined to two parties has these international effects. Similarly, any significant alteration of Earth's climate has an impact on the entire international system. Because acting on one part of the global environment necessarily has implications for the rest of the environment, one State's decision or a decision taken by a group of States to geoengineer can have direct consequences for other States. Regardless of whether these consequences are harmful or salutary, the larger point is that they have material consequences for all States, but one State or a small number of States (or even substate actors) may be able to undertake them without consulting other international actors.

Unlike geoengineering, however, war is the subject of centuries of treaty-based and customary international law. Specific questions about conducting warfare, from expectations about existing treaty obligations to the acceptable range of weapons to the treatment of prisoners of war, have been discussed at length by States, and the product of those discussions annually fills many volumes of legal documents. Among these are international agreements that address the environmental impact of war. Two notable examples are the 1976 Environmental Modification Convention (ENMOD) and Protocol I (1977) to the 1949 Geneva Conventions on the Laws of War.[6] However, both of these treaties are generally understood to pertain to environmental alteration during wartime and neither they nor any other agreements specify

---

[5] See the two reports from the U.S. National Academy of Sciences, *Climate Intervention: Carbon Dioxide Removal and Reliable Sequestration* and *Climate Intervention: Reflecting Sunlight to Cool Earth*, both published in 2015. They are available online at https://nas-sites.org/americasclimatechoices/other-reports-on-climate-change/climate-intervention-reports/

[6] For the full text of the *Convention on the prohibition of military or any other hostile use of environmental modification techniques (ENMOD)*, see United Nations, *Treaty Series*, vol. 1108, p. 151 and depositary notification C.N.263.1978.TREATIES-12 of 27 October 1978 (rectification of the English text); For the full text of Protocol I to the 1949 Geneva Conventions on the Laws of War, see International Committee of the Red Cross (ICRC), *Protocol Additional to the Geneva Conventions of 12 August 1949, and relating to the Protection of Victims of International Armed Conflicts (Protocol I)*, 8 June 1977, 1125 UNTS 3, available at: https://www.refworld.org/docid/3ae6b36b4.html [accessed 12 January 2019]

an application to geoengineering.[7] Furthermore, ENMOD Article III allows environmental modification for peaceful purposes. Unless CBG is specifically initiated as an act of war, States could argue they are deploying peaceful CBG in accord with Article III.

In war, States on both sides choose to engage in use of violent force. This is true even when States face a genuine dilemma. States that would otherwise avoid war may be forced by circumstances to fight in self-defense. Such States have chosen to engage in war rather than accept some other undesirable outcome, such as being overrun or incurring the wrath of a protector. For comparison to geoengineering, though, the critical point is that in war there is always the possibility, however unlikely or disastrous, that any State may surrender or otherwise refuse to be a belligerent. No such option exists on the part of the environment. The Earth may be able to re-freeze a polar ice cap or lower sea levels via processes that are apart from human actions, but only on a time scale so long that it is unlikely to influence policy decisions. In war, States' specific intent is to coerce, regardless of whether they are on the offense or defense, exhibiting agency or receiving the consequences of another actor's agency. In geoengineering, the environment is permanently in a defensive posture. When speaking of the environment, there are no concepts that correspond to 'intent' and 'foresight' in human activity.

This distinction raises an important question. Is one's moral culpability different while interacting with something purposive as compared to interacting with something non-purposive? There certainly are functional differences between the two categories of interaction. For example, States grappling with the effects of climate change do not have a diplomatic option when interacting with the environment. Similarly, States have no opportunity to use the levers of foreign policy to sway nature's behavior. The usual ways that States seek to influence others in the system do not apply to influencing the global environment. This imbalance in intent may constrain some of a State's actions in the situation. It does not alter our ability to assess the moral significance of those actions, but it may alter the way we assess moral responsibility on the part of the purposive actor.

States may bear an equal responsibility to avoid harming the environment, as compared to their responsibility to other types of agents. The points of similarity between war and geoengineering are striking and of moral significance. In specific, the roles of unintended consequences, and of multiple effects of a single action

[7] For example, both the International Committee on the Red Cross (ICRC) and the U.S. Department of State (DOS) refer to ENMOD as an agreement about wartime behavior. See ICRC, "1976 Convention on the Prohibition of Military or any Hostile Use of Environmental Modification Techniques," 01/2003 available at https://www.icrc.org/en/doc/assets/files/other/1976_enmod.pdf; DOS Bureau of International Security and Nonproliferation, narrative introduction to the Convention on the Prohibition of Military or Any Other Hostile Use of Environmental Modification Techniques, available at https://www.state.gov/t/isn/4783.htm. In addition, the documents of the Second ENMOD Conference in 1992 indicate that States differed dramatically in their understandings of how (and whether) the treaty applied outside of wartime. See ENMOD/CONF.II/12, Part III, p35. Available at https://s3.amazonaws.com/unoda-web/documents/library/conf/ENMOD-CONF-11-12.pdf

more generally, are important connections between thinking about war and thinking about geoengineering. In the absence of international legal or moral frameworks for making decisions about CBG, we look to one of the significant moral frameworks for evaluating war – just war theory.

3 Just War Theory and Constraint in War

3.1 Practical and Legal Constraints

States have long debated and negotiated the appropriate parameters for limiting warfare. Such parameters are widely acknowledged by States through national military manuals, Articles 2(4) and 51 of the United Nations Charter, international agreements such as the 1899 and 1907 Hague Conventions and the 1949 Geneva Conventions, treaties restricting the development and use of weapons of mass destruction, and a variety of norms and expectations about appropriate conduct in wartime. While there are individual violations of these rules and norms, the more notable facts are that States have taken the time to articulate their expectations and that, on the whole, those expectations persist. This is the body of rules and norms that Michael Walzer calls the 'war convention,' and it includes "the set of articulated norms, customs, professional codes, legal precepts, religious and philosophical principles, and reciprocal arrangements that shape our judgments of military conduct."[8]

Given how much is at stake in wartime, it is notable that States subject themselves, however imperfectly, to such constraint in deciding whether and how to use violent force. There are at least four reasons for such behavior. The first is a pragmatic reciprocity. If States are going to interact with each other in any productive way, diplomats must be able to speak and act freely, without fear of retaliation or punishment by the sovereign hosting them. States that do not afford such protections to others' diplomats are unlikely to find their own emissaries enjoying immunity.[9] At least since the Sixteenth Century, European leaders have granted legal immunity to official representatives of other governments in order to guarantee like treatment for their own diplomats abroad.[10] Similarly, States may act with restraint in wartime with the expectation that other States will do the same.

Second, day-to-day international relations requires some basic norms and expectations about how States will interact with one another. A certain level of international cooperation is required if currency exchanges are going to take place, postal

[8] Michael Walzer, Just and Unjust Wars: A Moral Argument with Historical Illustrations 44 (3rd ed. 2000)

[9] Emer de Vattel. The Law of Nations 705-06 (Béla Kapossy and Richard Whatmore eds. 2008).

[10] Christopher A. Whytock, *Foreign State Immunity and the Right to Court Access* 93 Boston U. L. Rev. 2033, 2038 (2013).

service is to operate transnationally, and ships and aircraft are to have safe travel lanes and harbor. Wartime unpredictability can wreak havoc with such quotidian functions, and a State contemplating war needs to consider the opportunity cost of that disruption.

Third, even leaving aside treaty law, once norms have developed, most States do not want to be too far outside them. This can contribute to the development of legal custom. Patterns of behavior that become regular over time lead to States to regard such behavior as obligatory. However, the power of such expectations is not seen most clearly by looking at the most powerful States in the system. The vast majority of States in the global system are not strong enough to flout custom with impunity; rather, they need to work in concert with others to have significant international impact. As a consequence, a State in that position will need to think twice before doing something that very clearly labels them as a rule-breaker, convention-flouter, or rogue.

Finally, the seat of moral decision-making is the human conscience. States, as corporate entities, do not have moral preferences. When States behave in concert with or opposition to moral dictates, they do so because of the choices of individual humans. Those individuals have consciences shaped by the standards they embrace as morally authoritative. As Mark Douglas observes,

> Just warriors ought not to be thought of simply as rule-followers – though they are also that – but also as people whose convictions about the way the world is lead them to pursue specific patterns of moral discernment in the world.[11]

There is a substantial literature on bureaucratic politics that explores the influence of an individual's organizational role on the preferences they express.[12] For our purposes, though, it is unimportant whether 'seats determine stands' or individual decision makers are otherwise 'captured' by their organizations. Regardless of the source of their standards of right and wrong, individual decision makers are applying some sort of standard. As long as political decisions are made by humans, those decisions will be the product of some kind of ethical assessment. For any risky behavior, decision makers will assess their options, including the implications of resisting systemic constraints. To the extent that decision-makers care whether they are behaving ethically, regardless of the moral metrics they apply, they will also care whether they can justify high-risk, destructive, and possibly irrevocable actions such as war or climate geoengineering. The just war tradition has developed within this environment of openness to restraint in wartime.

---

[11] Mark Douglas. *Changing the Rules: Just War Theory in the Twenty-First Century* 59(4) THEOLOGY TODAY 529, 538 (2003).

[12] Perhaps the best-known explanation of the bureaucratic politics model comes from Graham Allison, *Conceptual Models and the Cuban Missile Crisis* 63(3) AM. POL. SCI. REV. 689 (1969).

3.2 *Just War Theory*

In the West, just war theory has origins in the ancient Mediterranean region, including in the thought of Cicero and the early Christian writings of Augustine.[13] These early versions offered very limited criteria for making decisions about war and encouraged case-by-case consideration. Unlike some contemporary articulations of just war criteria, those of Cicero and Augustine did not assume that war was to be an absolute last resort, avoided at almost any cost. These thinkers also emphasized that the purpose of a just war was to establish a just peace – a properly ordered world. Particularly for Augustine, the goal of a just war was to bring humans back into accord with God's ordering of Creation. He writes that, "peace between men is an ordered agreement of mind with mind… [T]he peace of the whole universe is the tranquility of order."[14]

In the Thirteenth Century, Thomas Aquinas drew on the work of both Augustine and Aristotle to develop a meticulous and systematic theological approach to the question of war's justice.[15] Subsequent Scholastic philosophers (e.g., Francisco Vitoria, Francisco Suarez) adjusted, but maintained, the link between moral virtue and Catholic dogma in just war theorizing. Hugo Grotius was the first major post-Thomist to reach back to Cicero and avoid referring to the Catholic Church as the final authority on questions of justice and war. From that point, there are clear statements – from both Catholic and non-Catholic contexts – that the basic moral principles about how people treat one another have a bearing on decisions about war. Contemporary just war theorists disagree as to whether the morality of States and the morality of persons differ and lead to divergent imperatives for justice in wartime.[16]

While a consciously secular branch of just war theory developed from the works of Hugo Grotius and Emer de Vattel, in recent decades the Catholic Church in the United States has prominently returned to the discussion. One of the most popular and widely-cited contemporary discussions of just war can be found in *The Challenge of Peace: God's Promise and Our Response*, the U.S. Catholic Bishops' 1983 statement about the permissibility of war in the nuclear age. That document lays out a set of just war criteria that are restated and refined in *The Harvest of Justice is Sown in Peace,* the bishops' tenth anniversary reflection on *The Challenge*

[13] For a concise overview of Cicero's place in the just war tradition, *see* Gavin Stewart, "Cicero," in JUST WAR THINKERS: FROM CICERO TO THE 21ST CENTURY 8–20 (Daniel R. Brunstetter and Cian O'Driscoll, eds. 2018).

[14] AUGUSTINE, CITY OF GOD (Henry Bettenson trans, 1984) Book XIX, Ch. 13, at 870.

[15] THOMAS AQUINAS addresses the topic of just war specifically in SUMMA THEOLOGIAE, II-II, Q. 40.

[16] For example, compare the Christian realism of JEAN BETHKE ELSHTAIN, WOMEN AND WAR (1987); the moral historicism of JAMES TURNER JOHNSON, JUST WAR TRADITION AND THE RESTRAINT OF WAR: A MORAL AND HISTORICAL INQUIRY (1981); and the reductive individualism of JEFF MCMAHAN, KILLING IN WAR (2009).

of Peace.[17] The criteria are divided into two categories. *Jus ad bellum* criteria set the standards by which to assess the justice of going to war. *Jus in bello* criteria speak to the conduct requirements for fighting justly once hostilities have begun. The two broad categories overlap in important ways, as do some of the criteria, and the *jus ad bellum* criteria remain salient for strategic decision-making even after fighting has begun. The language of the criteria is quite general so that the standards may serve as guidelines in the widest possible set of circumstances. *The Harvest of Justice is Sown in Peace* lays out the just war criteria as follows:

Jus ad bellum

- *Just Cause*: force may be used only to correct a grave, public evil, i.e., aggression or massive violation of the basic rights of whole populations;
- *Comparative Justice*: while there may be rights and wrongs on all sides of a conflict, to override the presumption against the use of force the injustice suffered by one party must significantly outweigh that suffered by the other;
- *Legitimate Authority*: only duly constituted public authorities may use deadly force or wage war;
- *Right Intention*: force may be used only in a truly just cause and solely for that purpose;
- *Probability of Success*: arms may not be used in a futile cause or in a case where disproportionate measures are required to achieve success;
- *Proportionality*: the overall destruction expected from the use of force must be outweighed by the good to be achieved;
- *Last Resort*: force may be used only after all peaceful alternatives have been seriously tried and exhausted.

Jus in bello

- *Noncombatant Immunity*: civilians may not be the object of direct attack, and military personnel must take due care to avoid and minimize indirect harm to civilians;
- *Proportionality*: in the conduct of hostilities, efforts must be made to attain military objectives with no more force than is militarily necessary and to avoid disproportionate collateral damage to civilian life and property;
- *Right Intention*: even in the midst of conflict, the aim of political and military leaders must be peace with justice, so that acts of vengeance and indiscriminate violence, whether by individuals, military units or governments, are forbidden.[18]

---

[17] National Conference of Catholic Bishops, The Challenge of Peace: God's Promise and Our Response, nos. 85–110 (1983); United States Conference of Catholic Bishops, The Harvest of Justice is Sown in Peace (1993).

[18] U.S. Conference of Catholic Bishops. The criteria in The Harvest of Justice are identical to those mentioned in The Challenge of Peace, with one exception. The 1983 statement gives the first three criteria, in order, as just cause, legitimate authority, and comparative justice. The 1993 statement alters the ordering, so that the first three criteria read 'just cause, comparative justice, and legitimate authority.' It is doubtful this was an accidental change. The presumption against use of force is one of the more controversial components of this formulation. The earliest Christian proponents of just war did not stipulate a presumption against fighting. A main argument

Contemporary just war theorists have addressed in depth the parameters, meaning, and validity of these criteria. The full extent of such discussion is beyond the scope of this chapter.[19] We base our own work on the bishops' formulation because it has widely been treated as morally authoritative and has engendered significant discussion.

In addition, the changing nature of warfare has meant that new dimensions of conflict must be part of any discussion of limiting war. Most wars no longer involve regular armies meeting on the battlefield after sovereigns have formally declared a state of war. The realities of terrorism, drone warfare, cybersecurity threats, and guerrilla insurgents, among others, mean that theories of just war need to be flexible in their assumptions about conflict in order to stay relevant.[20] We believe one piece of that rethinking should be addressing non-traditional security threats such as environmental concerns.

4 Just War and the Doctrine of Double Effect

The just war tradition treats unintended consequences at length, often in terms of double effect, a problem that arises when both morally permissible and morally impermissible results can be foreseen for a particular action. Provided they have met the pertinent just war criteria, actions against military targets are morally permissible. Actions against non-military targets are morally prohibited. Non-military casualties of either type of attack are collateral damage.

An archetypal instance of double effect is the outcome of bombing a military objective that cannot be targeted discretely from a civilian facility. For example, a school near a military airfield may be close enough that a bomb dropped on the airfield (morally permissible, since it targets combatants) could easily damage the school (morally impermissible, since the schoolchildren are noncombatants), as well. Is it then morally permissible to attack the airfield? The doctrine of double effect (DDE) provides a method for thinking through this question. In sum, the

---

of THE CHALLENGE OF PEACE and the bishops' statements more broadly is that the advent of nuclear technology must fundamentally change how States understand their moral options. For arguments against the moral worth of this this changed perspective, *see* James Turner Johnson, *The Just War Idea: The State of the Question*, 23(1) SOCIAL PHILOSOPHY AND POLICY 180 (2006); William V. O'Brien, *The Challenge of War: a Christian Realist Perspective* in JUST WAR THEORY 169–176 (Jean Bethke Elshtain, ed. 1992) 169–176.

[19] Thoughtful examples include Eamon Aloyo, *Just War Theory and the Last of Last Resort,* 29(1) ETHICS & INTERNATIONAL AFFAIRS 187 (2015); and Ian Halliday, *When is a Cause Just?* 28 REV. INT'L STUDIES 557 (2002).

[20] For example, Jovana Davidovic argues that the changing nature of warfare decreases the distinction between the morality of States and the morality of persons. *See* J. Davidovic, *Should the Changing Character of War Affect Our Theories of War?* 19 ETHICAL THEORY AND MORAL PRACTICE 603 (2016).

DDE says it is morally permissible to undertake an act that will cause a double effect if

1. The act is good in itself;
2. Only the good effect is intended;
3. The bad effect is not the means to the good effect; and
4. The magnitude of the good done is at least as great as the magnitude of harm done.[21]

By this reasoning, bombing the airfield may be a morally permissible means of conducting a just war, provided the action is undertaken with appropriate intent and thoughtful assessment of causality.

The DDE is helpful for thinking through the moral acceptability of certain types of geoengineering. It is reasonable to assume that any CBG large enough to have an impact on the climatic condition will have an impact sufficient to alter environmental conditions globally. In that case, it is safest also to assume (and modeling suggests) that the environmental changes resulting from CBG will not be identical at all points on Earth and that some of the changes will be unintended. This increases the likelihood of a double effect and makes the DDE relevant.

Imagine that State Z is considering engaging in CBG to address its environmental concerns. By thinking about the global environment and the State(s) affected by State Z's action as the two 'targets' of Z's geoengineering, we can use the concept of double effect to explore the moral justifiability of actions that are equivocally (or at least discernibly) both good and bad. Presumably, these cases are the most complex morally, since the other possible outcomes are either unequivocally harmful or unequivocally salutary for all entities that experience the effects.

If we consider the double effect criteria in turn, we raise important moral questions about the practice of CBG. First, we must consider whether there is sufficient evidence that CBG is good in itself. That requires weighing the evidence both that geoengineering technologies will be able to deliver the outcomes they promise and that the changes are likely to have sufficiently widespread salutary effects to justify manipulating the environment at the planetary level. The potential moral hazard is real. Geoengineering cannot be a license to continue or increase current levels of environmental degradation. The burden of proof here should be high. Human interventions caused the current disruptions in the natural environment. Therefore, there must be exceptionally strong and clear evidence that further environmental interventions will improve the situation in more than just the immediate term.

The second moral question about CBG is how to assess costs and benefits should the technologies be deployed. This will require deployers to acknowledge the scope and nature of stakeholders (e.g., States, geographic regions, social and cultural groups, economic actors, individual persons, etc.) and to be ready to respond to the range of interested parties. In cases of conflicting interests, whose should be given

---

[21] Brian Orend, War and International Justice: a Kantian Perspective 164 (2000).

priority? On what grounds?[22] Given the complexity of the system being reshaped by CBG, it is unlikely that a set of precise, one-size-fits-all decision rules will pertain from deployment opportunity to deployment opportunity. In any case, decision makers will need to think broadly and flexibly about the actors potentially concerned.

The DDE has been particularly relevant for thinkers in the just war tradition. This suggests that other aspects of just war thinking may be valuable guidelines for States considering deploying CBG and that the just war tradition may therefore provide us with a decision-making framework that can be modified for consideration of geoengineering.

5 Just Geoengineering and the Doctrine of Double Effect

Noting the parallels between war and geoengineering, we have argued elsewhere that a modified version of the just war criteria can usefully be applied to decisions about geoengineering.[23] As with the just war criteria, the 'just geoengineering criteria' can be divided into those guiding decisions about the resort to using CBG ('jus ad climate' criteria) and those guiding decisions about conduct once CBG technologies have been deployed ('jus in climate' criteria.) As with the *jus ad bellum* and *jus in bello* criteria, the jus ad climate and jus in climate criteria are not intended as wholly discrete categories. That is, jus ad climate criteria may require reconsideration even after CBG technologies have been deployed and there may be overlap among some of the criteria. Here we offer an overview of these just geoengineering criteria as the preface to a more specific discussion of double effect and geoengineering.

Jus ad climate

1. Just cause/last resort: We combine these two criteria. CBG should be used only in response to a major climate emergency and only when other measures have been shown to be inadequate to the task. This will require that States have target thresholds in mind for what level of environmental disruption is too much. While global variation and technological uncertainties make it impossible to generate a universal set of criteria, States should take it upon themselves to think through their level of tolerance to climate change before a disaster requires crisis decision-making. They are also obliged to maximize mitigation efforts before undertaking geoengineering.

---

[22] For example, in their writings on climate change and the environment more generally, the U.S. Catholic bishops have repeatedly stated that any mitigation efforts must prioritize the needs and well-being of the poor and weak. *See* U.S. Conference of Catholic Bishops, Global Climate Change: A Plea for Dialogue, Prudence, and the Common Good (2001).

[23] Elizabeth L. Chalecki & Lisa L. Ferrari, *A New Security Framework for Geoengineering* 12(2) Strategic Studies Q. 82 (2018).

2. Comparative justice: The fact that CBG deployment could have some negative consequences is not, in itself, an argument against such geoengineering. However, the nature and magnitude of foreseeable consequences for both acting and not acting need to be compared. States must choose options that improve global circumstances, rather than benefitting some States while harming other, particularly less powerful, States. Given the current uncertainty about the effectiveness of CBG, and to minimize unnecessary negative effects, States should be prepared to cease or alter their CBG intervention in a timely manner if negative effects in one region outstrip positive ones in another. Time frames may need to be calculated case-by-case, but the overarching principle should be that States should err on the side of stopping harm as quickly as possible.
3. Legitimate authority: National governments must be responsible for decisions affecting their States. However, national decisions must be subject to international oversight. Because CBG technologies uniquely combine global range and the potential to affect the livability of Earth in a fundamental way, decisions about CBG should not be made by a single State or in isolation. For deployment of geoengineering technologies to meet the criterion of legitimate authority, States would need their actions to be authorized by the U.N. Security Council, assuming no new institution is formed specifically to address such environmental security issues.
4. Right intention: The legitimate goal of geoengineering is to promote the long-term health and livability of the planet. Efforts to manipulate the climate instrumentally, such as to create unfavorable environmental conditions for an adversary, are not morally permissible.
5. Probability of success: Since most CBG technologies with global reach cannot be tested at scale, this is an especially speculative criterion. However, States and the international institutions they consult with need to use the best scientific and economic information and analyses available.
6. Proportionality: To some extent, this criterion is similar to that of just cause. If CBG is permissible only under extreme circumstances, then one threshold for measuring proportionality is addressed when determining whether a State has just cause to act. In another sense, proportionality could refer to the amount of environmental manipulation to be undertaken to address the crisis at hand. States need to keep in mind that the scope of their geoengineering should be sufficient to address the current problem, without acting to generate undue environmental change.

Jus in climate

1. Discrimination: We have argued that the concepts of double effect and collateral damage have critical relevance to assessing CBG. Since CBG will affect disparate parts of the world, States that geoengineer must take care that their actions have minimal negative impact on life in other States.
2. Proportionality: As with the decision to commence geoengineering, the decision to continue with actions that have begun must show positive effects in a timely manner and globally, not just for one or two States. As with comparative justice,

some States must be ready to cease geoengineering in a timely manner, even if they are reaping benefits, should the effects be harmful to others. Ongoing use of such technologies could cause further environmental degradation and a false sense of security among decision-makers who need to address the problem of climate change more effectively.

3. Right intention: Even after geoengineering technologies have been deployed, the deploying State needs to continue monitoring its own goals in acting to be sure that ongoing action remains focused on the global good rather than on a narrower interest.

These just geoengineering criteria are intended to motivate States' ongoing inquiry into the moral permissibility of their actions and motivations, rather than to dictate specific answers. As such, much of the terminology in the criteria is deliberately general, as are statements of the just war criteria. The salient moral question is whether a State's manipulation of the environment meets the moral standards set by the just geoengineering criteria and with the caveats introduced in our discussion of the DDE.

6 The Role of Intent

Not all moral standards concern themselves with an agent's intent; however, the just war criteria clearly do. Right intention is one of the longest-standing *jus ad bellum* criteria. Moral assessment of politics raises two kinds of questions about actors. First, how can we know what they intend? Second, what is the relationship between foreseeability and intent? If one is assessing one's own inclination to fight, it is relatively simple to identify and judge one's own intentions. As an outside observer to a conflict, though, assessing the intentions of belligerents can be much more difficult. Warring parties may not be transparent or credible in stating their goals and motivations.

This difficulty may not be insurmountable. Some scholars argue persuasively that actors' intent can be evaluated, to a significant extent, by looking at behaviors and outcomes, rather than just by attempting to discern the interior state of an actor at the time of decision-making.[24] For example, if the outcome of an action is not consistent with the actor's stated intention in acting, observers are justified in questioning the actor's intentions. Clearly, many actions have unintended consequences, but labeling something 'unintended' becomes more problematic if the unintended consequence was nonetheless foreseeable.

If we assume that States are rational actors and consider multiple possible outcomes of their actions, we can find some indication of what the State regarded as a

[24] Darrell Cole, *War and Intention* 10(3) J.MILITARY ETHICS 174 (2011); Rosemary B. Kellison *Impure Agency and the Just War: a Feminist Reading of Right Intention* 43(2) J. RELIGIOUS ETHICS 317 (2015).

morally acceptable outcome by looking at what the State actually did. Unless States are utterly unable to predict the outcomes of their actions, most outcomes of most actions can be assumed to fall within the bounds of acceptable risk for a State. That information about risk can help us extrapolate information about intent. Two examples illustrate this point.

The first example is Iraq's invasion of Kuwait in 1990. Presumably, Iraq did not attack primarily in the hope of drawing States beyond Kuwait into a regional ground war. However, Iraq must have recognized that such a war was possible and nonetheless chose to attack. Whether the invasion of Kuwait was specifically intended to start a larger war, it is fair to say that Iraq recognized the possibility of a regional military response, particularly from Saudi Arabia and Iran, and nonetheless decided to launch the attack.[25] Given outsiders' incomplete knowledge of what States intend, we can still say that States must intend one or more of the foreseeable results of an action. Identifying foreseeable outcomes does not allow us to pinpoint a specific intent, but it does provide information about the range of outcomes a State may have intended. That, in turn, tells us something important about a State's intention. Of course, States cannot anticipate every implication of a policy choice, but neither are they blind to all possibilities except the one they settle on. In other words, we know that a State's intended outcome is one of the foreseeable outcomes of its action. In that sense, we can treat foreseeability as a guide to recognizing intent.

The second example is Egypt's construction of the Aswan High Dam during the 1960s. The dam was intended to generate significant hydroelectric power while shielding Nile Valley agriculture from the vagaries of irregular river flooding. The dam has achieved both of these goals. It has also created a number of adverse effects, both socio-economic (resulting from the resettling of thousands of Nubians whose land was flooded upstream from the dam and from the increased production costs for Nile Valley farmers who had to invest in new fertilizers) and environmental (including increased pollution in the Nile Delta and soil erosion that is no longer compensated for by silt from flood waters). While the Egyptian government had high hopes for the benefits of the Aswan High Dam, a number of environmental experts warned about the negative effects of further regulating the flow of the Nile.[26] The government discouraged this line of argument and proceeded with construction of the new dam. As in the case of Iraq invading Kuwait, Egypt was aware of the potential costs of its action and determined that the benefits outweighed those costs.[27]

[25] Much has been made of U.S. Ambassador April Glaspie's remarks to Saddam Hussein that the U.S. had no stake in disputes between Arab States. Yet many States besides the U.S. had a stake in whether Iraq annexed Kuwait, and would have regarded such an action as problematic. *See*, for example, the response of the Saudi king: Fahd bin Abdel Aziz, *Defending the Kingdom* 56(2) Vital Speeches of the Day, 675 (1990). For discussion of external influences on Iraq's strategic decisions, *see* Abdulkhaleq Abdulla, *Gulf War: The Socio-Political Background* 16(3) Arab Studies Q. 1–13 (1994).

[26] An earlier dam, now generally referred to as the Aswan Low Dam, was erected in 1903 and expanded in 1913 and 1933.

[27] For a retrospective assessment of the dam's impact, *see* Hesham Abd-El Monsef, Scot E. Smith, and Kamal Darwish, *Impacts of the Aswan High Dam After 50 Years* 29 Water Resources Mgmt. 1873 (2015).

Of course, we have argued here that geoengineering is morally problematic precisely because it is not possible to anticipate its full range of outcomes. So, can the logic of Kuwait and Aswan apply to geoengineering, too? In other words, when it is difficult or impossible to have confidence that one understands the most likely set of outcomes of a policy decision, is it still fair to say that looking at behaviors and outcomes tells us something important about intentions? We argue that it does. For example, in 1961, President John F. Kennedy began considering use of defoliant herbicides to assist in the Vietnam war effort. Agent Orange could be sprayed on the jungle canopy in Vietnam to make guerrilla fighters easier to spot. It could also be used to destroy crops that were feeding supporters of North Vietnam. However, several of Kennedy's advisors, notably Secretary of State Dean Rusk, warned of the negative consequences of spraying Agent Orange. Some argued that spraying herbicide would be abhorred by the international community as engaging in chemical warfare. Rusk was particularly concerned that guerrilla war could be won only with popular support and that destroying crops would alienate the people of Vietnam. The administrations of both Kennedy and Lyndon Johnson were further criticized by Americans who saw the use of Agent Orange as a tactic of starvation. Nonetheless, Kennedy approved Agent Orange for crop destruction in 1962 and Johnson continued the practice as he escalated the war, because the U.S. had political goals that warranted risking international censure and loss of critical domestic support in Vietnam.[28]

States motivated to employ some form of CBG will calculate costs and benefits and then act, even in the face of limited information about the possible consequences. If States find themselves in circumstances so dire that any prospect of worsening them can be discounted, we can expect rational actors to take any action available, including extreme and untested ones like deploying CBG. That is not to say such behavior constitutes good or morally acceptable decision making.[29] It just means that the opacity of one or more consequences is not enough to put off a desperate State from choosing that option; and that States' intentions may sometimes be 'whatever comes of this action.' The moral distinction between intended and unintended outcomes holds, but States may need to broaden their understanding of foreseen consequences and further problematize the distinction between foresight and intention.

[28] For a concise overview of the factors influencing Kennedy's initial use of Agent Orange, *see* Edwin A. Martini, *Hearts, Minds, and Herbicides: The Politics of Chemical War in Vietnam,* 37(1) Diplomatic History 58–84 (3024).

[29] For example, there is growing, though not absolute, consensus among just war scholars that the bombings of Hiroshima and Nagasaki were not morally justified. *See* Jean Bethke Elshtain, Just War against Terror: the Burden of American Power in a Violent World 62 (2003).

7 Conclusion

In this chapter we have considered the practical and moral reasons for restraint in deploying CBG technologies and offer a modified version of just war criteria as a model for assessing the permissibility of such deployment and the moral implications of double effect. This line of reasoning suggests that the current development and discourse of geoengineering is a critical moment in international relations. There are neither international agreements nor international organizations with the specific task of monitoring or restraining States' actions concerning CBG. Meanwhile, climate change is progressing and the urgency of addressing it is growing. If desperate States will take high risks to survive, the international system is reaching a point where the untested nature of CBG would not be a deterrent to deployment. Given the stakes of such action—changing the environment in potentially irrevocable ways—States would benefit from a set of ethical guidelines or a moral framework to support responsible policy decisions about geoengineering. We argue that a set of just geoengineering criteria, based on the just war criteria, would serve that purpose well.

References

1. Chalecki, E.L., Lisa, L.: Ferrari: A new security framework for geoengineering. Strategic Studies Q. **12**(2), 82 (2018)
2. Christopher, A.: Whytock: Foreign State Immunity and the Right to Court Access. Boston U. L. Rev. **93**(2033), 2038 (2013)
3. Douglas, M.: Changing the rules: Just war theory in the twenty-first century. Theology Today. **59**(4), 529 (2003)
4. Orend, B.: War and International Justice: A Kantian perspective. Wilfrid Laurier University Press, Waterloo (2000)
5. Robock, A., et al.: A test for geoengineering? Science. **327**(530) (2010)
6. Strong, A.L., Cullen, J.J., Chisholm, S.W.: Ocean fertilization: Science, policy, and commerce. Oceanography. **22**(3), 236–261 (2009)

A Human Rights Framework for Climate Engineering: A Response to the Limits of Cost-Benefit Analysis

Brian Citro and Patrick Taylor Smith

1 Introduction

Imagine a world where specially designed aircraft and dirigibles launch particles into the atmosphere at regular intervals throughout the year. These particles reflect a certain amount of sunlight back into space, thereby slowing, stopping, or even reversing the previously inexorable rise of global temperatures.[1] However, they also create their own climate disruptions, dramatically altering patterns of precipitation across the globe.[2] Different people will suffer and die if we continue with business as usual, if we mitigate and adapt, if we engineer the climate, or if we adopt a complicated suite of responses that includes all three.[3] The time where we could have prevented climate change at essentially no cost is long gone[4]—if it ever existed—and so "we" will need to decide how these costs will be distributed. How should that be done and by whom? What values should inform our reasoning? The purpose of this chapter is to argue that the moral assumptions that have undergirded much of the debate about climate engineering (CE), and solar radiation management (SRM) in particular,[5] are problematic and have been accepted because no real alternatives

[1] Crutzen [1].

[2] Tilmes et al. [2].

[3] *See* Keith [3].

[4] Jamieson [4].

[5] For a dated but generally accurate survey of these techniques, *see The Royal Society's* Geoengineering the Climate: Science, Governance, and Uncertainty, September 2009.

B. Citro (✉)
Human Rights Lawyer & Independent Researcher, Chicago, IL, USA

P. Taylor Smith
Assistant Professor of Philosophy, University of Twente, Enschede, the Netherlands

© Springer Nature Switzerland AG 2021

W. Burns et al. (eds.), *Climate Geoengineering: Science, Law and Governance*, AESS Interdisciplinary Environmental Studies and Sciences Series,
https://doi.org/10.1007/978-3-030-72372-9_5

have been offered. We aim here to present an improved set of values that can be productively applied to the issue.

Serious consideration of SRM began with claims concerning its "incredible economics."[6] In this view, the primary benefit for SRM—in comparison to less radical suites of mitigation and adaptation responses—is its cost.[7] This type of argument is typical of the dominant form of public policy theorizing about climate change, which can be described broadly as economically-minded cost benefit analysis (CBA).[8] Climate change impacts and mitigation and adaptation measures are analyzed in economic terms, and one tries to maximize the aggregate economic value of the system over the long run.[9] When advocates or opponents contend that CE's economic potential is "incredible" or that it is "cheap," this is what they mean: the economic costs of albedo modification are small in comparison to the economic benefits it generates and in comparison to the costs of other responses.

While many are dissatisfied with the dominant position of CBA in the analysis of climate change,[10] it is surprisingly difficult to present an alternative that is comparably useful and policy-guiding. In this chapter, we suggest a way to incorporate the normative considerations of human rights law into a less impoverished evaluative framework for CE, and SRM in particular. We first characterize the broad features of CBA and explain its problematic presuppositions. We then describe the features of a human rights framework for CE and how it responds to these problems. Next, however, we demonstrate that human rights discourse, as it is currently understood, may fall short of usefully guiding action in the context of CE. Finally, we offer a revised human rights framework more readily applicable to CE.

2 The Limits of Cost-Benefit Analysis

Let us consider a CBA of a regulatory action to rein in air pollution. We would need to *compare* the benefits of the regulation (e.g., fewer cases of asthma) to the costs (e.g., job losses). But this, one might think, is nothing more than a platitude. Of course, one should weigh benefits and costs when making decisions. The important questions, however, are *what* counts as a benefit or a cost and *how* to compare them. It is the answer to these questions that makes CBA a distinctive view of how to approach public policy.

---

[6] Barrett [5].

[7] Other purported benefits of SRM, especially its speed, can ultimately be understood in terms of cost considerations. The justification for not engaging in immediate, total decarbonization, on the other hand, is that it would be excessively costly in terms of human welfare.

[8] For the general structure and popularity of CBA, *see* the introduction of M. Adler and E. Posner [6].

[9] For explanation and utilization of an approach that attempts to maximize the aggregate economic value of a system, *see, e.g.*, Adler and Posner, *ibid.*; Crutzen, *supra* note 1; Barrett, *supra* note 6.

[10] *See* Richardson [7]; R. Frank, 'State and Federal Regulatory Reform: A Comparative Analysis' in Adler and Posner, *supra* note 8; Kelman [8].

CBA, properly understood as a specific account of how to characterize and compare costs and benefits, is predicated on three broad theoretical commitments.[11] First, the fundamental goal of public policy is to produce the greatest overall aggregate of the relevant values that make up the common good.[12] That is, the goal is not to ensure that each person has an adequate amount of a certain value, or to distribute a value *fairly,* but only to produce the greatest net *amount* of the relevant values. Second, the values can be accurately tracked via (or are actually constituted by) the preferences of the constituent members of the relevant community.[13] Third, we can concretize and capture the weightiness of various preferences through an economic welfare function that can be understood in terms of currency.[14] In other words, CBA is—essentially—utilitarianism applied to public policy questions, combined with claims about value and how that value is evaluated, compared, and discovered. Namely, CBA reduces value to preference satisfaction that can then be understood in terms of a willingness to pay and compensate in a shared medium of exchange. Returning to regulation to curb air pollution, CBA theorists would argue that one way to compare the costs and benefits is to first calculate the economic effects of asthma cases produced by pollution (manifested in lost work productivity, medical costs, etc.) and then determine whether that number is greater than the economic costs of the regulation.

If one suggests that SRM is an attractive policy response to climate change by referencing how *cheap* it is in comparison to mitigation and adaptation, one implicitly accepts a CBA normative framework.[15] That is, the economic costs of albedo modification are small in comparison to the economic benefits it generates, and in comparison to the costs of other responses. Consider how one might argue for injecting sulfates into the atmosphere. By slowing or eliminating average global temperature increases, SRM could prevent many of the significant economic costs of climate change, make adaptation less expensive by slowing the rate of change, and allow for faster economic growth as the world economy would not have to engage in costly efforts to eliminate externalities or reduce dependency on fossil fuels.[16] However, these interventions to modify the planet's albedo will have their own consequences, distributed unevenly across the world.[17] These consequences, like CE's benefits, can be translated into economic costs. The CBA argument in favor of CE is that the economic benefits of its deployment are greater—in terms of dollars in aggregate—than its costs.[18] But it is not obvious, for example, that a small

---

[11] For a description of the basic structure of the CBA, *see* Schmidtz [9].

[12] Railton [10].

[13] *See* Orr [11].

[14] *See* Schmidtz, *supra supra* note 11; Grob [12].

[15] This should not be taken as conceding that SRM will be cheap. Perhaps it will, perhaps it will not. Yet, judgments about the appropriateness of SRM that depend on its economic cost are the target of these objections.

[16] *See* Keith, *supra* note 3.

[17] *See* Tilmes, *supra* note 2.

[18] *See* Crutzen, *supra* note 1; Barret, *supra* note 6.

per capita benefit for a large number of people ought to outweigh a smaller overall cost that is distributed amongst a much smaller number of people.[19] Moreover, it should also be noted that aggregation under CBA is indifferent to what people deserve and to individual responsibility.[20] In the case of CE, this means groups that are more responsible for climate change may benefit from CE, allowing them to avoid incurring the costs associated with their behavior, while groups that did not contribute to rising temperatures may disproportionately bear the costs of CE.

Beyond worries about distribution, the underlying theory of value for CBA is suspect. First, preference satisfaction is a problematic theory of human interests. Imagine that I deliberately inject you with an addictive drug that generates a preference in you every morning.[21] Every morning I give you the drug, satisfying your desire. The overall number of your desires or preferences that are satisfied is increased by one per day. On a preference satisfaction view, it seems like your life has gotten better. If you are willing to pay for the drug, we could imagine scenarios where such addictions contributed to economic growth, contributing to social welfare. This is implausible; it seems like a preference satisfaction view needs an account of the "right" kinds of preferences that *ought* to be satisfied. This is to say that preference satisfaction cannot be a full theory of value.

Second, preference-satisfaction accounts of value do not properly attend to the mechanics of preference formation. They fail to take into account the dynamics of "adaptive preferences."[22] Individuals who are consistently faced with the prospect having their projects, desires, and preferences foiled and left unsatisfied will often shape or re-shape their preferences in a less ambitious direction. Worse, a sufficiently oppressive set of social dynamics might prevent individuals from even considering projects or interests that might make them much happier. The failure of economic, social, and political systems to provide opportunities to their citizens could be a severe injustice, particularly when those opportunities (and the lack of them) are distributed according to morally arbitrary factors, such as race and gender. Yet, if a woman adapts her preferences to her limited opportunities, CBA will have a difficult time explaining what is unjust about her plight.[23] After all, her adapted preferences are being satisfied while the preferences she would have had in a more egalitarian system would be left unsatisfied—indeed, they do not exist. Of course, one might respond by arguing that preventing women from effectively participating in the economy might lower economic productivity and, eventually, overall preference satisfaction. It seems odd, though, to say that what is wrong with gender discrimination is that it prevents women from effectively working towards the satisfaction of the preferences of their oppressors. It is more plausible that forcing

[19] *See* Kelman, *supra* note 10.

[20] This is a conceptual claim: if your key principle is "maximize X," then you must be indifferent to all other values, except insofar as they maximize X.

[21] *See* Parfit [13].

[22] *See* Nussbaum [14]; Sen [15].

[23] *See* Matthew D. Adler & Eric Posner, 'Implementing Cost-Benefit Analysis When Preferences are Distorted' in Adler and Posner, *supra* note 8.

an agent to adapt and limit her preferences is a deeply unjust way of treating that person regardless of its effect on others.

Finally, CBA assumes that we can characterize the relevant goods and interests in terms of economic value.[24] This means that the value of preventing a case of asthma or in protecting a national park must be understood in pecuniary terms. The value of the Grand Canyon, for example, might lie in the willingness of human beings to pay for a particular kind of environmental experience that is irreducible to its economic value.[25] One can offer internal and external criticisms of the "economic value" assumption of the CBA. Internally, the concern is that economics may not track the underlying theory of value—preference satisfaction or welfare—particularly well. Willingness (or ability) to pay does not obviously track the importance or weightiness of one's preferences in all circumstances. Externally, one might worry that willingness to pay is not a good proxy for understanding the full value of various objects and experiences. A painting might have aesthetic value even if no one wishes to look at it and a forest may have environmental value even if no one wishes to protect it. In addition, individuals can have *mistaken preferences*, driven by bias, ignorance, or plain moral viciousness.[26] For example, reproductive labor may be undervalued due to patriarchal social structures. Or one might worry that some preferences—such as those of animals—have been entirely excluded from the economic analysis because they cannot participate in the market.

In sum, CBA makes some philosophical assumptions about the nature of value and these assumptions are, at best, controversial. As such, the dominant position of CBA-type thinking in debates about climate change and CE is problematic. In what follows, we will sketch out an alternative. This alternative is not meant to replace CBA in public policy analysis. CBA is not without its strengths, and it is so deeply entrenched that—from a practical, political perspective—trying to eliminate it would be quixotic. Rather, the point of working towards a practicable framework built on human rights is to offer a complementary set of tools that will allow policymakers and the public to incorporate normative factors into their analysis of CE as a response to rising global temperatures.

3 A Human Rights Framework for Climate Engineering

In light of the limitations of CBA, we consider a human rights framework for CE. Human rights provide a well-established legal and normative basis on which to examine the procedural, distributive, and consequential concerns associated with

---

[24] *See* Schmidtz, *supra* note 11.

[25] *See* Ackerman and Heinzerling [16].

[26] For the idea that people can be "environmentally vicious," *see* Sandler [17].

CE.[27] The Paris Agreement within the United Nations Framework Convention on Climate Change acknowledges the importance of human rights for actions taken to address climate change:

> "Parties should, when taking action to address climate change, respect, promote and consider their respective obligations on human rights, the right to health, the rights of indigenous peoples, local communities, migrants, children, persons with disabilities and people in vulnerable situations and the right to development, as well as gender equality, empowerment of women and intergenerational equity."[28]

The international human rights regime has been in place since 1948, when the Universal Declaration on Human Rights (UDHR) was adopted following World War II.[29] Since then, international and regional human rights instruments have proliferated. These include, most importantly, the International Covenant on Civil and Political Rights (ICCPR) and the International Covenant on Economic, Social, and Cultural Rights (ICESCR) at the international level.[30] Along with the UDHR, these instruments form what is commonly referred to as the International Bill of Human Rights.[31] Human rights have also been enshrined in numerous constitutions throughout the world. The UDHR has served as a model for many national constitutions.[32] More recently, constitutions have explicitly adopted language from international instruments.[33] Judges in domestic courts also look to the human rights in these

[27] For a similar analysis of CE governance in the context of the Paris Agreement using a human rights framework, *see* Centre for International Governance Innovation, 'The Paris Agreement and Climate Geoengineering Governance: The Need for a Human Rights-Based Component (2016) CIGI Papers No. 111 (by William C.G. Burns).

[28] Paris Agreement to the United Nations Framework Convention on Climate Change, Paris (France), 12 Dec. 2015, in force 5 Oct. 2016.

[29] UN GA Resolution A/RES/217(III), of 10 Dec. 1948, on Universal Declaration of Human Rights.

[30] International Covenant on Civil and Political Rights (ICCPR), New York (United States of America), 16 Dec. 1966, in force 23 Mar. 1976; International Covenant on Economic, Social and Cultural Rights (ICESCR), New York (United States of America), 16 Dec. 1966, in force 3 Jan. 1976.

[31] Human rights treaties on specific issues have also been adopted at the international level. *See, e.g.*, International Convention on the Elimination of All Forms of Racial Discrimination (ICERD), New York (United States of America), 7 Mar. 1966, in force 4 Jan. 1969; Convention on the Elimination of All Forms of Discrimination Against Women (CEDAW), New York (United States of America), 18 Dec. 1979, in force 3 Sept. 1981; Convention on the Rights of the Child (CRC), New York (United States of America), 20 Nov. 1989, in force 2 Sept. 1990; Convention on the Rights of Persons with Disabilities (CRPD), New York (United States of America) 30 Mar. 2007, in force 3 May 2008. Human rights treaties have also been adopted at the regional level. *See, e.g.*, American Declaration of the Rights and Duties of Man (ADRDM), Bogota (Colombia), 2 May 1948, in force 2 May 1948; European Convention on Human Rights (ECHR), Rome (Italy), 4 Nov. 1950, in force 3 Sept. 1953; American Convention on Human Rights (ACHR), San José (Costa Rica), 22 Nov. 1969, in force 18 Jul. 1978; African Charter on Human and Peoples' Rights (ACHPR), Nairobi (Kenya), 27 Jun. 1981, in force 21 Oct. 1986.

[32] Hannum [18].

[33] *See, e.g.*, Fundamental Law Of Hungary, art. IV(1) (2011, rev. 2013) (adopting language from ICCPR); Const. of Kenya, art. 43 (2010) (adopting language from ICESCR); Const. of Zimbabwe,

instruments—and the elaboration of the rights' content and scope by human rights experts—to inform their interpretation of the rights in their constitutions.[34]

Whether or not we develop *ex ante* a comprehensive human rights framework for CE, individuals and groups that are harmed by environmental impacts resulting from CE deployment will likely make legal claims based on human rights in courts, as people have done for other environmental harms.[35] In this sense, in addition to framing positive and negative CE impacts, a human rights framework can be understood as a descriptive theory, anticipating legal and political reactions to negative deployment impacts. Thus, we aim to use the internal resources of human rights law and theory to evaluate their usefulness in the context of CE because it is likely that people who are negatively impacted by CE interventions or those who believe they stand to gain from them will make use of human rights law in the courts and political advocacy. This approach is in line with Dworkin's "constructive interpretation," in that it attempts to use the resources internal to human rights discourse and practice to offer an interpretation in the CE context that best fits with the internal logic of the practice.[36]

---

art. 80 (2013) (adopting language similar to that in CEDAW); Const. of the Republic of Uganda, art. 28(1) (1995, rev. 2005) (adopting language similar to that in ICCPR).

[34] *See, e.g.*, *Roper v. Simmons*, 543 U.S. 551 (2005) (United States of America) (referring to article 37 of CRC in juvenile death penalty case); *Ochieng v. Attorney Gen.*, Pet. No. 409 (2009) (H.C.K.) (Kenya) (referring to article 12(1) of ICESCR and an amicus curiae brief submitted by the UN Special Rapporteur on the right to health in a case involving access to HIV drugs); *Shatrughan Chauhan v. Union of India*, 3 S.C.C. 1 (2014) (India) (referring to prohibitions on cruel and degrading treatment or punishment in UDHR and ICCPR and a report of the UN Special Rapporteur on extrajudicial, summary or arbitrary executions in a death penalty case involving mentally ill persons).

[35] *See, e.g.*, *State of the Netherlands v. Urgenda Foundation*, Case 200.178.245/01, The Hague Court of Appeal (2018) (Netherlands). (holding government responsible for controlling country's levels of emissions, and finding government must do more to avert negative effects of climate change, because of its duty under the European Convention on Human Rights to protect and promote rights to life and respect for private and family life); *Mossville Envtl. Action Now v. United States*, Inter-Am. Comm'n H.R., Report No. 43/10, (2010) (Inter-American Commission on Human Rights agreed to hear a discrimination claim involving a chemical-producing industrial facility accused of contaminating the environment and producing ill-health effects that predominantly affected African-American households); *Roche v. United Kingdom*, Eur. Ct. H.R. (2005) (holding government violated Article 8 of ECHR and had to make available to claimant all information related to health risks from military mustard and nerve gas tests); *Taskin v. Turkey*, 2004-X Eur. Ct. H.R. 1149 (2004) (holding government violated Article 8 of ECHR for failing to provide claimant with information about risks to health due to living next to mining site).

[36] Dworkin [19].

3.1 Crosscutting Human Rights Principles

Human rights provide a widely accepted normative framework, rich with crosscutting principles that inform the content and application of specific rights. We define crosscutting principles as those that cut across and inform the application of various human rights at the governance stage and in consideration of possible CE deployment impacts.[37] These include: a focus on vulnerable or marginalized groups,[38] prioritization of the principle of nondiscrimination,[39] a requirement that affected communities participate in decision-making processes that impact their lives,[40] and the assignment of duties, accountability, and remedies for human rights violations.[41]

As an evaluative tool, human rights place the focus on vulnerable or marginalized populations, understood as those with the least capacity to cope with potential negative CE impacts, but also those facing severe risks from rising temperatures who are likely to benefit most from climate change mitigation. A human rights approach also prioritizes the principle of nondiscrimination at all levels of

---

[37] *See, e.g.*, Meier and Chakrabarti [20]. (employing a similar concept of crosscutting principles, including equality and nondiscrimination, participation and accountability, in examining Bhutan's health system through lens of the right to health).

[38] *See, e.g.*, UN CESCR, *Gen. Comment No. 3, The Nature of States Parties' Obligations*, para. 12, UN Doc. E/1991/23, 14 Dec. 1990; UN CESCR, *Gen. Comment No. 14, The Right to the Highest Attainable Standard of Health*, paras. 12(b)(i, ii), 35, 37, 40, 43(a, f), 52, 62, 65, UN Doc. E/C./12/2000/4, 11 Aug. 2000; Audrey R Chapman & Benjamin Carbonetti, 'Human Rights Protections for Vulnerable And Disadvantaged Groups: the Contributions of the UN Committee on Economic, Social and Cultural Rights' (2011) 33 *Human Rights Quarterly*, pp. 682–732; Lourdes Peroni & Alexandra Timmer, 'Vulnerable groups: The promise of an emerging concept in European Rights Convention law' (2013) 11(4) *International Journal of Constitutional Law*, pp. 1056–85.

[39] Prohibitions against discrimination appear in every major human rights instrument at the international and regional level. *See, e.g.*, UDHR, *supra* note 29, art. 7; ICCPR, *supra* note 30, art. 26; ICESCR, *supra* note 30, art. 2(2); ICERD, *supra* note 31; CEDAW, *supra* note 31; CRC, *supra* note 31, art. 2; CRPD, *supra* note 31, art. 5; ECHR, *supra* note 31, art. 14; ACHR, *supra* note 31, 1; ACHPR, *supra* note 31, 2. *See also*, UN HRC, *Gen. Comment 18, Non-discrimination*, UN Doc. HRI/GEN/1/Rev.1 26, 1994, 10 Nov. 1989; UN CESCR, *Gen. Comment No. 20, Non-discrimination in Economic, Social and Cultural Rights*, UN Doc. /C.12/GC/20, 2 Jul. 2009.

[40] *See, e.g.*, ICCPR, *supra* note 30, 25; UN HRC, *Gen. Comment No. 25*, UN Doc. CCPR/C/21/Rev.1/Add.7, 27 Aug. 1996; UN OHCHR, Promotion, protection and implementation of the right to participate in public affairs in the context of the existing human rights law: best practices, experiences, challenges and ways to overcome them, UN Doc. A/HRC/30/26, 23 Jul. 2015; Fabienne Peter, 'Human Right to Political Participation' (2013) 7(2) *Journal of Ethics & Social Philosophy*, (arguing that the right to participate in political affairs is necessary for human rights to secure political legitimacy).

[41] *See, e.g.*, UDHR, *supra* note 29, art. 8; ICCPR, *supra* note 30, art. 2(3); ICERD, *supra* note 31, art. 6; *Gen. Comment No. 3*, *supra* note 38; ACHR, *supra* note 31, art. 25(1); UN OHCHR, The OHCHR Accountability and Remedy Project, *Illustrative examples for guidance to improve corporate accountability and access to judicial remedy for business-related human rights abuse*, (July 5, 2016), http://ohchr.org/EN/Issues/Business/Pages/OHCHRstudyondomesticlawremedies.aspx; Jon M. Van Dyke, *Promoting Accountability for Human Rights Abuses*, 8 Chap. L. Rev. 153 (2005).

governance and in evaluating the distribution of the potential impacts of CE. Together, these principles highlight the role human rights can play in considering the distribution of CE interventions, rather than simply their aggregate global impact. The participation of individuals and communities in decision-making that impacts their lives, or threaten to do so, is also key to a human rights framework and undergirds the procedural aspects of the approach.

Finally, a human rights framework provides a structure through which to assign duties, determine accountability, and provide remedies for potential negative impacts of CE, understood as substantive human rights violations. Examining States' duties under international human rights law also allows us to consider whether CE deployment may assist States in meeting their duties to mitigate human rights impacts due to climate change. Framing the consequences of CE interventions as human rights violations or human rights promotion requires us to first determine who bears the legal duties, what their content is, what remedies should be available when rights are violated, and what system will provide these remedies and enforce accountability.

3.2 *Procedural Rights*

Human rights include robust protections for procedural rights meant broadly to facilitate fair, inclusive, and transparent legal and governmental processes. These rights are directly relevant to concerns about potential CE governance: How do we design a fair and inclusive governance regime to consider CE research or deployment? What decision-making procedures should be in place in such a regime? How do we ensure information is available to allow for the meaningful participation of all affected parties? Who should have a seat at the table during negotiations and at what level should the regime operate?[42]

We will not focus on CE research governance here, but we note that efforts to ensure CE research is transparent and inclusive (if conducted) are well underway.[43] A variety of mechanisms have been suggested toward this end, including public research registries, such as those used for nuclear power research, and online information clearinghouses or other information databases to facilitate disclosure and

[42] For a related discussion of the role of procedural rights in addressing climate change, *see* Kravchenko [21].

[43] *See, e.g.*, Science and Tech. Comm., *The Regulation of Geoengineering: Fifth Report of the Session* (2009–10), H.C. 221, p. 18 (U.K.) (Comments of John Virgoe to House of Commons committee discussing geoengineering governance, calling for 'principles around openness, transparency in research, [and] notifying a neighboring country or countries which might be affected'); Royal Soc'y, Geoengineering the Climate: Science, *Governance and Uncertainty* (2009), pp. 41–43; Carr et al. [22].

transparency.[44] These mechanisms would be made accessible to scientists, governmental regulatory regimes, and the public. Environmental Impact Assessments, although not tailored specifically for CE research or deployment, could be used and made publicly available to promote transparent and inclusive CE research.[45]

Procedural rights relevant to CE include, among others, the rights to public participation, access to justice, and information, as well as the rights to nondiscrimination and self-determination. These rights appear in various forms in the international and regional instruments discussed above and in constitutions around the world.[46] Principle 10 of the Rio Declaration on Environment and Development—a non-binding declaration signed by more than 170 countries—establishes the importance of public participation in the realm of the environment, noting the related rights of access to information and access to justice:

> "Environmental issues are best handled with the participation of all concerned citizens, at the relevant level. At the national level, each individual shall have appropriate access to information concerning the environment that is held by public authorities … and the opportunity to participate in decision-making processes. States shall facilitate and encourage public awareness and participation by making information widely available. Effective access to judicial and administrative proceedings, including redress and remedy, shall be provided."[47]

The Aarhus Convention on Access to Information, Public Participation in Decision-Making and Access to Justice in Environmental Matters (Aarhus Convention) of the United Nations Economic Commission for Europe provides an additional example. The Convention, ratified by 46 countries and the European Union, declares that "adequate protection of the environment is essential to human well-being and the enjoyment of basic human rights."[48] It establishes the "rights of access to information, public participation in decision-making, and access to justice in environmental matters" and promotes "accountability of and transparency in

---

[44] *See* Nigel Moore et al., 'Procedural Governance of Field Experiments in Solar Radiation Management' (2014) IASS Working Paper, pp. 10–11.

[45] *See* Convention on Environmental Impact Assessment in a Transboundary Context, Espoo (Finland), 25 Feb. 1991, in force 10 Sept. 1997.

[46] For international and reginal instruments, *see, e.g.*, UDHR, *supra* note 29, arts. 10, 11, 19, 21; ICCPR, *supra* note 30, arts. 1, 14, 19, 25; ECHR, *supra* note 31, arts. 6, 10; ACHR, *supra* note 31, arts. 8, 13, 23, 25; ACHPR, *supra* note 31, arts. 7, 9, 13, 20; Convention on Access to Information, Public Participation in Decision-Making and Access to Justice in Environmental Matters, (Aarhus, Denmark), 25 Jun. 1998, in force 30 Oct. 2001 [hereinafter Aarhus Convention]; *supra* note 39 for prohibitions against discrimination in regional and international instruments. For constitutions, *see, e.g.*, Const. of United States of America, amends. V, VI, XIV; Const. of Federative Rep. of Brazil of 1988, arts. 4(III), 5(XIV, XXXIII, LIV, LXXVIII), 14; Const. of Rep. of Ghana of 1992, secs. 17, 19, 21(1(f), 3); Const. of India, secs. 15, 16, 21, 22.

[47] Rio Declaration on Environment and Development, principle 10, UN Doc. A/CONF.151/26/Rev.1 (Vol. I), Annex I, 12 Aug. 1992.

[48] UN Treaty Collection, Status of Ratification, Chapter XXVII, Environment, 13. Convention on Access to Information, Public Participation in Decision-Making and Access to Justice in Environmental Matters; Aarhus Convention, *supra* note 46, pmbl.

decision-making" processes.[49] The Aarhus Convention also highlights the role of civil society in decision-making processes at local, national, and regional levels and establishes its own compliance review mechanism.[50] The Convention's principles have influenced European Union environmental policies through a series of Directives.[51] Finally, the 2018 Regional Agreement on Access to Information, Public Participation and Justice in Environmental Matters in Latin America and the Caribbean (Escazú Agreement)—a binding agreement coming from the United Nations Conference on Sustainable Development (Rio + 20)—establishes the rights of access to environmental information, public participation in environmental decision-making processes, and access to justice in environmental matters.[52] The Escazú Agreement is not yet in force, though. Sixteen countries have signed the treaty, but 11 must ratify it before it enters into force.[53]

Procedural human rights are grounded in basic democratic principles and notions of procedural fairness.[54] The principle of transparency is also closely associated with these rights and their underlying norms, particularly as relates to good governance.[55] The right to public participation requires that all parties likely to be affected by CE, both positively and negatively, have a right to be at the table when decisions are made.[56] This means that representatives of governments likely to experience the impacts of CE be included—and given an equal voice—in intergovernmental processes and that affected communities must be included in decision-making about CE deployment.[57] In line with the principle of transparency, deliberations and decisions at all levels should be transparent and made publicly accessible. Participation and transparency, in turn, implicate the right to information—in this case, the right of people to access information about all aspects of CE, including its likely impacts,

---

[49] Aarhus Convention, *supra* note 46, pmbl, art. 1.

[50] Aarhus Convention, *supra* note 46, pmbl., arts. 3, 6, 15.

[51] *See, e.g.*, European Comm., *The EU & the Aarhus Convention: in the EU Member States, in the Community Institutions and Bodies*, 8 Nov. 2015 (listing directives that have adopted provisions of the Aarhus Convention).

[52] Regional Agreement on Access to Information, Public Participation and Justice in Environmental Matters in Latin America and the Caribbean (Escazú, Costa Rica), 4 Mar. 2018, opened for signature 9 Apr. 2018.

[53] UN Treaty Collection, Status of Ratification, Chapter XXVII, Environment, 18. Regional Agreement on Access to Information, Public Participation and Justice in Environmental Matters in Latin America and the Caribbean.

[54] *See, e.g.*, Universal Declaration on Democracy, Cairo (Egypt), Inter-Parliamentary Council, 16 Sept. 1997; UN OHCHR, *Good Governance and Human Rights*, http://www.ohchr.org/EN/Issues/Development/GoodGovernance/Pages/GoodGovernanceIndex.aspx

[55] *See, e.g.*, *Good Governance and Human Rights*, *ibid*; UN OHCHR, *Human Rights and Anti-corruption*, http://www.ohchr.org/EN/Issues/Development/GoodGovernance/Pages/AntiCorruption.aspx

[56] *See, e.g.*, Aarhaus Convention, *supra* note 46, pmbl., arts. 6, 7, 8.

[57] *Gen. Comment No. 25*, *supra* note 40, para. 6.

both positive and negative, and the nature and operation of decision-making processes and governance mechanisms.[58]

Human Rights Impacts Assessments (HRIAs) may be useful in promoting transparency and the rights to information and participation related to potential CE deployment. HRIAs are used to identify, predict, and assess impacts on human rights resulting from a variety of interventions, including trade agreements, business operations, and health policies.[59] They may be used prior to, during, or after an intervention. Their "essential elements" include a normative human rights framework, public participation, equality and nondiscrimination, transparency and access to information, and accountability.[60] In order to strengthen procedural rights in the context of CE, HRIAs could be conducted by governments or private actors during *ex ante* deliberations concerning deployment.

Recognition and protection of procedural rights will promote the cross-cutting principle of non-discrimination—also a freestanding right—in CE governance mechanisms.[61] Together, the right to non-discrimination and the right to self-determination require that the interests of groups likely to be impacted by CE deployment be at the center of deliberations at the international and national levels. The idea is that, if the effects of CE interventions such as SRM are projected to disproportionately impact particular regions of the world, it may amount to discrimination if deployment occurs without the involvement of communities likely to be affected in decision-making processes. This claim is based on the principle that discriminatory intent is not necessary to prove discrimination, but rather only a discriminatory impact.[62] A commitment to ensure CE interventions are not discriminatory provides further support for the claim, under the right to public participation, that all affected parties must be at the table when decisions are made. This does not necessarily mean that affected communities should be granted a veto during decision-making processes, but it does require they are given an equal voice if they are likely to experience disproportionately negative or positive impacts.

Indigenous groups may benefit from particular CE interventions, while others may experience significant deprivations, resulting in the degradation of their natural environments.[63] The enjoyment of natural resources and traditional cultural prac-

---

[58] *See, e.g.*, Aarhaus Convention, *supra* note 46, art. 4; Steve Rayner et al., *The Oxford Principles*, Climate Geoengineering Governance Working Paper Series: No. 1, 1 May 2013, 27–31.

[59] *See, e.g.*, UN OHCHR, *Guiding principles on human rights impact assessments of trade and investment agreements*, UN Doc. A/HRC/19/59/Add.5, 19 Dec. 2011; World Bank & Nordic Trust Fund, *Human Rights Impact Assessments: A Review of the Literature, Differences with other forms of Assessments and Relevance for Development*, Feb. 2013; MacNaughton [23].

[60] World Bank & Nordic Trust Fund, *ibid*, p. xi.

[61] *See supra* note 39.

[62] *See, e.g.*, *Gen. Comment 18*, *supra* note 39, para. 7; ICERD, *supra* note 31, art. 1; CEDAW, *supra* note 31, art. 1; *Mossville*, *supra* note 35 (Inter-American Commission on Human Rights agreed to hear discrimination claim on chemical-producing industrial facility accused of contaminating the environment and producing ill-health effects that predominantly affected African-American households).

[63] *See, e.g.*, Whyte [24, 25].

tices is key to realizing the right to self-determination for indigenous groups, understood as a group's right to "freely determine their political status and freely pursue their economic, social and cultural development."[64] If indigenous groups are negatively impacted by CE deployment, under-inclusive decision-making may infringe their right to self-determination. Recognizing the right to self-determination would grant these groups a seat at the negotiating table and ensure their interests are given appropriate consideration. This, in turn, promotes the crosscutting principle calling for a focus on vulnerable or marginalized groups.

3.3 *Substantive Rights*

Human rights also offer a meaningful way to frame and address the potential consequences—both positive and negative—of CE deployment. While there is no certainty or universal consensus regarding the precise nature or magnitude of the impacts of CE, deployment of certain technologies may implicate, both positively and negatively, the substantive rights to life, health, housing, food, and water. Substantive human rights, like procedural rights, are established at the international, regional, and national levels in treaties and constitutions. The right to life is recognized in virtually every human rights instrument and in most constitutions.[65] The other rights, commonly referred to as social and economic rights, derive primarily from the UDHR and ICESCR at the international level, but are also enshrined in various forms in each of the specific international human rights conventions, such as ICERD, CEDAW, and the CRC.[66] At the regional level, these rights are provided at least some degree of protection in the human rights conventions of the Americas and in the African Charter.[67] Many national constitutions also recognize and protect economic, social, and cultural rights.[68]

Examining the possible positive and negative consequences of various forms of CE through the lens of substantive human rights requires us to acknowledge that the

---

[64] ICCPR, *supra* note 30, art. 1; *see also*, UN HRC, *Ominayak and Lubicon Lake Band v. Canada*, Comm. No. 167/1984, UN Doc. CCPR/C/38/D/167/1984, 26 Mar. 1990 (Human Rights Committee found a violation of the right of indigenous group to determine its own culture in a case involving environmental changes that prevented indigenous groups from hunting on traditional lands).

[65] *See, e.g.*, UDHR, *supra* note 29, art. 3; ICCPR, *supra* note 30, art. 6; ECHR, *supra* note 31, art. 2; ACHR, *supra* note 31, art. 4. The right to life appears in 117 constitutions, *see* Constitute Project, https://www.constituteproject.org/search?lang=en&q=Right%20to%20life

[66] *See, e.g.*, CRC, *supra* note 31, pmbl.; CEDAW, *supra* note 31, art. 13; ICERD, *supra* note 31, arts. 1, 2, 5.

[67] *See* Additional Protocol to the American Convention on Human Rights in the Area of Economic, Social and Cultural Rights (Protocol of San Salvador), San Salvador (El Salvador), 17 Nov. 1988, in force 16 Nov. 1999; ACHPR, *supra* note 31, art. 4; ACHR, *supra* note 31, ch. 3; ADRDM, *supra* note 31, art. XXII.

[68] *See* Jung et al. [26] ('[n]early all new democracies, and several established ones, have included some form of [economic and social rights] in their constitutions').

rights of some may be promoted while the rights of others may be infringed. What this tells us about the utility of the human rights framework is the subject of the next section, which focuses on the framework's limitations. We also acknowledge that it may be difficult or impossible to determine with certainty the chain of causality resulting in human rights violations, particularly in a world where multiple CE interventions are occurring, along with climate change. This poses a problem from the accountability standpoint, and potentially in providing remedies for deprivations, but it does not reduce the value of a human rights framework in identifying and evaluating consequences when they can be attributed to CE deployment.

We will largely focus here on the value a human rights framework offers in considering negative consequences of CE interventions. The basic argument is that the substantive rights of individuals that experience negative environmental effects resulting from CE deployment may be violated. The particular content of each right—the kinds of actions it prohibits and the particular entitlements it establishes under law—provide unique evaluative tools with which to understand the nature and magnitude of these impacts. Considering the distribution of the impacts allows us to determine whether the affected communities are vulnerable or marginalized within their larger societies or as a matter of relative geopolitical power, and what capacity they have to respond effectively. For instance, droughts resulting from a reduction or changes in precipitation patterns may disrupt the production and distribution of food or reduce access to water in certain communities.[69] This, in turn, may impact the health of individuals within the community and may further burden health systems in countries already struggling to provide access to good quality health services for all.

Let us consider how the specific content of the right to water in Article 11(1) of ICESCR would help in framing the impacts of a hypothetical deployment scenario. Again, we note that there is no universal consensus on the impact of particular CE deployment scenarios, but we nonetheless take the liberty to consider this hypothetical because it is useful from an analytical perspective, regardless of whether it is likely to occur. Imagine deployment of sulfur aerosol injection—a form of SRM—resulted in an overall reduction of temperatures in India, including in parts of the countries that experience extreme heat at certain times of the year.[70] This may mean some communities have increased access to water at particular times of the year.[71] However, the same deployment scenario may result in a reduction of monsoon rains in the south of the country.[72] At least in the short term, this could mean

[69] *See, e.g.*, Robock et al. [27]; Burns [28]; Russel et al. [29]; Crook et al. [30].

[70] For an explanation and discussion of sulfur aerosol injection, *see, e.g.*, National Research Council et al., *Climate Intervention: Reflecting Sunlight to Cool the Earth* (2015), pp. 66–101; Burns, *supra* note 69, 289–92.

[71] For the possibility that sulfur aerosol injection may increase availability of water, *see* Russel et al., *supra* note 69, p. 360.

[72] *See, e.g.*, Robock et al., *supra* note 69; Russel et al., *supra* note 69, 356; Tilmes et al., *supra* note 2; *Climate Intervention: Reflecting Sunlight to Cool the Earth, ibid.*, 56–57, 83, 85.

increased access to water for some, but reduced access for others. Let us consider two possible outcomes here.

If, in the aggregate, India experienced a reduction in drinking water as a result of reduced monsoon rains, this would almost certainly reduce access in the aggregate and amount to a violation of the right to water. If on the other hand, the reduction in monsoon rains in the south of the country was offset by an increase in precipitation in other areas, under a cost-benefit analysis approach we might declare that overall the deployment had a positive impact on enjoyment of the right to water. However, the right to water requires not only that water is available in the country, which would be met by an aggregate approach, but also that it is physically and financially accessible to all.[73] An increase in the aggregate water supply in the country at the cost of reduced physical or financial accessibility for certain communities would therefore constitute a violation of the right to water for those that lacked access. This may also be at odds with the principle requiring focus on the vulnerable or marginalized groups, depending on the communities impacted.

3.4 *Duties, Accountability and Remedies for Substantive Human Rights Violations*

3.4.1 Duties and Accountability

States have three distinct kinds of duties under international human rights law—to respect, protect, and fulfill.[74] The duty to respect requires governments to refrain from violating the rights of people within their territories through their own acts or

[73] ICESCR, *supra* note 30, art. 11; UN CESCR, *Gen. Comment No. 15, The Right to Water*, para. 12, UN Doc. E/C.12/2002/11, 20 Jan. 2003.

[74] *See, e.g.*, *Masstricht Guidelines on Violations of Economic, Social and Cultural Rights*, para. 6, UN Doc. E/C.12/2000/13, 2 Oct. 2000 [hereinafter Maastricht Guidelines] ('Like civil and political rights, economic, social and cultural rights impose three different types of obligations on States: the obligations to respect, protect and fulfil'); UN HRC, *Gen. Comment No. 31, The Nature of the General Legal Obligation Imposed on States Parties to the Covenant*, paras. 3, 7, 8, UN Doc. CCPR/C/21/Rev.1/Add. 13, 26 May 2004; UN CESCR, *Gen. Comment No. 12, The Right to Adequate Food*, para. 15, UN Doc. E/C.12/1999/5, 12 May 1999 ('The right to adequate food, like any other human right, imposes three types or levels of obligations on States parties: the obligations to respect, to protect and to fulfil'); UN CESCR, *Gen. Comment No. 13, The Right to Education*, para. 46, UN Doc. E/C.12/1999/10, 8 Dec. 1999 ('The right to education, like all human rights, imposes three types or levels of obligations on States parties: the obligations to respect, protect and fulfil'); *Gen. Comment No. 14*, *supra* note 38, para. 33 ('The right to health, like all human rights, imposes three types or levels of obligations on States parties: the obligations to respect, protect and fulfil'); UN HRC, *S. I. D. et al v. Bulgaria*, Comm. No. 1926/2010, para. 3.8, UN Doc. CCPR/C/111/D/1926/2010, 29 Sept. 2014; *Social and Economic Rights Action Centre and the Centre for Economic and Social Rights v. Nigeria*, Comm. No. 155/96, paras. 44–48, (2001) AHRLR 60. *See also* De Schutter [31]; Knox [32].

omissions.[75] It is thus the duty of a government not to deploy CE technologies, if it would result in human rights violations within its borders. States may also have an extraterritorial obligation not to deploy CE technologies—either from within or outside their borders—if doing so would violate or interfere with the enjoyment of human rights of people in other countries.[76]

The duty to protect requires governments to take steps to prevent human rights violations committed by private, non-governmental actors within their borders.[77] This would include individuals, corporations, and non-governmental organizations capable of CE interventions. Governments could act to ensure private entities do not deploy CE technologies likely to violate human rights within their borders through use of the law and regulations. Fulfilling the duty to protect vis-à-vis other governments, however, would be more difficult and may involve an obligation to advocate against deployment on the international stage or in the context of a transnational governance mechanism. As with the duty to respect, states may have extraterritorial obligations to protect against human rights violations outside their borders resulting from CE deployment by non-governmental actors, including private individuals and organizations and transnational corporations, if a State holds influence or is "in a position to regulate" the actor.[78] This would include deployment occurring both within and outside the State's borders that impacts people in other counties.

The duty to fulfill in international law requires States to take positive action to facilitate the enjoyment of human rights.[79] In some cases, deployment of CE technologies by a government may constitute a positive action to mitigate human rights violations caused by rising temperatures, promoting enjoyment of certain rights in line with the duty to fulfill. States may also have an obligation under the duty to fulfill to make information publicly available on all aspects of CE deployment or to engage in environmental or human rights impact assessments prior to deployment.[80] As with the duties to respect and protect, states may have extraterritorial obligations to fulfill the human rights of people outside their borders, which could include CE deployment, if deployment promotes enjoyment of human rights by people in other

---

[75] *See, e.g.*, Maastricht Guidelines, *ibid.*; *Gen. Comment No. 31*, *ibid*, paras. 6, 10; *Gen. Comment No. 13*, *ibid*, para. 47; *Gen. Comment No. 14.*, *supra* note 38, para. 33.

[76] *Maastricht Principles on Extraterritorial Obligations of States in the area of Economic, Social and Cultural Rights*, paras. 3, 8(a), 19, 20, 28 Sept. 2011, reprinted in (2012) 34 *Human Rights Quarterly*, pp. 1084–1169 [hereinafter Maastricht Principles]; UN HCRC, *Report of the Office of the United Nations High Commissioner for Human Rights on the relationship between climate change and human rights*, para. 86, UN Doc. A/HRC/10/61, 15 Jan. 2009 [hereinafter OHCHR Report on Climate Change and Human Rights].

[77] *See, e.g.*, Maastricht Guidelines, *supra* note 74; *Gen. Comment No. 31*, *supra* note 74, para. 8; *Gen. Comment No. 13*, *supra* note 74, para. 47; *Gen. Comment No. 14.*, *supra* note 38, para. 33.

[78] Maastricht Principles, *supra* note 76, paras. 3, 23–26; OHCHR Report on Climate Change and Human Rights, *supra* note 76, para. 86.

[79] *See, e.g.*, Maastricht Guidelines, *supra* note 74; *Gen. Comment No. 31*, *supra* note 74, para. 7; *Gen. Comment No. 13*, *supra* note 74, para. 47; *Gen. Comment No. 14.*, *supra* note 38, para. 33.

[80] For the relationship between the duty to fulfill and the right to information, *see, e.g.*, Mariela Belski, *Access to Information: An Instrumental Right for Empowerment* (2007), pp. 15–16.

countries.[81] This duty would not be unlimited, but rather contingent, among other things, upon a State's "economic, technical and technological capacities, available resources, and influence in international decision-making processes," and could involve cooperation with other States.[82]

3.4.2 Remedies

The right to an effective remedy for human rights violations is integral to a human rights framework.[83] Article 2 of the ICCPR requires that victims of human rights violations have their claims determined by competent authorities—judicial, administrative, legislative, or otherwise—and that effective remedies are available and enforceable through these authorities.[84] Human rights law recognizes several varieties of remedies. These include restitution, compensation, rehabilitation, satisfaction, and guarantees of non-repetition.[85] Restitution involves, for example, the restoration of liberty, of enjoyment of particular human rights, of citizenship, or return to one's place of residence.[86] Compensation requires financial compensation for financially assessable damages.[87] Rehabilitation requires provision of medical, psychological, legal, or social services to address harms resulting from rights violations.[88] Satisfaction includes injunctions and judicial or administrative sanctions against perpetrators.[89] Guarantees of non-repetition comprise investigations, prosecution and sanctioning of perpetrators, as well as human rights education.[90]

Remedies for human rights violations take distinct forms, depending on the nature of the violation, but the critical point is that accessing an effective and enforceable remedy through a competent authority is itself a human right.[91]

---

[81] Maastricht Principles, *supra* note 76, paras. 3, 28; OHCHR Report on Climate Change and Human Rights, *supra* note 76, para. 86.

[82] Maastricht Principles, *supra* note 76, paras. 30, 31; OHCHR Report on Climate Change and Human Rights, *supra* note 76, para. 86..

[83] *See, e.g.*, UDHR, *supra* note 29, art. 8; UN GA Resolution A/RES/60/147, 16 Dec. 2005, on Basic Principles and Guidelines on the Right to a Remedy and Reparation for Victims of Gross Violations of International Human Rights Law and Serious Violations of International Humanitarian Law; UN OHCHR, *Accountability and Remedy Project: improving accountability and access to remedy in cases of business involvement in human rights abuses*, ('The right to a remedy is a core tenet of the international human rights system ...'); Australian Human Rights Commission, *Right to an effective remedy*, ('The right to an effective remedy is an essential component of human rights under the ICCPR and other human rights instruments').

[84] ICCPR, *supra* note 30, art. 2.

[85] *See* UN GA Resolution 60/147, *supra* note 83.

[86] *Ibid.* para. 19.

[87] *Ibid.* para. 20.

[88] *Ibid.* para. 21.

[89] *Ibid.* para. 22.

[90] *Ibid.* para. 23.

[91] *See supra* notes 83 and 84.

Moreover, ensuring access to a remedy is a crosscutting principle, along with accountability, because it applies to violations across the spectrum of human rights, including both substantive and procedural rights. Conceiving of remedies as rights is a core aspect of what makes a human rights framework unique and valuable. In the CE context, it requires that individuals or groups who experience human rights violations resulting from deployment of CE technologies have access to effective remedies. In a sense, it is the final step in applying a human rights framework: procedural rights inform and constrain governance and decision-making mechanisms; substantive rights provide evaluative legal and normative tools to identify potential and actual CE impacts; duty-bearers are identified and held accountable; and effective remedies are provided to victims if violations occur.

The system by which accountability is determined and remedies are distributed could take many forms. Existing national, regional, and international courts and other bodies that adjudicate human rights claims under national constitutions and international and regional instruments could be used. Alternatively, the international community could develop a new body, possibly linked to CE governance mechanisms, that anticipates and prepares—financially and otherwise—to provide remedies for human rights violations resulting from CE deployment.[92] The particular nature and authority of such a mechanism, however, is outside the scope of this paper.

4 The Limitations of a Human Rights Framework

As the previous section demonstrates, a human rights framework has many beneficial features. It focuses our attention on distributive issues, especially impacts on the poor and marginalized, while offering a more plausible account of the relevant human interests at stake. Yet, the practical applicability of the framework to CE, including SRM and carbon dioxide removal (CDR), is not obvious. It can tell us what morally salient human interests are affected by a particular policy, but it has far greater difficulty in providing practical guidance for policy-makers. In other words, the human rights framework may be much better at diagnosis than at treatment.

The radically non-ideal nature of climate change policy is the essential feature that burdens the application of the human rights framework.[93] If we faced a policy choice where one option violated human rights and the other did not, then it seems plausible that the human rights framework could provide decisive policy guidance.

---

[92] For a discussion of CE compensation concerns and possible regimes, *see, e.g.*, Svoboda and Irvine [33]; Horton [34]; Valdivia [35].

[93] By radically non-ideal, we mean to refer to two features of climate change policy. First, we are dealing with a public policy problem that has been generated by individuals and groups acting in a seriously unjust way. Second, the nature of climate change and its potential responses make it impossible for us to act in an ideally just way. That is, we have done wrong and someone will unjustly suffer as a consequence. The possibility of acting a way that does no wrong has been foreclosed.

Human rights might serve as a trump in that case. However, that is not our choice in responding to climate change. As Dale Jameison has pointed out, the time for preventing climate change impacts has long passed.[94] Even radical mitigation that is implemented immediately will still result in seriously burdensome impacts, especially on the poorest and most marginalized populations on the planet.

But are these impacts human rights violations? In short, yes, we believe they are. However, we acknowledge that there is less than universal consensus around whether the deleterious impacts of climate change constitute infringements of human rights or whether States have human rights obligations to stop or mitigate these impacts.[95] Nonetheless, a growing body of lawsuits and court decisions, United Nations resolutions, expert opinions, and scholarship support the claim that some climate change impacts constitute human rights violations, with corresponding State obligations.[96]

For example, in 2017, the Inter-American Court of Human Rights issued an Advisory Opinion on the Environment and Human Rights that recognized an

---

[94] *See* Jamieson, *supra* note 4.

[95] *See, e.g.*, Gordon [36] (Discussing the Inter-American Commission on Human Rights' rejection of an Inuit petition seeking relief from human rights violations resulting from global warming caused by acts and omissions of the United States. The Commission indicated in a letter to the petitioners that the information provided in the petition 'was insufficient for making a determination'); Posner, [37] (Arguing that international human rights litigation will not lead to a desirable outcome for victims of climatic change).

[96] *See, e.g.*, *Male' Declaration on the Human Dimension of Global Climate Change*, 14 Nov. 2007 (Representatives of 'Small Island Developing States' expressing concern that 'climate change has clear and immediate implications for the full enjoyment of human rights'); UN HRC Resolution 7/23, 28 Mar. 2008, on Human Rights and Climate Change (Declaring concern that 'climate change poses an immediate and far-reaching threat to people and communities around the world and has implications for the full enjoyment of human rights' and calling on the UN OHCHR to study the relationship between climate change and human rights); OHCHR Report on Climate Change and Human Rights, *supra* note 76; UN HRC Resolution 10/4, 25 March 2009, on Human Rights and Climate Change; UN HRC Resolution 18/22, 17 October 2011, on Human Rights and Climate Change; UN HRC Resolution 26/27, 15 July 2014, on Human Rights and Climate Change; UN OHCHR, *Mapping Human Rights Obligations Relating to the Enjoyment of a Safe, Clean, Healthy and Sustainable Environment: Focus report on human rights and climate change* (2014); UN HRC, *Summary report of the Office of the United Nations High Commissioner for Human Rights on the outcome of the full-day discussion on specific themes relating to human rights and climate change*, UN Doc. A/HRC/29/19, 1 May 2015; UN HRC Resolution 29/15, 30 June 3015, on Human Rights and Climate Change; Zeid Ra'ad Al Hussein, UN High Commissioner for Human Rights, *Burning Down the House*, Speech delivered at Paris Climate Change Conference, 3 Dec. 2015 (Stating that 'international human rights law imposes affirmative legal obligations on all states to take the necessary steps in law, policy, institutions, and public budgets to protect human rights from [] harms' due to climate change'); UN HRC, *Report of the Special Rapporteur on the issue of human rights obligations relating to the enjoyment of a safe, clean, healthy and sustainable environment on the human rights obligations relating to climate change*, UN Doc. A/HRC/31/52, 1 Feb. 2016. *See also*, Doudda et al. [38]; International Council on Human Rights Policy, *Climate Change and Human Rights: A Rough Guide* (2008); John H. Knox, *supra* note 74; Limon [39]; Knox [40]; Humphreys [41]; Bodansky [42]; Allard [43]; Quirico and Boumghar (eds.) [44].

autonomous right to a healthy environment under the American Convention on Human Rights, as well as extraterritorial obligations for States to prevent transboundary environmental harms originating from their own territories that impair the rights of persons outside their territories.[97] In its opinion, the Court emphasized the "adverse effects of climate change [on] the effective enjoyment of human rights."[98] In 2018, a Dutch Court of Appeal held that there is a real and imminent danger that climate change will infringe the right to life and the right to private and family life in the European Convention on Human Rights and that the Dutch government has an obligation to take protective action.[99] In another case, the Supreme Court of Justice of Colombia held that the fundamental rights to life, health, minimum subsistence, freedom and human dignity are "substantially linked and determined by the environment and the ecosystem."[100] The Court acknowledged the imminent and irreversible dangers of climate change and humans' responsibility for the situation, and it ordered the Government of Colombia to adopt measures to reduce greenhouse gas emissions, to implement strategies towards climate change adaptation, and to formulate plans to counteract the deforestation rate in the Amazon to address the effects of climate change.[101] In yet another 2018 case, plaintiffs from seven countries, including children, brought suit in the European Union General Court to compel the European Union to take more stringent actions to reduce greenhouse gas emissions. They argued that climate change threatens their fundamental rights of life, health, occupation, and property under the Charter of Fundamental Rights of the European Union.[102]

In light of all this, we proceed based on the presumption that allowing climate change to continue unabated—i.e., doing nothing—will result in human rights violations that States have an obligation to prevent.

The translation of different atmospheric manipulations into environmental impacts and the generation of humanly relevant impacts from those environmental consequences is mediated through a series of complex systems that will likely generate subtly different social outputs with even small differences in atmospheric states.[103] As a result, every meaningfully different suite of climate change responses—including those that include large-scale CE deployment and more mod-

---

[97] The Environment and Human Rights (State Obligations in Relation to the Environment in the Context of the Protection and Guarantee of the Rights to Life and to Personal Integrity – Interpretation and Scope of Articles 4(1) and 5(1) of the American Convention on Human Rights), Advisory Opinion OC-23/18, Inter-American Court of Human Rights (2017).

[98] *Ibid.* para. 47

[99] *State of the Netherlands v. Urgenda Foundation, supra* note 35.

[100] *Future Generations v. Ministry of the Environment and Others*, STC4360–2018, p. 13, Supreme Court of Justice (2018) (Colombia).

[101] *Ibid.*

[102] *Armando Ferrão Carvalho and Others v. The European Parliament and the Council*, Case T-330/18, General Court (EGC) (2018).

[103] What is more, these differences will frequently cut across standard political, social, economic, or geographic groups: global south/north, rich/poor, etc.

erate forms of CE—will result in different individuals being subject to relevantly different impacts. Similarly, anthropogenic climate change will violate the human rights of particular people, while benefiting others, at least up to certain temperature thresholds. Thus, no response will prevent all the violations or generate only benefits, and every response will produce different combinations of both violations and benefits.[104] Even some forms of radical decarbonization, by slowing economic growth and undermining the development of countries where the global poor are concentrated, may contribute to human rights violations. In other words, doing nothing, ideal mitigation and adaptation, radical decarbonization, CE, and any combination of these responses are likely to lead to human rights violations. Moreover, they will all lead to human rights violations—both now and over time—of different groups of people.

It is this last point—the different violations and benefits caused by each response—that makes the application of the human rights framework especially difficult. If the same population was subject to roughly the same human rights violations only to a lesser or greater degree, then it might be uncontroversial to suggest that we should simply minimize rights violations. Yet, this is not the scenario we are faced with; different responses will impose different violations and positive human rights impacts on different people. Even the best suite of responses might generate a scenario that looks like this:

	No CE Deployment	**CE Deployment**
Group A	Human rights violations	Positive human rights impacts
Group B	Status quo	Human rights violations

In some ways, this may understate the problem as regards climate change. Even within most social groups, different people may experience benefits from emitting additional carbon while others will suffer dire consequences as a result of our response to climate change. For example, in large developing countries, carbon emissions may contribute to economic growth that lifts people out of poverty and provides a larger tax base for social services. While it is likely true that some of the extreme climate change scenarios will be human rights worse for everyone, the choice among some plausible emissions pathways, at least in the short to medium term, will involve making some worse off than they would have been under other scenarios.

Will we accept a 1.5, 2 or 3-degree increase? To what extent will we emphasize adaptation over mitigation? How much CE is acceptable? Each answer will generate different benefits and costs, different improvements and human rights violations, for different people. So, we need a set of distributive principles that will guide

[104] It is uncontroversial in the philosophical literature that anthropogenic climate change will very likely cause egregious human rights violations. *See, e.g.*, Caney [45].

us in the various tradeoffs between individuals and between different sets of violations and positive impacts. And, unfortunately, procedural rights are not immediately helpful in this regard. After all, climate change is already leading to violations of procedural rights that are as egregious as they would be in the case of CE: the people of Bangladesh get as little say in Chinese or American emissions behavior as they are likely to get in the decision to deploy SRM.

To put it another way, standard human rights discourse has assumed that there is always a human rights-dominant strategy for social reform or that there is a Pareto-optimal action in terms of human rights.[105] That is, much of the time where human rights discourse is employed, there is an action that is human rights-superior for some and human rights-inferior for none. This is fairly plausible in paradigmatic cases of human rights violations. When one demands that arbitrarily imprisoned journalists critical of an authoritarian regime be released, it seems reasonable to think that the only human rights effects of that release will be positive. Yet, this is not likely to be true when evaluating CE policy. Any suite of climate responses will be human rights-inferior for some and human rights-superior for others. So, we need principles to help us decide between non-optimal policies.[106] SRM may stop or reduce the rate of global temperature increase, reducing human rights violating impacts while also disrupting precipitation patterns in ways that will also undermine human rights. CDR techniques, such as bio-energy with carbon capture and storage (BECCS), will reduce climate change impacts while also consuming arable land and freshwater.[107] There is no clear human rights-dominant policy in the context of climate change or CE.

Unfortunately, the human rights framework is not especially well-suited to developing principles for adjudicating tradeoffs. The reasons for this are both theoretical and political. Theoretically, we can imagine human rights as playing two different roles. First, human rights might have especially heavy *pro tanto* weight because they protect especially urgent interests.[108] If that is how we conceptualize human rights, then the framework lacks the key element that would allow us to apply it to decisions under non-ideal conditions: a public and uncontroversial set of principles for weighting and comparing various human rights violations. Even if we acknowledge that human rights are the kinds of moral protections that could be overridden, we still do not know how to make the comparison. And, as we shall see, there are rhetorical and political reasons why human rights theorists have been unwilling to develop them.

---

[105] We mean pareto-optimal here in the broad, metaphorical sense of improvement in terms of value X for at least one person Y and no decreases in X for anyone else.

[106] It is also far from obvious that the best response to unavoidable human rights violations is to try to minimize the number of violations. Even setting aside questions of responsibility, there remains the question of distribution. *See* Held [46] (arguing that—when faced with unavoidable violations—we should try to more *equitably* distribute human rights violations before minimizing).

[107] For a more detailed discussion of these claims, see the last section of this paper.

[108] *See* Shue [47]; Beitz [48].

The concern is that some will be asked to accept decreased human rights protection in the name of greater protection for others. Even if we grant that human rights can be traded off, we would likely want to subscribe to James Griffin's conclusion they should be highly resistant to being outweighed.[109] We would need good reasons—and reasons of a specific type—to justify imposing inadequate human rights protection on others. Unfortunately, it is not obvious—and is in fact deeply contested—what sacrifices can be rightly demanded. Should we simply attempt to aggregate human rights protection as some sort of proportionality or balancing test may demand? Or should we prioritize the least well off? Guarantee a sufficient level of protection for everyone at the cost of higher overall protection globally? The point we wish to emphasize here is that simple aggregation is not obviously correct—even if we could develop a measure of commensurability that would allow us to make aggregative judgments and if we grant that human rights can be balanced in the first place. What is more, balancing and proportionality tests in the real world acknowledge this point. Balancing is usually not simply aggregation, and thus is quite different from CBA. So, we reject the idea that a CBA-style principle of simple aggregation is the most appropriate in human rights tradeoff cases. Even if we include aggregative elements, they will need to be placed in the context of—and be constrained by—other principles and values.

Examining proportionality tests in a bit more detail is illustrative because they show how we can have comparative, balancing judgments without collapsing our analysis to CBA. Proportionality tests, as currently practiced by courts including and especially at the constitutional level, contain many constraints that are inconsistent with CBA. First, proportionality tests usually only include specific interests as being relevant to the test, such as law of armed conflict proportionality tests that weigh civilian lives more than military ones. Second, there is typically some claim that governmental action needs to be narrowly tailored or necessary for achieving the constitutionally acceptable goal. This is quite unlike the CBA, which purports to offer an account where all values are commensurable and included in the analysis. Finally, once these considerations have been sufficiently considered, the courts can then engage in a narrow balancing test. Yet, even this test can include values—such as equality, fairness, or due process—that imply balancing judgments that favor the least well off, the oppressed, or the powerless.[110] So, it is not obvious that the aggregative language of proportionality is much more than a metaphor, as opposed to the much stricter decision-procedure of CBA. Adopting a proportionality test in the context of *pro tanto* human rights does not necessarily direct policymakers as to how to make the relevant tradeoffs in novel policy situations such as climate change and CE. In other words, there might come to be a time where a simple aggregation or narrow balancing test will be appropriate in the context of CE, but that will be at the conclusion of a human rights analysis and not its beginning.

[109] Griffin, [49].

[110] A Stone Sweet and J. Matthews [50].

On the other hand, we might adopt an absolutist conception of human rights by which rights are treated as universal trumps or side constraints.[111] That is, human rights might be a set of protections that are never acceptable to violate in the name of some other value. If that is the relevant conception, then the tradeoffs and comparisons that policymakers will need to engage in are impossible. Again, this provides little guidance for cases when human rights must necessarily be violated. It is otiose. So, the relevant question, for our purposes, will be how to develop distributive principles for the *pro tanto* conception of human rights.

There are political and rhetorical considerations that have made human rights theorists and advocates unwilling to develop non-ideal, distributive principles. After all, one might be forgiven for thinking that the point of developing an account of human rights—as opposed to an account of fundamental human interests or needs—is precisely to create a set of normative considerations that are immune from normal tradeoffs. After all, it is a standard rationalization of those who wish to bypass or ignore human rights that they are outweighed by other considerations such as national security or economic growth. One might think that the political value of human rights—particularly as a foundation for criticizing regimes—lies in an absolutist conception of those considerations. Rhetorically, adopting a non-absolutist conception would complicate international criticism and open up possibilities for bad actors to offer new justifications for their behavior.

One way forward is to develop a set of distributive principles that are *internal* to the human rights framework. On this view, the only justification that can be offered for trading off one human right is the effect that action will have on other human rights. For example, parliamentary rules of debate constrain and burden some individuals' free speech rights. However, parliamentary rules of debate are justified in the name of ensuring that everyone is in a position to enjoy rights to free speech and participation. So, we can only justify parliamentary rules of debate if they allow for the fairer enjoyment of human rights. Conversely, this internal view would not justify parliamentary rules of debate if those rules were used to silence a particular group in the name of generating outcomes that produced greater welfare. Welfare would be an external consideration; protecting the capacity to exercise free speech rights are, conversely, internal to the framework. So, in the next section, we will describe some features of the human rights framework that can be used to generate an internal account of tradeoffs between—and only between—the normative considerations described by the framework itself. This framework will almost certainly require some kind of comparative, balancing judgment where protecting some human rights for some classes of people will justify violating—or failing to protect—the human rights of other people. This is unavoidable in non-Pareto-optimal human rights scenarios. However, since we do not accept simple aggregation as the appropriate principle for making these tradeoffs and because we limit our analysis only to certain interests as described by human rights discourse, this balancing test does not collapse into CBA. The fundamental dilemma, then, is how to develop

---

[111] *See* Nozick [51]; Dworkin [52].

human rights discourse in a way that can be used more productively in non-Pareto situations without collapsing into CBA. It is to that question we now turn.[112]

5 A Revised Human Rights Framework

We have established that human rights are well-suited to diagnose and frame expected CE impacts, both positive and negative, in legal and normative terms. It is less clear, however, whether a human rights framework provides us with the tools to prioritize competing claims and interests stemming from these impacts. If human rights are urgent moral claims, how do we determine which claims are the most urgent? Some will win and some will lose as a result of CE deployment, just as some will lose and some will win, at least in the short to medium term, if we adopt any of the available mitigation or adaptation pathways and allow the current climate change trajectories to continue without extensive CE. We may think that CE offers a relatively quick fix to address known, ongoing human rights violations due to climate change—violations that are likely to increase as temperatures continue to rise—whereas the potential negative impacts of CE deployment are only prospective. If so, we are faced with the decision whether to act to address ongoing, known harms or take a cautious approach to avoid prospective harms. In either event, if we decide to act and deploy CE, we will need to balance the benefits and harms of the winners and losers.

There are several ways to approach this within human rights discourse. First, we will consider whether there is a hierarchy of human rights that allows us to rank positive and negative CE impacts accordingly. Second, we will consider the value in examining the core and periphery obligations of human rights. Third, we will revisit the crosscutting principle that calls for a focus on vulnerable or marginalized groups, and consider both their capacity to respond to negative impacts and what they might gain from positive impacts. Finally, we will consider what role the principle of non-retrogression might play in weighing CE impacts.

[112] Some—*see, e.g.*, Morrow and Svboda [53]—have suggested that radically non-ideal nature of climate policy-making, where we do wrong no matter we do, can be resolved through use of a 'clinical' moral theory that compares feasible alternatives that produce the 'least' injustice. We welcome these contributions, but our project is somewhat distinct, in two ways. First, these views often rely on 'intuitive' notions of proportionality by which many incommensurable goods and values can be compared. So, our analysis, to an extent, starts up where theirs leaves off. Second, our view is an attempt to see what internal resources the *human rights discourse* has to deal with these kinds of tradeoffs. In that sense, our view is interpretative as well as normative.

5.1 A Hierarchy of Rights: The Right to Life as a Trump

The first thing to consider is whether there is a hierarchy of rights.[113] If so, we simply need to determine which rights are likely to be promoted or fulfilled and which may be violated by CE deployment, consult the hierarchy, and determine where things stand based on the relative position of the implicated rights. Unfortunately, there is no well-accepted hierarchy of human rights, certainly not one that is explicitly built into the international human rights regime that allows us to say with certainty whether one right is more important than another.[114]

Despite the lack of an accepted hierarchy of rights, CE impacts that implicate the right to life—either in eliminating or increasing threats to it—might be weighed more heavily than those that implicate other rights. In this view, the right to life would be something of a trump: if CE deployment will eliminate threats to the right to life by mitigating the impacts of climate change, even if it is likely to result in violations of other rights, we should deploy. On the other hand, if deployment is likely to result in violations of the right to life, even if doing nothing allows temperatures to continue to rise, resulting in violations of other rights, we should not deploy. We find this approach attractive and in line with the centrality of the right to life in human rights law, most importantly that the right to life is non-derogable, meaning it may not be restricted at any time, for any reason.[115] In addition, social and economic rights, including the rights to health, water, and food, can be understood as broadly concerned with the protection and promotion of life.[116]

What if, however, the right to life is implicated both if we deploy and if we do not deploy? In this case, we are left balancing competing claims involving the

---

[113] For a discussion of a hierarchy of human rights, *see, e.g.*, Scheinin [54] (suggesting that human rights 'hierarchies could [] be relied upon, for instance, in resolving conflicts between human rights, by giving primacy to the hierarchically superior right').

[114] *See* UN OHCHR, *Frequently Asked Questions on a Human Rights-Based Approach to Development Cooperation*, UN Doc. HR/PUB/06/8 (2006) (asking 'Is there any hierarchy among human rights?' and answering, no, 'all human rights are equally important' and declaring human rights 'all have equal status as rights, and cannot be ranked, a priori, in a hierarchical order'); World Conference on Human Rights, *Vienna Declaration and Programme of Action*, para. 5, UN Doc. A/CONF.157/23, 12 Jul. 1993.

[115] ICCPR, *supra* note 30, art. 4(2) (declaring States Parties may not derogate the right to life, among others, even during a 'time of public emergency which threatens the life of the nation and the existence of which is officially proclaimed,' as described in article 4(1)); *See also, e.g.*, Scheinin *supra* note 112, p. 2 ('The right to life and the prohibition against torture, and violations of human dignity … are strong candidates for [] special status'); Popovic [55] (stating the right to life represents the most basic human right and 'figures prominently in all basic international human rights instruments').

[116] *See generally* UDHR, *supra* note 29, art. 25(1) ('Everyone has the right to a standard of living adequate for the health of himself and of his family, including food … and medical care and necessary social services'); *Gen. Comment No. 15*, *supra* note 73, para. 1 ('Water is a limited natural resource and a public good fundamental for life and health'); ICESCR, *supra* note 30, art. 11(1) (recognizing 'the right of everyone to an adequate standard of living for himself and his family, including adequate food …').

right to life. The trump card does not work here. We might ask how many people are involved on each side and make some kind of consequentialist determination that acting to save N people is justified if the alternative is allowing the right to life of N + X people to be threatened—or vice versa. We may also consider the difference between direct acts and acts of omission. When the right to life is involved, is a direct act that threatens to violate the right worse than a failure to act that allows a state of affairs to continue in which the right is threatened? Notwithstanding these challenges, we believe employing the right to life as a trump is useful in weighing competing interests, at least when the right is not implicated on both sides of a decision.

5.2 *Core and Periphery Obligations and Values*

Human rights have cores and, by extension, peripheries with corresponding degrees of obligation.[117] The Committee on Economic, Social, and Cultural Rights has declared that governments "have a core obligation to ensure the satisfaction of, at the very least, minimum essential levels of each of the rights enunciated in the [ICESCR]."[118] Core obligations correspond to those aspects of the right that are most central to its content and scope. Duty-bearers are meant to prioritize the fulfillment of these obligations ahead of others. Obligations associated with less central aspects of a right could be said to be on the periphery, or at least not in the core.

For example, the right to health in Article 12 of ICESCR has six core obligations:

1. To ensure the right of access to health facilities, goods and services on a non-discriminatory basis ...;
2. To ensure access to the minimum essential food ...;
3. To ensure access to basic shelter, housing and sanitation, and an adequate supply of safe and potable water;
4. To provide essential drugs ...;
5. To ensure equitable distribution of all health facilities, goods and services;
6. To adopt and implement a national public health strategy ...[119]

In contrast, the duties to provide immunization against major infectious diseases and to provide appropriate training for health personnel are not core obligations.[120] In weighing potential CE impacts, priority could be granted to those that promote or prevent duty-bearers in upholding core human rights obligations. For instance, if changes in precipitation patterns resulting from CE deployment reduce access to

[117] Scheinin, *supra* note 112, pp. 5–10 (arguing that 'every human right contains a core with the quality of a rule' and the 'inviolability of the essential core of any human right is an important step in the assessment of permissible limitations to the broader human right surrounding that core').

[118] *Gen. Comment No. 3, supra* note 38, para. 10.

[119] *Gen. Comment No. 14,* supra note 38, para. 43.

[120] *Gen. Comment No. 14,* supra note 38, para. 43.

minimum essential food or if health systems are disrupted, significantly burdening vulnerable or marginalized groups' access to health services, we may weigh these violations more heavily than benefits in the form of minor improvements in health for other, better off populations. In this sense, the consequences of CE that negatively impact core human rights obligations should be given greater weight than positive consequences that only implicate the periphery of those rights. On the other hand, CE impacts that promote fulfillment of core aspects of human rights should be weighed more heavily than those that negatively impact the periphery of rights.

This approach provides some guidance in balancing positive and negative CE impacts, but is not likely to resolve all competing human rights claims. There may be disagreement on what constitutes the core and the periphery for particular rights. If so, the indeterminacy of the content and scope of the core and the periphery may render this approach nugatory. Even if we agree on what constitutes the core, it is possible that core human rights obligations will be implicated both positively and negatively in certain deployment scenarios, leaving us with a zero-sum game. Moreover, promoting realization of a periphery human rights obligation for a particularly vulnerable group through CE deployment might be weighed more heavily than a marginal decrease in the enjoyment of even a core aspect of another right for a wealthier, less vulnerable group. This leads us to revisit the cross-cutting principle that calls for a focus on vulnerable or marginalized groups.

5.3 *Vulnerable or Marginalized Groups and Relative Capacities*

Focusing on the potential positive and negative impacts of CE deployment on vulnerable or marginalized groups may assist in balancing competing interests. The basic idea is that we should weigh potential positive and negative impacts more heavily if they are likely to affect groups what are worse off than others—i.e., vulnerable or marginalized in some meaningful way. Positive impacts on vulnerable or marginalized groups should be given more weight than similar impacts on better off groups. On the other hand, negative impacts on vulnerable or marginalized groups should be given greater consideration than similar impacts on groups in a better relative position.

We define vulnerable or marginalized groups as those that lack political, economic, or social power relative to other groups within a particular country or across borders. Such groups may be defined, among other things, along lines of race, nationality, ethnicity, income, religion, citizenship, health status, or gender. The key feature of these groups is that they lack certain kinds of capacities—political, social, financial, or otherwise—relative to better off groups. As a result, they are likely to benefit more meaningfully from marginal increases in the enjoyment of human rights and experience greater deprivations, with less capacity to recover, from human rights violations.

There are at least three reasons to support this approach. First, weaker groups, particularly those that are politically marginalized, are likely to be underrepresented in global, regional, or national CE governance mechanisms. Granting greater concern to potential human rights impacts on these groups may, in part, account for their lack of participation in decision-making processes. This should not, however, substitute for meaningful participation during governance processes.

Second, positive impacts resulting in greater enjoyment of human rights for groups that traditionally experience a lack of protections and fulfillment of their rights should be weighed more heavily than marginal increases (or decreases) in the enjoyment of rights for groups that normally enjoy robust protection and fulfillment of their rights. This allows for the recognition that, in most cases, the enjoyment of a right is not simply a good to be maximized, with each marginal increase equivalent to the next. Rather, rights set baselines, below which urgent moral claims exist for governments to act to address deprivations, but above which we are less concerned with marginal fluctuations in the enjoyment of the right.[121]

Third, negative impacts on vulnerable or marginalized groups should be weighed more heavily than both negative and positive impacts on better off groups because the former lack capacity to address deprivations and recover from human rights violations. Wealthy or politically powerful groups may respond quickly and effectively to address violations of their rights through judicial or political means in order to access remedies and ensure further violations do not occur. Groups without political or economic power may lack the capacity to respond effectively and may suffer greater and longer lasting harms from human rights violations as a result.

We may face scenarios in which vulnerable or marginalized groups are likely to experience positive impacts while others may be impacted negatively from CE deployment. We again find ourselves facing a zero-sum game dilemma. To resolve such dilemmas, we may identify meaningful distinctions in the relative capacities of each group, in the sheer number of people involved on each side, in the kinds of rights implicated—e.g., is the right to life affected—or in the nature of the impacts on the rights—i.e., do they implicate core or periphery aspects. In this way, we can begin to develop an approach, discussed more fully below, that combines the balancing mechanisms introduced in this section.

This priority placed on marginalized and vulnerable populations is an essential way in which our account differs from CBA. Unlike an aggregative principle that is indifferent to distribution, this sort of prioritarian principle would not permit violations of the rights of the worse off in order to improve net human rights protection or enjoyment. So, we suggest that a key feature—one that adds considerable value to public policy analysis—of human rights discourse is this orientation towards balancing principles that are explicitly concerned with distribution in general and the status of the least well-off in particular.

---

[121] *See, e.g.*, Narula [56] (arguing rights-based approach sets baseline for protecting rights of land users); Wilson [57] (discussing rights that set the floor of the Inter-American system).

5.4 Principle of Nonretrogression

Finally, the principle of nonretrogression provides a tool by which to distinguish and rank impacts on human rights likely to result from CE deployment. The principle establishes a strong presumption against implementation of measures that result in backwards steps in the enjoyment of human rights, i.e., retrogressive measures.[122] The strong form of the principle holds that adoption of deliberately retrogressive measures resulting in decreased enjoyment of social or economic rights constitutes a *prima facie* human rights violation.[123] In the context of CE, if deployment is likely to result in reduced access to water, health, or food for certain groups, the principle of nonretrogression would prohibit governments from deploying, even if deployment would likely fulfill other human rights. Under the obligation to protect—requiring governments to protect people from human rights violations committed by non-State actors—governments would also have a duty to ensure private actors did not deploy either.

This approach seems useful to the extent it is possible to identify with certainty acts that are likely to result in a decrease in the enjoyment of human rights. Unlike the other balancing mechanisms discussed here, the principle provides a bright-line rule that avoids zero-sum game dilemmas. That is, if we are able to identify retrogressive measures, we simply may not implement them regardless of the positive impacts they may have; there is no balancing against such measures. However, this assumes only intentional, direct acts may constitute retrogressive measures. If we consider the regressive effects of acts of omission—i.e., failures to act when there is a duty to do so—[124] such as failing to act to mitigate climate change, we may again face a zero-sum game dilemma, wherein both acting and not acting result in prohibited regressive impacts on the enjoyment of human rights. There may also be disagreement on what constitutes a "regressive" measure resulting in a decrease in the enjoyment of a right below an accepted baseline, as opposed to one that results in a marginal decrease in enjoyment above the baseline.

Finally, abiding by the principle of nonretrogression may simply amount to an entrenchment of the status quo, in which better-off groups maintain their relative advantage over the vulnerable or marginalized. If the principle is interpreted as simply establishing a floor based on current levels of rights protections, below which governments may not sink, then entrenchment and stasis is a concern.

[122] *See, e.g.*, UN OHCHR, *Fact Sheet No. 33: Frequently Asked Questions on Economic, Social and Cultural Rights* (2008), p. 16; *Gen. Comment No. 3, supra* note 38, para. 9; UN CESCR, *Gen. Comment No. 4, The Right to Adequate Housing*, para. 11, UN Doc. E/1992/23, 13 Dec. 1991; *Gen. Comment No. 12, supra* note 74, para. 19; *Gen. Comment No. 13, supra* note 74, paras. 45, 49; *Gen. Comment No. 14, supra* note 38, paras. 32, 48; *Gen. Comment No. 15, supra* note 73, paras. 19, 21, 42.

[123] UN OHCHR, *Report on austerity measures and economic and social rights*, paras. 40–43, UN Doc. E/2013/82, 7 May 2013 ('the adoption of deliberately retrogressive measures constitutes a prima facie violation of the [ICESCR]').

[124] *See, e.g.*, Bohlen [58]; *Stovin v. Wise* [1996] UKHL 1, 24 July 1996.

However, employed properly, the principle should act only as a bulwark against regression, requiring that, at a bare minimum, existing rights are maintained. If so, it will not prevent implementation of new policies or expansion of existing policies that increase enjoyment of human rights.

5.5 *What Is Left? A Combined Approach*

Considered in isolation, none of the balancing approaches internal to human rights discourse discussed here provide definitive guidance for all competing human rights claims likely to result from CE deployment. Each mechanism is vulnerable to indeterminacy regarding the content and scope of the terms involved and subject to zero-sum game dilemmas, wherein we find claims meant to act as trumps on both sides of the fence. Nonetheless, we believe each approach offers a unique and useful principle by which to begin weighing competing human rights claims, both negative and positive. More importantly, when considered together, these principles constitute a more robust framework for balancing claims than when applied independently. For instance, if the right to life of a vulnerable group is threatened by a particular CE deployment, which is also likely to result in a decrease in the enjoyment of core aspects of other rights, we should weigh this more heavily than positive impacts on periphery aspects of the rights of a better off group. In this way, a combined approach offers guidance in confronting competing human rights claims and avoiding narrowly construed zero-sum games.

6 Conclusion

The popularity of CBA is driven, in part, by its usefulness in adjudicating policy and values conflicts in cases where tradeoffs are unavoidable. Human rights discourse, on the other hand, is often limited by an assumption that there exists a singular human rights policy answer in most issue areas. In the context of CE, this assumption is false: any policy we advance to deploy currently feasible CE technologies at meaningful scale will result in negative and positive impacts on human rights. To confront this challenge, we have set forth and critically examined a human rights framework for CE. In doing so, we have proposed several means for establishing priority relations amongst human rights claims and comparing human rights impacts on both sides of decisions around climate change. These principles help clarify and reduce irremediable human rights conflicts. This makes human rights-oriented policymaking more tractable and action-guiding: we now have clarity over what to prioritize in comparing likely outcomes. Furthermore, these principles address the conflicts that are most theoretically and practically intransigent: tradeoffs of non-derogable human rights, or core components of rights, especially between members of vulnerable or marginalized groups.

Our analysis has moved the debate forward in two ways. We have clarified the nature of human rights conflicts and argued for a set of prioritarian principles derived from international human rights law. Unlike many versions of CBA, these principles are not purely aggregative. Instead, they argue that some human rights impacts must be specially-weighted in our deliberations—violations of the right to life or core components of other rights, retrogressive measures, and impacts on vulnerable or marginalized groups. Yet, these principles do not fully resolve the conflicts likely to be generated by deployment of CE technologies: definitional and measurement challenges exist and some human rights conflicts will remain even after the principles are applied. However, these limitations of the human rights framework are mitigated by its robust set of procedural rights. These rights provide normative, legal guidance to assist in developing fair and inclusive processes with meaningful participation by groups likely to be affected by CE deployment. Finally, duties, accountability, and remedies established by human rights law applied in the context of CE further strengthen the framework in providing both *ex ante* and *ex post* standards with which to consider and address the impacts of deployment.

References

1. Crutzen, P.: Albedo enhancement by stratospheric sulfur injections: A contribution to resolve a policy dilemma. Clim. Chang. **77**, 211–220 (2016)
2. Tilmes, S., et al.: The hydrological impact of geoengineering in the Geoengineering Model Intercomparison Project (GeoMIP). J. Geophys. Res. **118**(19), 11036–11058 (2013)
3. Keith, D.: A Case for Climate Engineering. The MIT Press, Cambridge, MA (2013)
4. Jamieson, D.: Adaptation, mitigation, and justice. In: Sinott-Armstrong, W., Howarth, R. (eds.) Perspectives on Climate Change: Science, Economics and Politics, pp. 217–248. Elsevier JAI, Amsterdam/San Diego (2005)
5. Barrett, S.: The incredible economics of geoengineering. Environ. Resour. Econ. **39**(1), 45–54 (2008)
6. Adler, M., Posner, E.: Cost-Benefit Analysis: Economic, Philosophical, and Legal Perspectives. University of Chicago Press, Chicago/London (2001)
7. Richardson, H.: The stupidity of the cost-benefit standard. J. Leg. Stud. **29**(2), 971 (2000)
8. Kelman S.: Cost-benefit analysis: An ethical critique. AEI J. Govern. Soc. Regul., pp. 33–40 (1981, Jan-Feb)
9. Schmidtz, D.: A place for cost-benefit analysis. Philos. Issues. **11**(1), 148–171 (2001)
10. Railton, P.: Costs and benefits of cost-benefit analysis: A response to Bantz to MacLean. In: 1982 PSA: Proceedings of Biennial Meeting of the Philosophy of Science Association, pp. 261–272 (1982)
11. Orr, S.: Values, preferences, and the citizen-consumer distinction in cost-benefit analysis. Politics Philos. Econ. **5**(3), 377–400 (2006)
12. Grob, G.F.: Providing recommendations, suggestions, and options for improvement. In: Wholey, J. (ed.) Hand book of Practical Program Evaluation (Wiley-Blackwell, 4th ed., 2010) 13. Parfit D.: Reasons and Persons. Oxford University Press, Oxford (1984)
14. Nussbaum, M.: Sex and Social Justice. Oxford University Press, New York (2001)
15. Sen, A.: Development as Freedom. Anchor, New York (1999)
16. Ackerman, F., Heinzerling, L.: Priceless: On Knowing the Price of Everything and the Value of Nothing. The New Press, New York (2005)

17. Sandler, R.: Character and Environment: A Virtue-Oriented Approach to Environmental Ethics. Columbia University Press, New York (2009)
18. Hannum, H.: The status of the universal declaration of human rights in national and international law. Ga. J. Int. Comp. Law. **25**, 287–297 (1996)
19. Dworkin, R.: Law's Empire. Belknap Press of Harvard University Press, Cambridge, MA (1986)
20. Meier, B.M., Chakrabarti, A.: The paradox of happiness: Health and human rights in the Kingdom of Bhutan. Health Hum. Rights. **18**(1), 193–208 (2016)
21. Kravchenko, S.: Procedural rights as a crucial tool to combat climate change symposium: International human rights and climate change. Ga. J. Int. Comp. Law. **38**(3), 613–638 (2010)
22. Carr, W., et al.: Public engagement on solar radiation management and why it needs to happen now. Clim. Chang. **121**(3), 567–577 (2013)
23. MacNaughton, G.: Human rights impact assessment: A method for healthy policymaking. Health Hum. Rights. **17**(1) (2015)
24. Whyte, K.P.: Now This! Indigenous sovereignty, political obliviousness and governance models for SRM research. Ethics Pol. Environ. **15**(2), 172–187 (2012)
25. Whyte, K.P.: Indigenous peoples, solar radiation management, and consent. In: Preston, C.J. (ed.) Engineering the Climate: The Ethics of Solar Radiation Management, pp. 65–76. Lexington Books (2012)
26. Jung, C., et al.: Economic and social rights in national constitutions. Am. J. Comp. Law. **62**(4), 1043–1098 (2014)
27. Robock, A., Oman, L., Stenchikov, G.L.: Regional climate responses to geoengineering with tropical and Arctic SO2 injections. J. Geophys. Res. **113**(D16101), 1–15 (2008)
28. Burns, W.: Geoengineering the climate: An overview of solar radiation management options. Tulsa Law Rev. **46**(2), 283–304, 290–91 (2010)
29. Russel, L.M., et al.: Ecosystem impacts of geoengineering: A review for developing a science plan. Ambio. **41**(4), 350–69, 358 (2012)
30. Crook, J., et al.: A comparison of temperature and precipitation responses to different Earth radiation management geoengineering schemes. J. Geophys. Res. **120**(18), 9352–9373 (2015)
31. De Schutter, O.: International Human Rights Law: Cases, Materials, Commentary, 2nd edn, pp. 280–291. Cambridge University Press (2014)
32. Knox, J.H.: Climate change and human rights law. Va. J. Int. Law. **50**(1), 163–218, 179–80 (2009)
33. Svoboda, T., Irvine, P.: Ethical and technical challenges in compensating for harm due to solar radiation management geoengineering. Ethics Pol. Environ. **17**(2), 157–174 (2014)
34. Horton, J.: Solar geoengineering: Reassessing benefits, costs, and compensation. Ethics Pol. Environ. **17**(2), 175–177 (2014)
35. Valdivia W.D.: The fair compensation problem of geoengineering. The Brookings Institution. (2016, 23 Feburary)
36. Gordon, J.: Inter-American commission to hold hearing after rejecting inuit climate change petition. Sustain. Dev. Law Pol. **7**(2), 55 (2007)
37. Posner, E.: Climate change and international human rights litigation: A critical appraisal. Univ. Pa. Law Rev. **155**, 1925–1945 (2007)
38. Doudda, J.V., Corkery, A., Chartres, R.: Human rights and climate change. Aust. Int. Law J. **14**(1), 161–184 (2007)
39. Limon, M.: Human rights and climate change: Constructing a case for political action. Harv. Environ. Law Rev. **33**, 439–476 (2009)
40. Knox, J.H.: Linking human rights and climate change at the united nations. Harv. Environ. Law Rev. **33**, 478–498 (2009)
41. Humphreys, S. (ed.): Human Rights and Climate Change. Cambridge University Press (2010)
42. Bodansky, D.: Introduction: Climate change and human rights: Unpacking the issues symposium: International human rights and climate change. Georgia J. Int. Comparative Law. **38**(3), 511–524 (2010)

43. Allard, K.: Lowenstein International Human Rights Clinic. Yale Law School, Human Rights and Climate Change Obligations, Draft Memorandum for the Experts. Group on Global Climate Obligations (2013)
44. Quirico, O., Boumghar, M. (eds.): Climate Change and Human Rights: An International and Comparative Law Perspective. Routledge (2016)
45. Caney, S.: Climate change, human rights, and moral thresholds. In: Humphreys, S. (ed.) Human Rights and Climate Change, pp. 69–90. Cambridge University Press (2010)
46. Held, V.: How Terrorism is Wrong: Morality and Political Violence. Oxford University Press, Oxford (2008)
47. Shue, H.: Basic Rights. Princeton University Press, Princeton (1980)
48. Beitz, C.: The Idea of Human Rights. Oxford University Press, Oxford (2009)
49. Griffin, J.: On Human Rights. Oxford University Press, Oxford (2008)
50. Stone Sweet, A., Matthews, J.: Proportionality balancing and global constitutionalism. Colum. J Transnatl. Law. **47**(72) (2009)
51. Nozick, R.: Anarchy, State, and Utopia. Basic Books, New York (1974)
52. Dworkin, R.: Rights as trumps. In: Waldron, J. (ed.) Theories of Rights. Oxford University Press, Oxford (1984)
53. Morrow, D., Svboda, T.: Geoengineering and non-ideal justice. Public Aff. Q. **30**(1), 85–104 (2016)
54. Scheinin, M.: Core rights and obligations. In: Shelton, D. (ed.) The Oxford Handbook of International Human Rights Law, pp. 527–540 (2013)
55. Popovic, N.A.F.: In pursuit of environmental human rights: Commentary on the draft declaration of principles on human rights and the environment. Colum. Hum. Rts. L. Rev. **27**, 487–603, 515 (1996)
56. Narula, S.: The global land rush: Markets, rights, and the politics of food. Stanf. J. Int. Law. **49**, 101–75, 139 (2013)
57. Wilson, R.J.: Supporting or thwarting the revolution? The Inter-American human rights system and criminal procedure Reform in Latin America. Southwest. J. Law Trade Am. **14**(2), 287–318, 299 (2008)
58. Bohlen, F.H.: The basis of affirmative obligations in the Law of Tort. Am. Law Reg. **53**(4), 209–239 (1905)

The Role of Human Rights in Implementing CDR Geoengineering Options in South Africa

Ademola Oluborode Jegede

1 Introduction

There is increasing discussion about the potential role of climate geoengineering options to address the rapidly escalating threat of climate change. For the purposes of this chapter, climate geoengineering is defined as "the deliberate large-scale manipulation of the planetary environment to counteract anthropogenic climate change",[1] as a climate intervention option. This is due to the reality that it is nearly impossible for conventional climate response measures alone to achieve net-zero carbon dioxide (CO_2) emissions and stabilize global temperatures below 2 °C.[2] This chapter is concerned with carbon dioxide removal (CDR) options.[3] CDR approaches, seek to remove and sequester carbon dioxide from the atmosphere through biological, geochemical, or chemical means.[4] Options under consideration include ocean fertilization to stimulate carbon-sequestering, afforestation, and bioenergy carbon capture and storage (BECCS) systems that convert biomass to heat, electricity, or

[1] Royal Society, Geoengineering the Climate: Science, Governance and Uncertainty (2009), at 15.

[2] Royal Society & Royal Academy of Engineering, Greenhouse Gas Removal (2018), at 7; T. Stocker et al., The Physical Science Basis. Contribution of Working Group I to the 5th Assessment Report of the Intergovernmental Panel on Climate Change (IPCC Summary for Policy Makers) (2013), at 25.

[3] Wil Burns & Simon Nicholson, *Bioenergy and Carbon Capture with Storage (BECCS): The Prospects and Challenges of an Emerging Climate Policy Response*, 7 J. Envtl. Studies and Sci. 527, at 527 (2017).

[4] *Id.* at 528; Phil Williamson, *Emissions Reduction: Scrutinize CO_2 Removal Methods*, 530 Nature 153, 155 (2016).

A. O. Jegede (✉)
School of Law, University of Venda, Thohoyandou, South Africa

© Springer Nature Switzerland AG 2021

W. Burns et al. (eds.), *Climate Geoengineering: Science, Law and Governance*, AESS Interdisciplinary Environmental Studies and Sciences Series,
https://doi.org/10.1007/978-3-030-72372-9_6

liquid or gas fuels.[5] CDR is not without uncertainties and risks that can potentially undermine the realization of a wide range of human rights in developing countries,[6] which inform the perennial questions raised by some authors as to whether the world is ready for it,[7] or whether its implementation will worsen or improve the climate in concrete terms.[8] However, if there has ever been any doubt regarding its potential role in the global climate agenda, this has been dispelled by the 2018 report of the IPCC on global warming of 1.5 °C which clarifies that the deployment of CDR geoengineering approaches at national level is necessary to achieve net zero carbon dioxide (CO_2) emissions and stabilize global temperatures below 2 °C.[9] This stark scientific assessment is reinforced by the recommendation of the United Kingdom's Royal Society and Royal Academy of Engineering in 2018 that implementing a global suite of CDR methods is required to meet the targets of the 2015 Paris Agreement.[10]

Implementing CDR options touches on a whole range of resources such as land, water, food, and energy that can affect human welfare and national development in developing countries,[11] a reality that makes the legal field important to its regulation. Is the human rights framework an appropriate response of law in fulfilling that role? A human rights framework, affirms Burns, would be critical in addressing anticipated justice issues around the deployment of CDR.[12] Other authors share a similar view that a human rights framework would serve to link inalienable rights

---

[5] William C.G. Burns, *Human Rights Dimensions of Bioenergy With Carbon Capture and Storage: A Framework for Climate Justice in the Realm of Climate Geoengineering*, in Climate Justice: Case Studies in Global and Regional Governance Challenges 150–170 (Randall Abate, ed. 2016).

[6] William C.G. Burns, *The Paris Agreement and Climate Geoengineering Governance: The Need for a Human Rights-Based Component* 111, Centre for International Governance Innovation Papers 1, at 16–17 (2016).

[7] David P. Keller, Ellias Y. Feng and Andreas Oschlies, *Potential Climate Engineering Effectiveness and Side Effects during a High Carbon Dioxide-Emission Scenario*, 5 Nature Commmunication 1, 9 (25 February 2014).

[8] Olof Corry, *The International Politics of Geoengineering: The Feasibility of Plan B for Tackling Climate Change,* 48(4) Security Dialogue 297, 299 (2017); Charles Q. Choi, *Geoengineering Ineffective Against Climate Change, Could Make Worse*, Live Science (25 February 2014).

[9] IPCC, *Summary for Policymakers*, in Global Warming of 1.5 °C. An IPCC Special Report on the impacts of global warming of 1.5 °C above pre-industrial levels and related global greenhouse gas emission pathways, in the context of strengthening the global response to the threat of climate change, sustainable development, and efforts to eradicate poverty 1–32 (Masson-Delmotte Valerie et al., eds. 2018) (hereafter, IPCC *Summary for Policymakers* 2018).

[10] Royal Society & Royal Academy of Engineering, *supra* note 2, at 10; Paris Agreement under the United Nations Framework Convention on Climate Change 2015, FCCC/CP/ 2015/L.9/Rev.1. (adopted by Conference of the Parties, 21st Session Paris, 30 November-11 December 2015.

[11] Neil A. Craik & William C.G. Burns, Climate Engineering under the Paris Agreement A Legal and Policy Primer, Centre for International Governance Innovation (2016), at 8.

[12] Burns, *supra* note 6

with duties and provide a normative framework and language of accountability for climate engineering interventions at both international and national levels.[13]

South Africa is exposed to climate vulnerabilities due to its socio-economic and environmental situation. It is among the countries that are simultaneously highly affected by carbon emissions, a leading CO_2 emitter in Africa that is ranked amongst the top 12 emitters in the world,[14] and with the potential for substantial deployment of CDR options. It is a State party to a range of international climate change and human rights instruments relevant to CDR options and human rights. It possesses a progressive Constitution encompassing a bill of rights and application of international law. However, whether and how human rights forming part of its existing corpus of law can respond to the issues around the suspected risks associated with CDR deployment, while harnessing the benefits, is not clear. Accordingly, this chapter clarifies the potential implications of benefits and uncertainties of CDR options in the context of human rights considerations, and demonstrates how human rights law can respond to the uncertainties around the implementation of CDR geoengineering options in South Africa.

2 The Significance of a Human Rights Framework to Climate Change Interventions

Authors have argued the meaning and legal basis for a human rights framework,[15] but the nexus between human rights and climate change interventions, that is, measures in response to climate change, in particular climate engineering, is only a subject of recent scholarly effort. In particular, after an analysis of literature and international instruments on human rights, Burns concludes that the implementation of climate engineering options could potentially undermine human rights under some circumstances.[16]

The legal basis of a human rights framework lies in a wide range of instruments under the aegis of the United Nations (UN), and to the developing states in Africa, instruments under the African Union (AU) which confer individual and collective rights and impose obligations on States. Under the rubric of the UN, key instruments include the Universal Declaration on Human Rights,[17] which though not a

[13] Toby Svoboda, Holly J. Buck and Pablo Suarez, *Climate Engineering and Human Rights*, 28(3) Envtl. Pol. 397 (2019).

[14] Patrick T. Sekoai & Michael O. Daramola, *Biohydrogen as a Potential Energy Fuel in South Africa*, 6 Biofuel Rsch. J. 223 (2015).

[15] Marie-Bénédicte Dembour, *What are Human Rights? Four Schools of Thought*, 32 Human Rights Q. 1 (2010); Jack Donelly, *International Human Rights Law: Universal, Relative, or Relatively Universal*, in International Human Rights Law: Six Decades after the UDHR and Beyond 31–48 (Manisuli Ssenyonjo & Mashood A. Baderin, eds. 2010).

[16] Burns, *supra* note 6.

[17] UN General Assembly, Universal Declaration of Human Rights, 217 A (III), (10 December 1948).

binding instrument, has provisions widely recognized as customary international law,[18] International Covenant on Civil and Political Rights (ICCPR),[19] International Covenant on Economic, Social and Cultural Rights (ICESCR),[20] International Convention on the Elimination of All Forms of Racial Discrimination (CERD),[21] Convention on the Elimination of All Forms of Discrimination Against Women (CEDAW),[22] Convention on the Rights of the Child (CRC),[23] and the Convention on the Right of Persons with Disabilities (CRPD).[24] For an African State such as South Africa, it is also subject to a range of AU instruments, such as the African Charter on Human and Peoples' Rights (African Charter),[25] the African Charter on the Rights and Welfare of the Child (ACRWC),[26] and the Protocol to the African Charter on Human and Peoples' Rights on the Rights of Women in Africa (Maputo Protocol).[27]

The relevance of the foregoing instruments as an assessment standard for rights and State obligations in the context of climate change and its interventions is discussed at the United Nations (UN) level, notably through the resolutions of United Nations Human Rights Council (UNHRC).[28] Following elaborate processes of discussions and consultation, the UNHRC, through Resolutions 7/23 of 2008, 10/4 of 2009, 18/22 of 2011, 26/33 of 2014 and 32/34 of 2016,[29] highlight the link of climate change interventions with human rights. Resolution 7/23 requires a detailed analytical study on the relationship between climate change and human rights.[30] The Office of the United Nations High Commissioner for Human Rights (OHCHR) Report, which responded to this request, describes the effect of climate change on a

---

[18] Vojin Dimitrijevic, *Customary Law as an Instrument for the Protection of Human Rights*, 7 ISPI Working Paper, 1 at 8–12 (2006).

[19] UN ICCPR, 999 UNTS 171, 19 December 1966 (entered into force 23 March 1976).

[20] UN ICESCR, 993 UNTS 3, 6 ILM 360, 16 December 1966 (entered into force 3 January 1976).

[21] UN ICERD, 660 UNTS 195, 21 December 1965 (entered into force 4 January 1969).

[22] UN CEDAW, 13 UNTS 1249, 18 December 1979 (entered into force 3 September 1981).

[23] UN CRC, 1577 UNTS 3, 28 ILM 1456, 20 November 1989 (entered into force 2 September 1990).

[24] UN CRPD, Doc A/RES/61/106, Annex 1, 13 December 2006 (entered into force 3 May 2008).

[25] OAU, African Charter on Human and Peoples' Rights, CAB/LEG/67/3 rev. 5, 21 ILM 58, 27 June 1981 (entered into force 21 October 1986).

[26] OAU, African Charter on the Rights and Welfare of the Child, 11 July 1990, CAB/LEG/24.9/49 (entered into force 29 November 29).

[27] AU, Protocol to the African Charter on Human and Peoples' Rights on the Rights of Women in Africa, adopted by the 2nd Ordinary Session of the Assembly of the Union Maputo (11 July 2003).

[28] Established by UNGA Resolution 60/251, Human Rights Council, A/RES/60/251 (15 March 2006).

[29] UN HRC, Human Rights and Climate Change, Res. 7/23, UN Doc. A/HRC/7/78 (2 July 2009); UN HRC, Human Rights and Climate Change, Res. 10/4, 41st meeting, A/HRC/RES10/4 (25 March 2009); UN HRC, Human Rights and Climate Change, A/HRC/RES/18/22 (24 March 2011); UN HRC, Human Rights and Climate Change, A/HRC/26/L.33 (23 June 2014); UN HRC, Climate Change and Human Rights, A/HRC/32/L.34 (30 June 2016).

[30] UNHRC Resolution 7/23, *supra* note 29.

range of rights, including right to life,[31] the right to adequate food,[32] the right to adequate water,[33] the right to health,[34] and the right to adequate housing.[35] This position is reinforced in UNHRC Resolution 10/4, which also affirms that human rights obligations and commitments have the potential to inform and strengthen international and national policy-making in the area of climate change.[36]Subsequently, the interface of climate change and human rights found expression in the work of treaty monitoring bodies, namely the Human Rights Committee,[37] Committee on Economic and Social Cultural Rights (CESCR),[38] the Committee on the Elimination of Discrimination Against Women (CEDAW),[39] and the Committee on the Rights of the Child (CRC),[40] and Resolutions 153,[41] 148,[42] 271[43] of the African Commission on Human and Peoples' Rights (African Commission), which is the treaty monitoring body of the African Charter.[44] Developments under the aegis of the United Nations Framework Convention on Climate Change (UNFCCC),[45] and its subsequent instruments, namely the Kyoto Protocol,[46] and the Paris Agreement,[47] is useful in clarifying the role of human rights. For instance, the Paris Agreement not only commits State parties to limit the global average temperature increase to "well

[31] UN HRC, Report of the Office of the United Nations High Commissioner for Human Rights on the Relationship between Climate Change and Human Rights, A/HRC/10/61 (15 January 2009) paras. 21–24

[32] *Id.* at paras. 25–27.

[33] *Id.* at paras. 28–30.

[34] *Id.* at paras. 31–34.

[35] *Id.* at paras. 31–34.

[36] UNHRC Resolution 10/4, *supra* note 29.

[37] Established under art. 28 (1) of the ICCPR, *supra* note 19.

[38] Established under the Economic and Social Council Resolution 1985/17, (28 May 1985).

[39] Established under art. 17 of CEDAW, *supra* note 22.

[40] Established under art. 43(1) of CRC, *supra* note 23.

[41] AU, African Commission of Human and Peoples' Rights, ACHPR/Res153 (XLVI) 09: Resolution on Climate Change and Human Rights and the Need to Study its Impact in Africa (25 November 2009).

[42] AU, African Commission of Human and Peoples' Rights, 148: Resolution on the Establishment of a Working Group on Extractive Industries, Environment and Human Rights Violations in Africa, adopted at 46th ordinary session held in Banjul, The Gambia (11–25 November 2009).

[43] AU, African Commission on Human and Peoples' Rights '271: Resolution on Climate Change in Africa', adopted at the 55th ordinary session of the African Commission on Human and Peoples' Rights held in Luanda, Angola (28 April–12 May 2014).

[44] Ademola O. Jegede, *Climate Change in the Work of the African Commission on Human and Peoples' Rights*, 31(2) Speculum Juris 136 (2017).

[45] United Nations Framework Convention on Climate Change (UNFCCC) ILM 851 (1992).

[46] United Nations Kyoto Protocol to the United Nations Framework Convention on Climate Change, 1998 (entered into force 16 February 2005).

[47] Paris Agreement, *supra* note 10.

below" 2 °C,[48] but also urges parties, when taking action to address climate change, to respect, promote and consider their respective obligations on human rights.[49]

The foregoing instruments recapture mostly the three levels of obligations under international human rights law, namely, to *respect*, to *protect*, and to *fulfil* human rights.[50] At the AU level, in *SERAC*,[51] the African Commission, in the context of environmental claims over the degradation of the land of Ogoni people, added the obligation to *promote* as a fourth layer of obligation.[52] With its reference in the Paris Agreement, the obligation to 'promote' has become a fourth layer of states' human rights obligation. As Burns has argued, a human rights framework as framed under these key instruments is applicable in the context of climate engineering.[53]

South Africa is a State party to human rights instruments which have shaped the link of climate change to human rights, and arguably climate change engineering, such as the ICCPR,[54] ICESCR,[55] CERD,[56] CEDAW,[57] CRC,[58] and the CRPD.[59] At the regional level, it is a State party to the African Charter,[60] the ACRWC,[61] and the

[48] *Id.* at art. 2.

[49] *Id.* at preamble.

[50] UN General Comment No. 31 [80] Nature of the General Legal Obligation Imposed on state parties to the Covenant, CCPR/C/21/Rev.1/Add.13 HRC (26 May 2004), paras. 5–6; UN General Comment No. 3: The Nature of States Parties' Obligations (Art. 2, Para. 1, of the Covenant), Fifth Session of the Committee on Economic, Social and Cultural Rights (adopted 14 December 1990).

[51] Communication 155/96, *Social and Economic Rights Action Center* (SERAC) *and Center for Economic and Social Rights (CESR) / Nigeria* (hereinafter *SERAC*).

[52] *Id.* at paras. 45–47.

[53] Burns, *supra* note 6.

[54] It became a state party on 10 December 1998, OHCHR https://tbinternet.ohchr.org/_layouts/15/TreatyBodyExternal/Treaty.aspx?CountryID=162&Lang=EN

[55] It became a state party on 12 January 2015, OHCHR https://tbinternet.ohchr.org/_layouts/15/TreatyBodyExternal/Treaty.aspx?CountryID=162&Lang=EN

[56] It became a state party on 10 December 1998, OHCHR https://tbinternet.ohchr.org/_layouts/15/TreatyBodyExternal/Treaty.aspx?CountryID=162&Lang=EN

[57] It became a state party on 15 December 1995, OHCHR https://tbinternet.ohchr.org/_layouts/15/TreatyBodyExternal/Treaty.aspx?CountryID=162&Lang=EN

[58] It became a state party on 16 June 1995, OHCHR https://tbinternet.ohchr.org/_layouts/15/TreatyBodyExternal/Treaty.aspx?CountryID=162&Lang=EN

[59] It became a state party on 30 November 2007, OHCHR https://tbinternet.ohchr.org/_layouts/15/TreatyBodyExternal/Treaty.aspx?CountryID=162&Lang=EN

[60] It became a state party on 7 September 1996, ACHPR, Ratification Table: African Charter on Human and Peoples' Rights, http://www.achpr.org/instruments/achpr/ratification/

[61] It became a state party on 7 January 2000, ACHPR, Ratification Table: African Charter on the Rights and Welfare of the Child, http://www.achpr.org/instruments/child/ratification/

Maputo Protocol.[62] It is a State party to the UNFCCC,[63] the Kyoto Protocol,[64] the 2015 Paris Agreement,[65] United Nations Sustainable Development Goals (SDGs),[66] Convention on Biological Diversity (CBD)[67] the Convention on the Prevention of Marine Pollution by Dumping of Wastes and Other Matter (CPMPD),[68] and the United Nations Convention of the Law of the Sea (UNCLOS).[69] The relevance of these instruments with or without domestication is not in doubt in South Africa in that by virtue of section 233 of the Constitution of the Republic of South Africa courts may prefer any reasonable interpretation of the legislation that is consistent with international law while interpreting any legislation.[70] Also key provisions in international instruments that are consistent with the Constitution or an Act of Parliament, may qualify as customary international law, which is applicable by virtue of section 232 of the Constitution. Similarly, the Constitution consists of a bill of rights and obligations of State which are akin to those conferred at the international level in the context of climate change and its interventions. The Constitution guarantees the right to life,[71] the right to have access to sufficient food and water,[72] the right to have access to health care services,[73] the right to property,[74] the right to adequate housing,[75] and environmental rights,[76] which are crucial to the interface of

[62] It became a state party on 17 December 2014, ACHPR, Ratification Table: Protocol to the African Charter on Human and Peoples' Rights on the Rights of Women in Africa http://www.achpr.org/instruments/women-protocol/ratification/17/12/2004

[63] It became a state party on 29 August 1999, UNFCCC, http://unfccc.int/essential_background/convention/status_of_ratification/items/2631.php

[64] It became a state party on 31 July 2002, UNFCCC http://unfccc.int/kyoto_protocol/status_of_ratification/items/2613.php

[65] It became a state party on 1 November 2016, UNFCCC http://unfccc.int/kyoto_protocol/status_of_ratification/items/2613.php

[66] SA News, SA committed to Sustainable Development Goals, http://www.sanews.gov.za/south-africa/sa-committed-sustainable-development-goals

[67] UNEP, Convention on Biological Diversity, UN Doc UNEP/CBD/COP/DEC/IX/16 (31 January 1996); *see* CBD, List of Parties, https://www.cbd.int/information/parties.shtml

[68] UN Convention on the Prevention of Marine Pollution by Dumping of Wastes and Other Matter, Resolution, LC-LP.2 (2010); South Africa became a state party on 7 August 1978, *see* https://www.fishbase.de/country/summaryconventions.php?ID=31#

[69] UNCLOS, South Africa became a state party on 23 December 1997, see UN, Chronological lists of ratifications of, accessions and successions to the Convention and the related Agreements https://www.un.org/depts/los/reference_files/chronological_lists_of_ratifications.htm

[70] Constitution of the Republic of South Africa, 1996.

[71] *Id.* at sec. 11.

[72] *Id.* at sec. 27 (1)(b).

[73] *Id.* at sec. 27 (1)(a).

[74] *Id.* at sec. 25.

[75] *Id.* at sec. 26.

[76] *Id.* at sec. 24

climate change and human rights of vulnerable groups.[77] In terms of section 7(2) of the Constitution, the State must respect, protect, promote and fulfil the rights in the Bills of Rights. Generally, courts must also regard international law while interpreting the Bill of rights by virtue of section 39 1(b) of the Constitution. The foregoing discussion suggests that a human rights framework applies, but its relevance to the deployment of CDR options in South Africa is not as settled.

3 CDR Engineering: A Potential Climate Intervention in South Africa?

With a focus on afforestation, BECCS, and ocean fertilization, the analysis below demonstrates that the ambivalent potential of CDR options in terms of its benefits and harm, and the uncertainty around its linkage with human rights, may influence their implementation in South Africa.

3.1 Ambivalent Potential

The potential harms and benefits of CDR options underlie the uncertain prospect of their implementation in South Africa. Afforestation involves planting or replanting forests over large areas for the purpose of absorbing carbon in both the trees and the soil as they grow.[78] As defined by the UNFCCC, afforestation is a human induced conversion of non-forested land to forested land through planting, seeding and/or promotion of natural seed sources.[79] Despite its potential for CDR, the position on afforestation in South Africa is in flux. Approximately 16% of South Africa's land, primarily the wetter eastern parts, is suited to afforestation.[80] This could be potentially beneficial in terms of ecosystems and human services,[81] and can mitigate environmental degradation and serve as a sink to carbon.[82] Yet, of the 24 projects registered under the Clean Development Program of the UNFCCC for South Africa,

[77] A.O. Jegede, The Climate Change Regulatory Framework and Indigenous Peoples' Lands in Africa: Human Rights Implications (2016).

[78] IUCN, *Afforestation and Reforestation for Climate Change Mitigation: Potentials for Pan-European Action* (2004).

[79] UNFCCC, Glossary of Climate Change Acronyms and Terms, https://unfccc.int/process-and-meetings/the-convention/glossary-of-climate-change-acronyms-and-terms#a

[80] Sarah Kiggundu, Afforestation in South Africa: Managing Forestry Resources Using Assessment Plans, https://www.polity.org.za/article/afforestation-in-south-africa-managing-forestry-resources-using-assessment-plans-2012-01-20

[81] IUCN, *supra* note 78, at 7.

[82] David J. Nowak, Robert Hoehn, and Daniel E. Crane, *Oxygen Production by Urban Trees in the United States*, 33(3) Arboriculture & Urban Forestry 220 (2007).

none are afforestation.[83] The major concern stalling afforestation lies in the fact that South Africa is a water-limited country with an average annual rainfall of 560 millimeters/year.[84] This concern is not unfounded as copious studies on South Africa have affirmed that forest plantations consume more water than the baseline vegetation, reducing water yield (streamflow) as a result.[85] In fact, afforestation utilizes water that would otherwise form catchment runoff.[86] Also afforestation requires largescale land use which carries a risk for local populations.[87] It can displace customary land title and uses of local populations living in rural areas whose populations depend on such resources for survival.[88]

BECCS involves the growing or collection and the processing of biomass for conversion to heat, electricity or liquid or gas fuels, capturing the resulting carbon, and storing it underground or in long-lasting products.[89] BECCS holds the prospect for "net negative" emissions.[90] With BECCS it is possible to capture 90% or more of the carbon dioxide released through biomass production.[91] Moreover, BECCS can help reverse reliance on coal-based energy production, and thus contribute to global reductions of carbon emissions and the adverse effects of coal production and burning on populations.

However, the implementation of BECCS in South Africa is uncertain despite its potential, Food crops such as maize are banned as an energy source while Jatropha is discouraged because it leaves behind a toxic seedcake.[92] Grain sorghum and soya beans have been recommended as the feedstocks for the manufacture of bioethanol and biodiesel respectively.[93] The two crops, in particular sorghum, gained favor because it is a drought-resistant and non-water intensive crop. However, major concerns around its mass cultivation remain: water limitations, land degradation and the

83 UNFCCC CDM, Project Search, https://cdm.unfccc.int/Projects/projsearch.html

84 Janine M. Albaugh, Peter J. Dye and John S. King, *Eucalyptus and Water Use in South Africa*, Int'l J. Forestry Research 1–12 (2013).

85 Mark B. Gush, *Modelling Streamflow Reductions resulting from Commercial Afforestation in South Africa: From Research to Application*, in Proceedings of the International Conference on Forest and Water, China (August 2006).

86 Jane K. Turpie et al., *The Ecological and Economic Consequences of Changing Land Use in the Southern Drakensberg Grasslands, South Africa*,10(4) South African Journal of Economic and Management Sciences 423 (2007).

87 Arttu Malkamäki et al., *A Systematic Review of the Socio-Economic Impacts of Large-Scale Tree Plantations, Worldwide*, 53 Global Envtl. Change 90 (2018).

88 Jegede, *supra* note 77, at 105; Lorenzo Cotula et al., *Testing Claims about Large Land Deals in Africa: Findings from a Multi-Country Study*, 50 (7) J. Development Studies 903 (2014).

89 Burns, *supra* note 6; Burns & Nicholson, *supra* note 4.

90 Preethy Thangaraj et al., *Fact Sheet: Bioenergy with Carbon Capture and Storage* (12 March 2018).

91 Joris Kornneeff et al., *Global Potential for Biomass and Carbon Dioxide Capture, Transport and Storage up to 2050*, 11 International Journal of Greenhouse Gas Control 117,118 (2012).

92 DoE, *Biofuels Industrial Strategy of the Republic of South Africa* (December 2007).

93 Department of Energy, The Draft Position Paper on the South African Biofuels Regulatory Framework, No. 37232 (15 January 2014), at 19.

effects of large-scale production on food security and employment creation.[94] The foregoing concerns around BECCS in South Africa reinforces the position of studies which have shown that large-scale deployment of BECCS could have negative implications for food security for some of the world's most vulnerable populations,[95] because it will require conversion of substantial area of land which have been wrongly classified as abandoned and marginal,[96] a development that can result in displacement of the poor from land.[97] Its implementation can endanger food security in developing countries,[98] affect adversely availability of water,[99] and exacerbate environmental degradation.[100]

Ocean iron fertilization is a CDR approach whereby iron is introduced into iron-poor areas of the ocean surface to stimulate phytoplankton production, which can enhance biological productivity and/or accelerate carbon dioxide sequestration from the atmosphere.[101] Proponents of this option contend that phytoplankton production is limited due to low concentrations of iron in the southern ocean, subarctic Pacific and eastern equatorial Pacific waters,[102] hence, adding iron artificially in these regions could stimulate phytoplankton production, thus enhancing carbon dioxide uptake.[103] The significance of ocean fertilization to carbon removal agenda has been contested, though. While previous studies demonstrate that ocean iron fertilization could reduce atmospheric carbon dioxide levels substantially,[104] other

[94] Willem Jonker et al., *Implications of Biofuel Production in the Western Cape Province, South Africa: A System Dynamics Modelling Approach*, 28(1) J. Energy in Southern Africa 1 (2017).

[95] Burns & Nicholson, *supra* note 3.

[96] Pete Smith et al., *Biophysical and Economic Limits to Negative CO_2 Emissions*, 6 Nature Climate Change 42, 46 (2016); Williamson, *supra* note 4, at 154.

[97] Sivan Kartha & Kate Dooley, *The Risks of Relying on Tomorrow's 'Negative Emissions' to guide Today's Mitigation Action*, SEI Working Paper No 2016–08 ((2016).

[98] Oliver De Schutter, *Note on the Impacts of the EU Biofuels Policy on the Right to Food* (23 April 2013). http://www.srfood.org/images/stories/pdf/otherdocuments/20130423_biofuelsstatement_en.pdf

[99] Pete Smith, *Soil Carbon Sequestration and Biochar as Negative Emission Technologies*, 22(3) Global Change Bio. 1315 (2016).

[100] Kartha & Dooley *supra* note 97, at 16.

[101] Richard S. Lampitt et al., *Ocean Fertilization: A Potential Means of Geoengineering?*, 366 Philosophical Transactions Royal Society A 3919, 3920 (2008).

[102] Sanjay K. Singh et al., *Response of Bacterioplankton to Iron Fertilization of the Southern Ocean, Antarctica,* Frontiers in Microbiology 1, 2 (2015).

[103] Matthew Hubbard, *Barometer Rising: The Cartagena Protocol on Biosafety as a Model for Holistic International Regulation of Ocean Fertilization Projects and Other Forms of Geoengineering,* 40 William & Mary Environmental Law and Policy Review 591, 598 (2016).

[104] Melissa V. Eick, *A Navigational System for Uncharted Waters: The London Convention and London Protocol's Assessment Framework on Ocean Iron Fertilization,* 46 Tulsa Law Review 351, 357 (2010).

findings conclude that it would result in minimal sequestration of atmospheric carbon.[105]

South Africa's experimentation with ocean fertilization has been controversial since the 2009 LOHAFEX expedition, which was co-sponsored by India and Germany. The field experiment introduced six tons of iron sulphate over an area of 300 square kilometers in the Southern Ocean.[106] Although that project was stopped due to environmental concerns, there is ongoing research, through the Southern Ocean Carbon & Climate Observatory (SOCCO),[107] on how iron impacts the Southern Ocean's biology and its relationship to climate and CO_2. Specifically, researchers are assessing whether it is feasible to promote the ocean's natural ability to take up CO_2.[108] There are environmental and other concerns in South Africa which resonate with general findings in studies about the potential risks of ocean fertilization to the ecosystems and humans who depend on ocean resources. Among other damaging consequences, it is feared that implementing ocean fertilization can block sunlight in deeper waters and overload bacterial decomposers that take up oxygen,[109] and increase mortality rates of critical prey species.[110] The foregoing benefits and harm potential of CDR options may have implications for the realization of human rights in South Africa.

3.2 *CDR Options as Human Rights' Enabler or Hindrance?*

Implementing CDR options has the dual potential to threaten and enhance human rights, in particular, the right to life, the right to have access to sufficient food and water, the right to have access to health care services, the right to property, the right to adequate housing, and the right to environment, all of which are guaranteed in the South African Constitution and international human rights instruments to which South Africa is a state party.

Section 11 of the Constitution guarantees as well as article 4 of the African Charter and article 6 of the ICCPR the right to life. In terms of section 27(5) of the

---

[105] Philippe Ciais et al., Carbon and other Biogeochemical Cycles, in Climate Change 2013: The Physical Science Basis Contribution of Working Group I to the First Assessment Report of the Intergovernmental Panel on Climate Change 465–570 (Thomas Stocker et al., eds. 2013) 549, at 551.

[106] Neil Overy, *State of Play? Geoengineering in South Africa*, 2018 https://za.boell.org/2018/06/05/state-play-geoengineering-south-africa

[107] *Id.*

[108] *Id.*

[109] Jennie Dean, *Iron Fertilization: A Scientific Review with International Policy Recommendations* 32(2) Environs 322, 330 (2009).

[110] John J. Cullen & Philip W. Boyd, *Predicting and Verifying the Intended and Unintended Consequences of Large-Scale Ocean Iron Fertilization*, 364 Marine Ecology Progress Series 295, 300 (2008).

Constitution, the right to life is entirely a non-derogable right. In addition to urging States to make efforts to avoid threats to human life, the UNHRC General Comment No. 6 warns that the right should not be narrowly interpreted.[111] The thinking at that level is reflected in *Mazibuko v City of Johannesburg*[112] *(Mazibuko* case) where the Constitutional Court noted that the constitutional protection of water rights in South Africa stems from the fact that water is critical for life.[113] The right to life can be both enhanced and hindered while implementing afforestation programs, for instance. On one hand, given that trees are central to the human respiratory process and the reduction of environmental degradation, they ultimately contribute to enhancement of the right to life. On the other hand, its heavy demand on water, a limited resource in South Africa, may undermine the right to life given the centrality of water to life.

The right of everyone to access to supplies of sufficient food is guaranteed under section 27(1)(b) of the Constitution and article 11 of the ICESCR. While examining the normative content of the right to food, the CESCR General Comment No 12 of 1999,[114] urges States parties to note that the normative elements of the right to food, that is the availability, accessibility, acceptability, and safety of food, can be adversely affected by climatic and ecological factors. It therefore urges States parties to adopt appropriate measures to ensure that climate change does not adversely affect the right to food.[115] In *Government of the Republic of South Africa and Others v Grootboom and Others,* the Constitutional Court endorses the position of CESCR in General Comment 3 that states have the minimum obligation to ensure that individuals are not deprived of essential foodstuffs.[116] In helping to stabilize the climate system, deploying all the CDR options is at least, indirectly, positive for the right to food as it will contribute to reducing extreme events and changes in temperature associated with continued carbon emissions that undermine land suitability and crop yields. However, the diversion of agricultural land for afforestation and BECCS projects can reduce the availability of food, thereby undermining the right to food in South Africa.

The right of everyone to access sufficient water is guaranteed under section 27(1)(b) of the Constitution and impliedly under articles 11 and 12 of the ICESCR. In delineating States' obligations in CESCR General Comment No 15 of 2002 on the

---

[111] UN Human Rights Committee (HRC), CCPR General Comment No. 6: Article 6 (Right to Life), adopted at the Sixteenth Session of the Human Rights Committee (30 April 1982) paras. 1 &2

[112] *Mazibuko v City of Johannesburg* 2010 4 SA 1 (CC) (the *Mazibuko* case).

[113] *Id.* at para 1

[114] UN Committee on Economic, Social and Cultural Rights, General Comment No. 12: The Right to Adequate Food (Art. 11), E/C.12/1999/5, adopted at the Twentieth Session of the Committee on Economic, Social and Cultural Rights (12 May 1999).

[115] *Id.* at paras 4 and 7.

[116] Government of the Republic of South Africa and Others v Grootboom and Others (CCT11/00) [2000] ZACC 19; 2001 (1) SA 46; 2000 (11) BCLR 1169 (4 October 2000) para. 29.

right to water,[117] States parties are urged to adopt strategies and programs that address developments such as climate change that may hamper the realization of the right to water.[118] In the *Mazibuko* case, the Constitutional Court noted that realizing the constitutional promise of access to sufficient water for all will require careful management of water, which is a scarce resource in South Africa.[119] On the one hand, the deployment of CDR options such as BECCS, ocean fertilization and afforestation, can help reverse the adverse effects of climate change such as flooding, which may undermine the availability and delivery of clean water to populations, and thereby enhance access to water. On the other hand, due to water demand and its limitation in South Africa, the options of afforestation and BECCS may threaten the right to water.

Equally, section 27(1)(a) guarantees the right of everyone to health in South Africa, an equivalent provision of which can be found in article 16 of the African Charter and article 12 of the ICESCR. While interpreting the right to life, CESCR General Comment No 14 identifies determinants of health, as including the prevention and reduction of exposure to harmful substances such as radiation and harmful chemicals or other detrimental environmental conditions that directly or indirectly impact upon human health.[120] The Constitutional Court in *Minister of Health and Others v Treatment Action Campaign and Others* maintained the view that the State must avoid unreasonable restrictions on the realization of the right to health.[121] The deployment of CDR options may help stabilize temperature, reduce extreme weather events, prevent the spread of associated epidemics and thereby contribute to the realization of the right to health. The potential impact of CDR options, for instance, the release of nitrous oxide associated with ocean fertilization may worsen pollution and thereby hinder the realization of the right to health in South Africa.

Section 25(5) of the Constitution safeguards against arbitrary deprivation of property and urges the State to ensure access to land on an equitable basis. Along similar lines, section 26 of the Constitution provides that everyone has the right to have access to adequate housing. Both rights are guaranteed respectively under article 14 of the African Charter and article 11 of the ICESCR. While setting out the obligations of States under General Comment No 4 on the right to adequate housing under the ICESCR, the CESCR links the right to housing to property by affirming that the legal tenure in the context of housing could take the form of 'accommodation, cooperative housing, lease, owner-occupation, emergency housing and

[117] UN Committee on Economic, Social and Cultural Rights, General Comment No. 15: The Right to Water (Arts. 11 and 12 of the Covenant), E/C.12/2002/11, adopted at the Twenty-ninth Session of the Committee on Economic, Social and Cultural Rights, on 20 January 2003.

[118] *Id.* at para. 28.

[119] Mazibuko, *supra* note 112, at para. 3.

[120] *Id.* at para. 15.

[121] Minister of Health and Others v Treatment Action Campaign and Others (No 2) (CCT8/02) [2002] ZACC 15; 2002 (5) SA 721; 2002 (10) BCLR 1033 (5 July 2002) paras. 4 and 136.

informal settlements, including occupation of land or property'.[122] In the context of climate change, the CESCR indicates that security of tenure, availability, accessibility, location, affordability, habitability, and cultural adequacy of housing may be adversely affected by climatic and ecological considerations.[123] In *Alexkor Limited and another v Richtersveld Community and Others,* the Constitutional Court noted that restoration of land and land rights of disadvantaged populations is an issue of supreme importance,[124] while in *Rahube v Rahube and Others*, it was held by the same Court that the right to housing is intrinsically linked to the dignity of the human person, which should not be undermined by any piece of legislation.[125] For its potential to undermine and dispossess local populations of their customary land tenure system and traditional use of land, the deployment of CDR options of afforestation and BECCS may affect adversely the rights of the population to property and adequate housing. Nonetheless, if implemented with due regard for the nature of land tenure of populations, afforestation and BECCS can accommodate the interest of local populations and boost the realization of property or adequate housing in South Africa.

Section 24 of the Constitution guarantees environmental rights, as does article 24 of the African Charter. The potential for afforestation to reduce water yield and increase soil erosion and the potential risks of ocean fertilization to the ecosystems may encourage ecological degradation and thereby threaten ecological sustainability in terms of section 24 (b)(i), as well as conservation and the right to protect the environment for present and future generations, according to section 24 (b). In particular, the potential risk of pollution that is associated with ocean fertilization contrasts with an important objective of the prevention of pollution under section 24 (b)(i). Its implementation appears inconsistent with key international instruments such as the CBD,[126] CPMPD,[127] and (UNCLOS),[128] insofar as they affect the biodiversity or pollute the sea. Over long stretches of time, deploying CDR options are not without positives in that it may help reduce if not eliminate developmental choices of the state that are inimical to sustainable environment. For instance, BECCS can constitute an alternative to reliance on coal- based energy systems by South Africa and its associated environmental effects, and thereby enhance the realization of the provisions of section 24. In *Earthlife Africa Johannesburg v Minister of Environmental Affairs and others,* the court noted that South Africa contributes to global GHG emissions because of coal-intensive energy systems, and that its population is

[122] UN Committee on Economic, Social and Cultural Rights, General Comment No. 4: The Right to Adequate Housing (Art. 11 (1) of the Covenant*)*, E/1992/23, adopted at the Sixth Session of the Committee on Economic, Social and Cultural Rights (13 December 1991), para. 8(a).

[123] *Id.* at para. 8.

[124] Alexkor Ltd and Another v Richtersveld Community and Others 2003 (12) BCLR 1301 para. 38.

[125] Rahube v Rahube and Others 2019 (1) BCLR 125 (CC) paras. 2 and 74.

[126] CBD *supra* note 67.

[127] CPMPD, *supra* note 68.

[128] UNCLOS, *supra* note 69.

vulnerable to the socio-economic and environmental impact of climate change.[129] Hence, administrative decisions on projects must take heed of climate change impact assessments.[130]

Overall, the large uncertainties associated with the potential benefits and risks in CDR options, including potential impacts on human rights, are a major impediment to implementation of such options in South Africa. How then can State human rights obligations be adequately scrutinized in response to such challenges?

4 Human Rights Obligations as a Response to Implementation Challenges

Arguably, the four layers of human rights obligations, to wit, the obligations to respect, p*rotect*, promote, and fulfill human rights may be engaged as a response to the challenges around the implementation of CDR options in South Africa. The obligation to respect mandates that States should not interfere in the enjoyment of human rights.[131] Also, it denotes that there should be respect on the part of the State for right-holders. In the context of socio-economic rights, this means that the State is obliged to respect the free use of resources owned individually or collectively.[132] In relation to a collective group, the obligation to respect entails that resources belonging to this group should be respected.[133] Accordingly, the application of the obligation to respect in the CDR context demands that the State should not deploy any option that endangers informal land tenure, the rights to life, property, housing, water, food, and environment of the populations. No doubt, it is almost impossible to implement afforestation and BECCS with a zero negative human rights impact on populations. This is not unexpected as generally, except for the right to life which is non-derogable, bill of rights may be limited in terms of section 36 (1)(e) insofar as such a limitation is reasonable and justifiable in an open and democratic society based on human dignity, equality, and freedom, and taking into consideration factors including the less restrictive means of achieving the purpose. However, in the case of a limited right, the State should not deploy measures which may intrusively undermine the rights of populations in South Africa.

The obligation to respect further requires that any legislation that is incompatible with the protection of rights should not be implemented. There are quite a number of pieces of legislation with provisions that may challenge the implementation of

[129] Earthlife Africa Johannesburg v Minister of Environmental Affairs and others (Thabametsi case) [2017] JOL 37526 (GP), paras. 25–27.

[130] *Id.* at para. 101.

[131] UN General Comment No. 31, *supra* note 50; UN General Comment No. 3, *supra* note 50.

[132] *SERAC*, *supra* note 51, at para. 45.

[133] *Id.*

CDR options in South Africa. These include the Water Act,[134] and the National Environmental Management Act,[135] Air Quality Act,[136] Conservation of Agricultural Resources,[137] Forestry Act,[138] and the Carbon Tax Bill.[139] For example, the rationing and management of water as provided for by the Water Act may challenge the implementation of afforestation and BECCS in South Africa.[140] Whether the term 'environmentally sound technology' in NEMA accommodates any of the CDR options is not certain.[141] The classification of nitrous oxide as an air pollutant makes the Carbon Tax Bill an unlikely instrument to support ocean fertilization,[142] in that nitrous oxide production is a likely side effect of ocean fertilization.[143] In fact, the idea behind the Bill, which is that polluters should pay for the wrongs they do to the environment is neither preventative nor remedial in the climate change context. It does not, for instance, discourage those who have the means to stop polluting, and its contribution to halting the dangerous progression to global warming beyond 1.5° is arguably negligible. Also the requirement for authorization to use water for afforestation and other land use under section 3 (1)(d) of the Forestry Act is problematic in both the context of afforestation and BECCS. For such an application to succeed, proposed projects must conserve natural resources, especially soil and water; and advance the status of disadvantaged populations.[144] The likely side effects of afforestation and BECCS on water and soil resources contrast with section 3 of the Conservation of Agricultural Resources Act, legislation established to protect and foster soil health, water sources, and vegetation.

The obligation to protect enjoins the State from adopting measures, including legislation, and provide effective remedies, to protect the interests of right-holders against infringement.[145] It requires the State to formulate an appropriate framework through a blend of laws and regulations so that beneficiaries of rights can achieve their rights. South Africa is a signatory to international environmental and human rights instruments relevant to CDR options. The absence of standalone legislation at the international level on CDR signifies that section 232 of the Constitution dealing with application of international law can only be applied with caution so that applicable international instruments can serve their appropriate ends. For instance, we

[134] National Water Act (NWA), No. 361998.

[135] National Environmental Management Act (NEMA)107,1998.

[136] National Environmental Management: Air Quality Act 39, 2004.

[137] Conservation of Agricultural Resources Act 43, 1983.

[138] National Forests Act, 1998.

[139] Carbon Tax Bill 2018.

[140] NWA, *supra* note 134, at section 6(3)(iv).

[141] NEMA, *supra* note 135, at section 24(1).

[142] Carbon Tax Bill, *supra* note 139, at preamble.

[143] Freestone D. Ray, *Ocean Iron Fertilization and International Law,* 364 (213–218), Marine Ecology Progress Series 227, 231 (2008).

[144] National Forests Act, *supra* note 138, at section 3(3)(vi) and (vii).

[145] UN General Comment No. 31, *supra* note 50; UN General Comment No. 3, *supra* note 50.

will need to acknowledge tradeoffs may be necessary to balance the potential impacts of CDR deployment and the potential ramifications of unchecked climate change. This analysis will need to be conducted under pertinent international instruments such as the CBD,[146] CPMPD,[147] and (UNCLOS).[148] Also insofar as the domestic legislation earlier examined challenges the implementation of CDR options, there is a need for its amendment to support implementation that enhances the realization of rights by State organs and non-State actors.

The obligation mandates that the State should create an environment that allows individuals to exercise their rights by promoting tolerance, raising awareness, and developing necessary infrastructures.[149] In the context of the deployment of CDR options, this is crucial, given that the position of the State on the potential benefits of CDR options for the future of climate system, and of course, the realization of rights is not yet clear. Hence, in line with the obligation to promote, one would expect the State to use its promotional organs of human rights to provide adequate and accessible information with the view of enhancing public awareness on the interface of human rights realization with CDR options in South Africa. In addition to improving the public perception of the measures, increased awareness may serve as a caution to prevent non-State actors from implementing initiatives in a manner that undermines the protection of human rights.

The obligation to fulfil requires the State to mobilize its machinery to foster realization of these rights.[150] While it is evident that the deployment of CDR options is not without side effects, it is the State's responsibility to ensure that the human rights benefits of the projects outweigh the negative impacts. In line with this obligation, the State may require all its decision-making organs regarding CDR to prioritize human rights assessment as a key consideration in the decision-making process. In particular, where the interest of local populations of South Africa is at stake, the State should ensure that the implementation of CDR options fulfill basic rights. Where such a demand requires resources beyond national capacity, States are obliged to seek, and be offered, international cooperation and assistance to fulfil this basic need in line with the provisions of article 2(1) of the ICESCR, which recognizes the need to seek international assistance for the progressive realization of rights. Such a route is also envisioned in the Paris Agreement, which requires developed States to financially assist developing countries with respect to both mitigation and adaptation in line with their existing obligations under the instrument.[151] Hence, in deserving circumstances, South Africa can request international assistance to address inevitable downsides to the deployment of CDR options.

---

[146] CBD, *supra* note 67.

[147] CPMPD, *supra* note 68.

[148] UNCLOS, *supra* note 69

[149] UN General Comment No. 31, *supra* note 50; UN General Comment No. 3, *supra* note 50.

[150] *Id.*

[151] Paris Agreement, *supra* note 10, at art. 9(1).

5 Conclusion

The application of a human rights framework to the deployment of CDR options, in particular afforestation, BECCS, and ocean fertilization in the context of South Africa, a State party to a range of international environmental law and human rights instruments, merits articulation. This chapter has set out to examine whether human rights can shape the feasibility of implementing CDR options, and if so, demonstrate how the concept can serve as a response to the uncertainties around implementation in South Africa.

The feasibility of implementing CDR options is unsettled in South Africa due to the ambivalence of its nature and the uncertainty around its potential for realization of human rights. This chapter suggests that the implementation of CDR options may result in both benefits and risks, which has implications for the realization and violation of human rights, especially, the rights to life, the right to have access to sufficient food and water, the right to access to health care services, the right to property, the right to adequate housing, and right to environment in South Africa. States' obligations to respect, protect, fulfill, and promote human rights, as this chapter has demonstrated, are a useful component of the human rights framework for responding to the challenges associated with the implementation of CDR options in South Africa.

References

1. Burns, W., Nicholson, S.: Bioenergy and Carbon Capture with Storage (BECCS): The prospects and challenges of an emerging climate policy response. J. Envtl. Studies and Sci. **7**, 527 (2017)
2. Corry, O.: The international politics of geoengineering: The feasibility of plan B for tackling climate change. Secur. Dialogue. **48**(4), 297–299 (2017)
3. Cotula, L., et al.: Testing claims about large land deals in Africa: Findings from a multi-country study. J. Develop. Stud. **50**(7), 903 (2014)
4. Cullen, J.J., Boyd, P.W.: Predicting and verifying the intended and unintended consequences of large-scale ocean iron fertilization. Mar. Ecol. Prog. Ser. **364**, 295–300 (2008)
5. Dean, J.: Iron fertilization: A scientific review with international policy recommendations. Environ. **32**(2), 322–330 (2009)
6. Dembour, M.-B.: What are human rights? Four Schools of Thought, Human Rights Q. **32**, 1 (2010)
7. Eick, M.V.: A navigational system for uncharted waters: The London convention and London protocol's assessment framework on ocean iron fertilization. Tulsa Law Review. **46**, 351–357 (2010)
8. Hubbard, M.: Barometer rising: The cartagena protocol on biosafety as a model for holistic international regulation of ocean fertilization projects and other forms of geoengineering. William & Mary Environ. Law & Policy Rev. **40**, 591–598 (2016)
9. Jegede, A.O.: Climate change in the work of the African commission on human and peoples' rights. Speculum Juris. **31**(2), 136 (2017)
10. Jonker, W., et al.: Implications of biofuel production in the Western Cape province, South Africa: A system dynamics modelling approach. J. Energy South Africa. **28**(1), 1 (2017)

11. Keller, D.P., Feng, E.Y., Oschlies, A.: Potential climate engineering effectiveness and side effects during a high carbon dioxide-emission scenario. Nat. Commmun. **5**, 1–9 (2014)
12. Kornneeff, J., et al.: Global potential for biomass and carbon dioxide capture, transport and storage up to 2050. Int. J. Greenhouse Gas Control. **11**, 117–118 (2012)
13. Lampitt, R.S., et al.: Ocean fertilization: A potential means of geoengineering? Philosophic. Trans. R. Soc. A. **366**, 3919–3920 (2008)
14. Malkamäki, A., et al.: A systematic review of the socio-economic impacts of large-scale tree plantations, worldwide. Glob. Envtl. Change. **53**, 90 (2018)
15. Nowak, D.J., Hoehn, R., Crane, D.E.: Oxygen Production by Urban Trees in the United States. Arboriculture & Urban Forestry. **33**(3), 220 (2007)
16. Sekoai, P.T., Daramola, M.O.: Biohydrogen as a potential energy fuel in South Africa. Biofuel Rsch. J. **6**, 223 (2015)
17. Svoboda, T., Buck, H.J., Suarez, P.: Climate engineering and human rights. Envtl. Pol. **28**(3), 397 (2019)
18. Turpie, J.K., et al.: The ecological and economic consequences of changing land use in the Southern Drakensberg Grasslands, South Africa. S. Afr. J. Econ. Manag. Sci. **10**(4), 423 (2007)
19. Williamson, P.: Emissions reduction: scrutinize CO_2 removal methods. Nature. **530**, 153–155 (2016)

Geoengineering and the Question of Weakened Resolve

David A. Dana

1 Introduction

One of the most-discussed topics in the social science literature regarding geoengineering is the inter-relationship between geoengineering and climate change mitigation. This literature has two distinct strands. What we might call Strand # 1 assumes that there may be some optimal mix of geoengineering and mitigation from a welfare economics perspective, and explores what an optimal mix might be and under what conditions it might obtain.[1] What we might call Strand #2 of the literature builds on the recognition that global mitigation efforts have been wildly sub-optimal and that a huge increase in mitigation efforts, as a normative matter, is needed for the sake of current and future generations. Strand # 2 recognizes, too, that mitigation is not easy, it is expensive, it requires individual and collective changes in behavior, it poses a threat to entrenched, powerful economic interests, and it implicates complicated questions as to who exactly should bear the costs of mitigation and in what measure. Strand # 2 is also understandably concerned with the following question:

[1] *See*,e.g., Jesse Reynolds, *A critical examination of the climate engineering moral hazard and risk compensation concern*, 2(2) THE ANTHROPOCENE REV. 174, 185 (2015) ("the simple economics of substitutes suggests that, to the extent climate engineering might actually reduce mitigation through substitution, that could be rational and beneficial"); Bjorn Lomborg, *Geoengineering – A Quick Clean Fix?*, TIME, Nov. 14, 2010 (arguing that geoengineering is justified by the high costs of proposed mitigation), available at http://content.time.com/time/magazine/article/0,9171,2030804,00.html;

Michael McCracken, *On the possible use of geoengineering to moderate specific climate change impacts*, 4(4) ENVTL. RES. LETTERS: 045107 (2009). (modelling scenarios in which geoengineering would enhance welfare).

D. A. Dana (✉)
Northwestern Pritzker School of Law, Chicago, IL, USA
e-mail: d-dana@law.northwestern.edu

© Springer Nature Switzerland AG 2021

W. Burns et al. (eds.), *Climate Geoengineering: Science, Law and Governance*, AESS Interdisciplinary Environmental Studies and Sciences Series,
https://doi.org/10.1007/978-3-030-72372-9_7

> Will publicity about and research into geo-engineering (and beyond that, actual deployment of geoengineering) weaken the (the already grossly insufficient) resolve to mitigate now and into the future?

To the extent the answer to this question is yes, geoengineering, even as an idea, even as what Corry calls "a sociotechnical imaginary,"[2] may pose a danger to social welfare. This Strand #2 question – the question of whether geoengineering weakens the resolve to mitigate -- has been variously dubbed a question of "moral hazard," "risk compensation," and "climate change moral hazard-risk compensation."[3] Moral hazard typically refers to situations where there is an incentive for one party to engage in a risky behavior because the party know the costs will be borne by others (e.g., people who invest in expensive homes in flood zones because underpriced federal flood insurance transfers the risk of flooding damage to the federal government). Risk compensation captures the idea that if there is a risk-reduction tool that reduces a risk to what one sees as an acceptable level, one may "compensate" for the risk reduction by engaging in the risky activity more than one otherwise would have in the first place (e.g., drivers who engage in speeding more once they know that their cars are equipped with air bags that may provide protection in the case of a crash).

Both moral hazard and risk compensation generally have negative connotations – moral hazard, because it entails a ducking of accountability for risk, and risk compensation because it is typically posited that the compensation for risk and the attendant increase in risky behavior is irrational on the level of individual and social welfare. The goal of seat belts and airbags is not to lead drivers to speed more so that driving is just as dangerous as before, but rather to make driving safer.

It is plausible to characterize geoengineering as either implicating moral hazard and/or risk compensation. If mitigation is reduced because of geoengineering, that can be seen as a moral hazard in that geoengineering (arguably, at least) is at best a temporary response to climate change, so reliance upon geoengineering by the current generation can be seen as a way of off-loading risk to future generations.[4] Alternatively, if mitigation is reduced because of geoengineering, that can be seen as (irrational) risk compensation because geoengineering is an inherently incomplete

---

[2] Olaf Corry, *The international politics of geoengineering: The feasibility of Plan B for tackling climate change*, 48(4) SECURITY DIALOGUE 297, 299 (2017).

[3] Keith and Lin both suggest that risk compensation is a more apt label for the concern than the most commonly-used term, moral hazard. *See* David Keith, A CASE FOR CLIMATE ENGINEERING 129-132 (2013); Albert Lin, *Does geoengineering present a moral hazard*, 40 ELQ 673, 688-690 (2013). Reynolds combines both ideas into "CE MH-RC," which stands for climate engineering moral hazard-risk compensation. Reynolds, *supra* note 1, at 178.

[4] To the extent that some geoengineering efforts do not actually reduce the build up of greenhouse gases in the atmosphere but rather only shield the surface from warming effects, the termination of geoengineering (intentional or not) could result in dramatic warming. *See* Andy Parker & Peter J. Irvine, *The Risk of Termination Shock From Solar Geoengineering*, Earth's Future, 11 March 2018, available at https://agupubs.onlinelibrary.wiley.com/doi/full/10.1002/2017EF000735 (describing the concerns surrounding SRM's "termination shock," but also suggesting ways this shock could be moderated).

response to climate change (solar radiation management will not by itself reduce ocean acidification, for example[5]) and (especially in the form of solar radiation management) is fraught with its own inherent risks.

The Strand #2 literature is grappling with is geoengineering's potential to weaken the resolve to take costly actions to address climate change through means other than geoengineering (namely, mitigation, but also adaptation, as discussed below). So, to highlight that core point implicated by all the commentators on the possible effect of geoengineering on mitigation, and to avoid the unhelpful conceptual questions introduced by "moral hazard" and "risk compensation," I suggest we re-label the Strand # 2 question, the "weakened-resolve" question (WRQ).

Behavioral psychology would tend to suggest that individuals, whether acting in their personal, business or political lives, will experience some weakened resolve in response to learning about geoengineering as an option to address the effects of climate change.[6] In general, people overestimate certain costs and underestimate the magnitude of uncertain or speculative costs. As Kahneman and Tversky formulate the phenomenon, "people overweigh outcomes that are considered certain, relative to outcomes which are merely probable."[7] For example, because insurance entails a sure loss in the form of a premium, people purchase what many observers believe to be inadequate levels of insurance against relatively small risks of large losses.[8]

Mitigation and adaptation both entail certain, upfront costs, in order to avoid the somewhat uncertain costs (in magnitude and incidence at least) of climate change. Geoengineering in some of its most-discussed forms, such as SRM, can have appeal – from a rational cost-benefit framework, *too much* appeal – because its certain costs (the direct costs of research and deployment) seem modest compared to the much larger, certain costs associated with mitigation and adaptation. The large but uncertain costs of geoengineering, such as disrupting weather patterns or even

[5] *See* David W. Keith and Douglas G. MacMartin, *A temporary, moderate and responsive scenario for solar geoengineering*, NATURE CLIMATE CHANGE, 16 February 2015, available at https://keith.seas.harvard.edu/files/tkg/files/174.keith_.macmartin atemporarymoderateandresponsivescenario-forsolargeoengineering.pdf (acknowledging that "[o]cean acidification is a risk of SRM," assuming SRM is "used as a substitute for emissions mitigation').

[6] As discussed below, the empirical support for the opposite proposition – that knowledge about geoengineering could increase support for mitigation efforts – comes primarily from small focus group studies in Europe, and that effect has not been established by any of the published studies from the United States. As noted below, there is a lack of empirical studies involving subjects outside of the United States, Western Europe and Australia

[7] Daniel Kahneman & Amos Tversky, *Prospect Theory: An Analysis of Decision Under Risk*, 47 ECONOMETRICA 263, 265 (1979).

[8] RICHARD THALER, THE PSYCHOLOGY OF CHOICE AND THE ASSUMPTIONS OF ECONOMICS, IN QUASI-RATIONAL ECONOMICS 137, 142 (Richard Thaler ed., 1991) (citing Kahneman & Tversky, *supra* note 22, 998. By contrast, where the loss at issue is small, but the probability of it occurring is so high as to make occurrence almost a sure thing, people are quite willing to purchase insurance. *See* Paul Slovic et al., *Accident Probabilities and Seat Belt Usage: A Psychological Perspective*, in THE PERCEPTION OF RISK 75-76 (Paul Slovic ed., 2000); Paul Slovic et al., *Preference for Insuring Against Probable Small Losses: Insurance Implications*, in THE PERCEPTION OF RISK, *id.*, at 51-72.

sparking terrorism and war, will tend to be underweighted precisely because they are uncertain.[9]

This Chapter first tries to frame the WRQ by asking: if we want to know whether geoengineering will weaken resolve, *whose* resolve is it that really matters? This question, in turn, requires us to ask, realistically, who is and will be determining how much of an effort to mitigate is made within any particular country or (as climate change is a global phenomenon where the action of major emitting nations is most relevant) within the group of countries that account for the bulk of annual GHG emissions? Moreover, because a great deal of resolve will be needed for meaningful adaptation efforts, the WRQ and pertinent research needs to expand to include adaptation: we also must ask, who will be making adaptation decisions?

The Chapter then summarizes the social science research to date, and includes some preliminary results from ongoing research which I have been undertaking with collaborators. Although there are a relatively large number of papers published on the WRQ, they provide an insufficient basis for postulating any answers. The existing literature provides thin support for the idea that geoengineering weakens resolve to mitigate, and equally thin support for the counter proposition – that geoengineering increases support for mitigation (a proposition that is sometimes called the moral galvanizing effect in the literature, and elsewhere referred to as a climate salience effect of geoengineering). The Chapter then argues that there are three areas (among many possible ones) that warrant more attention than they have received in the literature to date regarding the WRQ. These are:

- the WRQ as applied in cultural and political contexts outside of the United States and Northern Europe/EU, and especially China, India and other major non-Western emitters.
- the WRQ as applied to relevant elites (business, political, media, technological, political) as opposed to the general population;
- the WRQ in the context of the political polarization that has come to dominate United States politics (both popular and elite) and to an extent politics in England, Australia and elsewhere.

While more, broader-ranging social science to assess whether and how much geoengineering weakens resolve to mitigate is fully justified, there is a necessary limit to any answers social science can provide. For one thing, direct testing is infeasible. We cannot compare the path of mitigation efforts in a world where geoengineering is not discussed or researched, a world where it is discussed and researched,

[9] *See* David A. Dana, *A Behavioral Economic Defense of the Precautionary Principle*, 97 N NWU L. Rev. 1315, 1317 (2007) (arguing that the precautionary principle can function as a counterweight to heuristic biases in the context of climate change). I do not mean to suggest, however, that application of the precautionary principle to the geoengineering issue is straightforward, as one could make the argument that the precautionary principle counsels against geoengineering given its uncertain effects and also counsel in favor of geoengineering as insurance against climate change and the risk that mitigation will be inadequate. For a full discussion, see Professor Kalyani Robbins contribution to this book.

and a world where it is discussed, researched and deployed to some extent. "A major problem, especially for demonstrating the social and political forms of moral hazard, is the absence of a counterfactual."[10] Second, since geoengineering can be described in many different ways, is inherently complex, may take different forms in the future than we currently understand, and most people (even business and technology elites) have little or no relevant background knowledge, there is the possibility that social science will produce results that will not capture what people actually would think if they came to be introduced to geoengineering in a real, particular context over time.

Thus, at the end of the day, we may never know the magnitude of the risk of weakened resolve. In that sense, the WRQ is comparable to the question of what harms geoengineering, if deployed, would produce – we simply cannot know at this point. Given that, and given that there is at least private support for ongoing research into geoengineering and perhaps even an implicit recognition of geoengineering by the international community,[11] we also must ask: *assuming* geoengineering substantially weakens the resolve to mitigate, how can the weakening of the resolve itself be mitigated? Or, in the language used in the current academic literature, how can the moral hazard and risk compensation engendered by geoengineering itself be contained? Framing geoengineering as at best a risky, limited and temporary response to climate change may lessen its adverse impact on support for adaption and mitigation. The Chapter concludes with some ideas about how such a framing could be promoted.

2 Framing the Weakening Resolve Question: Who Actually Decides Mitigation and Adaptation Policy

In asking what effect the option of geoengineering will have on the resolve to address climate change through mitigation and adaptation, we cannot avoid the question of *who* decides on mitigation and adaptation. As an initial matter, it seems

[10] D. McLaren, *Mitigation deterrence and the "moral hazard" of solar radiation management*, 4 Earth's Future 596, 599 (2016).

[11] *See* Rona Fried, *Geoengineering solutions getting closer, Branson funding them*, May 12, 2012, available at http://www.sustainablebusiness.com/geoengineering-solutions-getting-closer-gates--branson-funding-them-50385/ ("Bill Gates, Sir Richard Branson, tar sands magnate Murray Edwards and Niklas Zennström, co-founder of Skype, and other wealthy individuals have financed official reports on the future use of geoengineering, raising concerns that wealthy people could have undue influence on policy. ") Gates analogizes geoengineering to heart surgery: "one of the complaints people have against that is that if it looks like an easy out, it'll reduce the political will to cut emissions. If that's the case, then, hey, we should take away heart surgery so that people know not to overeat." Bill Gates: The Rolling Stone Interview, March 13, 2014. The Paris Agreement could be read as implicitly acknowledging a need for geoengineering to some extent, as Daniel Farber has explained, http://blogs.berkeley.edu/2015/12/14/does-the-paris-agreement-open-the-door-to-geoengineering/

clear that the "who" question goes beyond the US, the EU, and other developed nations such as Canada and Australia. From both a mitigation and adaptation perspective, many other countries are as important. China has overtaken the United States as the largest annual emitter; India, Japan, Brazil, and Russia, contribute substantially to annual emissions.[12] (And of course all countries will have to contend with climate change adaptation.)

The political systems and cultures of these countries vary dramatically as among themselves, and may diverge in important ways from the United States, the EU, Australia or Canada. In China, for example, popular opinion may be less relevant in policy determination than in the United States because the political system there does not even purport to function as a representative democracy. Economic elites may matter less than political elites in China than in some other countries, and in particular Communist Party elites may matter the most (although there may be substantial overlap between party position and economic power). To the extent popular opinion does matter, public opinion too may vary greatly among countries with vastly different cultures, histories, and levels of economic wealth. Suggestively, a Pew survey of whether climate change "poses a major threat to my country" yielded percentages ranging from 35% (Russia) to 89% (Spain).[13]

Even if we choose to focus solely on the United States and the EU nations, the question of who makes policy is not straightforward. There is a tendency to believe (or maybe want to believe) that in representative democracy, popular opinion ultimately drives policy. But that is too simplistic a view.

Some political science literature suggests that the WRQ in the U.S. context may need to be explored in terms of whether geoengineering would weaken the resolve of elites, as opposed to the general public, to support mitigation and adaptation. Although the view that politics tracks "the median voter"[14] has a long pedigree in U.S. political science, the available evidence suggests a less than perfect relationship between median voter views and political outcomes. Rather, in large measure, votes in Congress and Executive action seem to track the preferences of economic elites. In 2014, Gilens and Page concluded, "Multivariate analysis indicates that economic elites and organized groups representing business interests have substantial independent impacts on U.S. government policy, while average citizens and mass-based interest groups have little or no independent influence."[15] More recently,

[12] Johanes Friedrich et al., Ap. 11, 2017, available at http://www.wri.org/blog/2017/04/interactive-chart-explains-worlds-top-10-emitters-and-how-theyve-changed

[13] Jacob Poushter and Dorothy Manevich, Aug. 1, 2017, available at http://www.pewglobal.org/2017/08/01/globally-people-point-to-isis-and-climate-change-as-leading-security-threats/

[14] See Tyler Cowen, *Why Politics Is Stuck in the Middle*, NY Times Feb. 6, 2010, available at https://www.nytimes.com/2010/02/07/business/economy/07view.html (discussing the median voter theorem, first proposed by Anthony Downs in 1957, that "there is a dynamic that pushes politicians to embrace the preferences of the typical or 'median' voter, who sits squarely in the middle of public opinion.").

[15] M. Gilens, M., & B. Page, *Testing Theories of American Politics: Elites, Interest Groups, and Average Citizens*, 12(3) Perspectives on Politics 564, 564.

Peter Temin concluded, "When the interests of the majority opposed those of the elites, they almost always lost out in political contests." In short, the Investment Theory of Politics – the theory that the extent of economic elites' investment in politics is a predictor of political outcomes - "is a far better predictor of political contests than the Median Voter Theorem."[16]

Moreover, whether we are concerned with elite (however we define the relevant elites[17]) or popular opinion in the US (as well as other nations such as the UK and Australia, although less clearly so), the growing literature on political polarization suggests that we should not focus on median or average opinion but rather the opinions held by elites and/or the lay public who adhere to polarized, political and cultural identities. From this vantage, the question becomes not whether geoengineering weakens the resolve of Americans or elite Americans, but whether it weakens the resolve of conservatives/Republicans and/or liberals/Democrats. Highly polarized segments of the population affiliated with highly polarized political leadership in the dominant parties has produced in the United States, a number of States in the US, and Australia a zig-zag pattern regarding climate change mitigation. This results in public policy shifts between treating climate change as a crisis requiring major regulatory initiatives on the one hand, and climate change denialism, or at best, climate change complacency on the other hand.[18] There is no *apriori* reason to anticipate that polarization will not characterize reactions to geoengineering and, in particular, the extent to which geoengineering will weaken resolve to mitigate and adapt.

The cultural cognition framework articulated by Dan Kahan and others also may have utility in studying the WRQ. The cultural cognition construct posits that the population is divided (polarized) not so much by political ideology per se, but by cultural orientation, with the two polar orientations being "egalitarian

---

[16] Peter Temin, The Vanishing Middle Class 74 (2017).

[17] Of course, economic elites are not monolithic. Conservative political elites and business interests tied to the fossil fuel industry have sought to undermine mitigation and adaptation initiatives within the federal government and in a number of state governments. *See, e.g.*, https://www.theguardian.com/us-news/2017/dec/12/big-oil-lobby-get-what-it-wants-epa-trump-pruitt. Other economic elites in the United States, including those in the finance and technology sector, seem to have no interest in climate change denialism. Thus, any assessment of elite opinion may have to take account of the fact that there are many elites that may be relevant in any policy domain, including the mitigation and adaptation domains. *See, e.g.*, http://money.cnn.com/2017/06/05/technology/business/businesses-paris-climate-agreement/index.html. The elites who may shape mitigation policy and adaptation policy, moreover, may be different. Both mitigation and adaptation require action at the level of the nation state, but many adaptation decisions will be made (or not) and implemented at the more local level. Thus, local elites may be as much the relevant audience for adaptation policy as national elites. *See generally* Hari M. Osofsky, *Polycentrism and climate change*, in Michael Faur (ed.), Elgar Encyclopedia of Environmental Law 324-36 (2016).

[18] For accounts of these shifts, *see, e.g.*, https://www.bna.com/australia-survived-climate-n73014452981/; http://www.businessinsider.com/trump-faces-limits-in-attempt-to-reverse-climate-change-policies-2017-10; https://www.politico.com/states/florida/story/2017/09/14/florida-governor-remains-unsure-about-climate-change-after-hurricane-irma-114498

communitarian" on one end and "hierarchical individualist" on the other. When sorted on the basis of cultural cognition, there are wide differences in views about climate change even among self-identified political moderates, at least in the United States.[19] To the extent that cultural cognition categories capture climate change polarization better, or at least differently, than conservative/liberal or Republican/Democratic categorizations, the cultural cognition categories might provide a useful framework for scrutinizing the WRQ as it applies to the U.S. population.

3 Overview of the Current Literature on Weakened Resolve

The empirical literature on the WRC to date in the context of climate geoengineering has employed on-line surveys using sophisticated social-psychological methods,[20] structured focus groups,[21] and reviews of statements in public debates.[22] The approach of the studies, with one notable exception, is to try to capture how people think about the possibility of weakened resolve for other people, and also how their own resolve may or may not weaken in light of the potential of geoengineering. Only one study, by Merk, seeks to ascertain if people change how they actually behave when they learn about geoengineering, assessing whether knowledge of geoengineering changes personal financial contributions toward mitigation.[23]

[19] Dan Kahan, June 21, 2012, available at http://www.culturalcognition.net/blog/2012/6/21/politically-nonpartisan-folks-are-culturally-polarized-on-cl.html

[20] *See, e.g.*, Adam Corner and Nick Pidgeon, *Geoengineering, climate change skepticism, and the "moral hazard" argument: an experimental study of UK public perceptions*, 372 PHIL. TRANS. ROYAL SOC'Y. 20,140,063 (2014); Victoria Campbell-Arvai et al., *The influence of learning about carbon dioxide removal on support for mitigation policies*, 143 CLIMACTIC CHANGE 321-336 (2017); Malcolm Fairbrother, *Geoengineering, moral hazard and trust in climate science: evidence from a survey experiment in Britain*, 139 CLIMACTIC CHANGE 477-489 (2016); Dan Kahan, et al., *Geoengineering and Climate Change Polarization: Testing a Two-Channel Model of Science Communication*, 658(1) ANNALS A. ACAD.POL.L& SOCIAL SCI. 199-222 (2015).

[21] *See, e.g.*, Victoria Wibeck et al., *Questioning the technological fix to climate change – Lay sense-making of geoengineering in Sweden*, 7 ENERGY RESEARCH & SOC. SCI. 23-30 (2015).

[22] See Duncan McLaren, *Public Conceptions of Justice in climate engineering: Evidence from secondary analyses of deliberations*, 41 GLOBAL ENT'L CHANGE 64-73 (2016).

[23] Merk et al. found German subjects were willing to offset their own emissions when they receive information about solar geoengineering. Subjects appeared to view solar geoengineering as a potential threat and appear willing to increase their investment in carbon offsets to help prevent a level of climate change that would make the deployment of solar aerosol injection more likely. Christine Merk et al., *Knowledge about aerosol injection does not reduce individual mitigation efforts*, 11 ENVT'L RESEARCH LETTERS (2016) 054009. For a general review of the weakened resolve literature, *see* Elizabeth T. Burns et al., *What do people think when they think about solar geoengineering? A review of empirical social science literature, and prospects for future research*, 4 Earth's Future 536–542 (2017), doi:https://doi.org/10.1002/2016EF00046; Corry, supra note 2, at 298.

The current literature is notable in its geographic scope (or lack of it). The studies are limited to wealthy, western nations – the US, Canada, the UK, Sweden and Germany. However, studies in even these few nations suggest potentially substantial differences among them in terms of the WRQ. Notably, while the studies from the US, UK, and Canada provide slim evidence for either weakened -resolve or heightened-resolve in light of the possibility of geoengineering, studies from Germany (Merk) and Sweden (Wibeck) more clearly suggest heightened resolve – that is, increased support for mitigation in the face of the possibility of deployment of geoengineering technologies. The studies from Germany and Sweden suggest, although certainly do not prove, that the WRQ may have a different answer in Sweden and Germany than it does in the US. Whether or not there are substantial differences between Germany and the US, it would be reasonable to assume large differences between the US and the population of major emitters in Asia and elsewhere around the globe.

The literature is also notable in that it is not trying to capture elite opinion, be it business/economic, political or technological. Rather, the effort is to capture representative samples of the population. Several studies have assessed whether socio-economic status and education within the representative example affect results. Wibeck, for example, segments the study population by education, and finds no meaningful differences. Corner and Pidgeon segment the study population by (among other things) socioeconomic status, and find socioeconomic status insignificant for some purposes, but mildly predictive as to whether subjects report that geoengineering might cause respondents to mitigate less in their own lives (with higher socioeconomic status being associated with a greater tendency to say geoengineering would lead to less mitigation in one's personal or private life.)

At least with respect to representative samples of the US and UK populations, the existing literature does seek to come to terms with how political and cultural polarization may be highly relevant to the WRQ. There were, at time of writing, five published papers that addresses the intersection of geoengineering, weakened resolve and polarization.[24] The papers differ in the categories they use to sort participants in terms of polarization, how they describe geoengineering for participants, and the questions they ask participants. Overall, however, the studies suggest a mixed view as to whether the introduction of geoengineering weakens resolve to mitigate among political and cultural conservatives to a greater extent than it does among non-conservatives.

Kahan (2015) highlights the potential value of cultural cognition categories in addressing the WRQ. Kahan surveyed a representative sample of 1500 U.S. residents and 1500 English residents. The subjects were measured with two worldview scales – hierarchical individualist and egalitarian communitarian – that had been

[24] See Corner and Pidgeon, supra note 20; Campbell-Arvai et al., supra note 20; Fairbrother, supra note 20; Kahan, et al., supra note 20; Kaitlin T. Raimi, Alexander Maki, David Dana & Michael P. Vandenbergh (2019) *Framing of Geoengineering Affects Support for Climate Change Mitigation, Environmental Communication*, 13:3, 300-319, DOI: https://doi.org/10.1080/17524032.2019.1575258

used in a number of studies of cultural cognition. One group was assigned what the authors called an antipollution prompt, which called for reductions in carbon dioxide emissions. Another group was assigned what the authors termed a geoengineering prompt, entitled "More Technology, Not More Limits Needed To Fight Climate Change,"[25] in which geoengineering is described as a relatively inexpensive, effective, desirable response to climate change that avoids the costs of cutting carbon dioxide emissions.

The subjects then were asked to evaluate the reliability of a scientific article describing the problem of climate change as real and harmful in a variety of concrete ways (e.g., rising sea levels, drought). The subjects also were asked to express their level of agreement with statements about climate change, such as "average global temperatures are increasing."

Kahan found that cultural polarization (that is, polarization between hierarchical-egalitarians and individualist-communitarians) over the validity of the science article decreased in the geoengineering condition as opposed to the antipollution conditions. The authors also found that hierarchical-individualists in the geoengineering condition expressed slightly more concern about climate change than those in the control condition.

Based on these effects, Kahan suggested that proponents of the weakened resolve argument have "things exactly backwards" – geoengineering (as in the Kahan study) makes hierarchical- individualists more open to accepting the science of climate change and hence more open to reasoned deliberation about what to do about climate change. As the authors explain, "[t]o overcome cultural resistance to sound scientific evidence that a problem exists, the two-channel communication strategy associated with the cultural-evaluator model," it is necessary that "people of diverse views must all be shown solutions that they find culturally congenial."

However, the Kahan study would have been far more illuminating regarding the WRQ had it asked subjects questions about their support for mitigation and adaptation after providing them with the control, antipollution, and geoengineering prompts. The results in Kahan are perfectly consistent with the view that after hierarchical-individualists are exposed to information about geoengineering, they will become more open to accepting climate change science, and hence more supportive of mitigation and adaptation. However, the results are equally consistent with the view that after hierarchical-individualists are exposed to information about geoengineering, and are then asked about their support for climate change mitigation, they will remain just as skeptical of mainstream climate change science.[26] Any increased embrace of climate science may be wholly contingent on subjects not being asked to consider concrete mitigation and adaptation proposals that entail

---

[25] Kahan, supra note 20, at 213.

[26] Indeed, the geoengineering prompt in Kahan (2015) depicts mitigation as essentially irrational comparted to geoengineering: "Land-based filters could remove excess CO2 from the air; high-altitude reflectors could be turned on and off to reduce solar heating . . . . 'geoengineering' technologies . . .would not only be more effective than enactment of emission restrictions, but also spare consumers and businesses [of] heavy costs . . . ." Kahan, *supra* note 20, at 213.

regulation and communal investment that run counter to a hierarchical-individualist worldview.

On the other hand, Fairbrother provides some support for what Kahan suggests regarding geoengineering and polarization. Fairbrother does not employ cultural cognition or personality categories, but rather addresses political polarization over climate change in the UK by sorting subjects as to whether they say they intend to vote Liberal, Labor or Conservative in the next election. With a pool of 167 subjects who identified as Conservative party supporters, Fairbrother found that Conservatives who are given a very brief statement about geoengineering (with no discussion of its risks[27]) are more supportive of a carbon tax to combat climate change than Conservatives who are not first provided the geoengineering prompt. Fairbrother acknowledges that his sample size for Conservatives is relatively small and that subjects were asked only about taxation in a general way, rather than being asked to consider a more particular tax proposal (or, for that matter, more overtly regulatory means of mitigation like emissions caps). That said, Fairbrother interprets his own results as supportive of the suggestions made by of Kahan.

There are three studies that, contrary to Fairbrother, suggest geoengineering information *increases* polarization over climate change and climate change mitigation. Crowder and Pidgeon ("Pidgeon") employed a nationally representative sample of 610 United Kingdom participants, who were classified based on their level of climate change skepticism and on a values scale designed to capture whether participants held "self-enhancing" or "self-transcending values." Pidgeon thus used a category dichotomy that is similar to Kahan's individualist/communitarian dichotomy.

The most notable finding in the Pidgeon study is that higher climate skepticism and identification with self-enhancing values were predictive of agreement with the statement, "Knowing geoengineering is a possibility makes me feel less inclined to make changes in my own behavior to tackle climate change." This finding suggests that, at least for personal mitigation efforts, geoengineering information does not strengthen "values conservatives" resolve to mitigate, but rather weakens it.

The Pidgeon study, however, did not ask participants to answer whether knowledge of geoengineering made them more or less likely to support mitigation polices as a political matter – as for example, by supporting a carbon tax or regulatory restrictions on CO2 emissions. More fundamentally, the basic design of the Pigeon study does not allow for the testing of whether information about geoengineering weakens resolve to mitigate among climate skeptics or "self-enhancers" because all the subjects in the study, including the control group, were given information about geoengineering. The experimental manipulation only involved adding one or another single paragraph to the substantial text on geoengineering that the control group received.

---

[27] *See* Fairbrother, supra note 20, at 481 (informing subjects that "to deal with global warming, scientists are developing ways of cooling the Earth's climate, such as by putting large mirrors in space to block some of the sun rays that heat the planet. Another technique they are researching is spraying particles in the atmosphere, to reflect some light from the sun back into space.").

The Campbell-Arvai study also provides some basis for questioning the view that geoengineering information reduces polarization. Campbell-Arvai surveyed 1114 residents of the United States, using an experimental manipulation that provides one group general information on climate change (the control group), a second group some general information on carbon dioxide removal ("CDR"), and others groups' information on particular forms of CDR, such as direct air capture and bioenergy with carbon capture and storage. Campbell-Arvai sorted subjects by political ideology, as reflected by subjects' self-assessment on a liberal-to-conservative scale, which may capture something akin to Kahan's worldviews or Pidgeon's values.

Campbell-Arvai found some evidence of weakened resolve for CDR generally and for all forms of CDR but reforestation; across the political spectrum, exposure to CDR information reduced subjects' perception of climate change as a threat and in turn lowered support for various mitigation policies. But reductions in the perceived threat of climate change in response to learning about CDR resulted in a more marked reduction in support for range of mitigation policies among conservatives, as compared to moderates or liberals.

A study of 781 U.S. subjects by myself, Rami, Vandenbergh and Alexander uses a similar approach as Rami's CDR paper but focuses on solar radiation management. The study results run counter to the suggestions of Kahan and Fairbrother about geoengineering, polarization, and weakened resolve, as its results indicates that conservatives (as well as moderates) become *less* worried about climate change after they are exposed to information about geoengineering. Less worry, in turn, reduces support for mitigation. The study also tests, and supports, the suggestion made by Lin that the framing of geoengineering as a major/permanent/riskless or partial/temporary/risky solution to climate change matters in terms of the WRQ. We found "the strength of moral hazard effects was strongest among conservatives in the major solution condition," the condition in which SRM is described as a comprehensive, low-cost solution. We conclude that the "[t]he Goldilocks approach—wherein geoengineering is described as a minor solution—may thus be the best way to frame geoengineering for all individuals if the goal is to educate the public about this technology without losing support for mitigation."

4 Limiting Weakened Resolve Through Framing

More study of weakened resolve could help clarify such questions as whether weakened resolve is a larger or lesser issue in different nations and cultural settings, the extent of weakened resolve in key elites with influence over policy (business, technological, political) as opposed to among the lay or general public, and how the phenomenon of weakened resolve interacts with the cultural/values and political polarization. To delve into these questions, researchers may need to adopt different methods than they have so far. For example, focus groups, on–line surveys and reviewing public records of debates may not yield sufficient information regarding elite opinion; instead, other methods, such as interviews, might be necessary to try to assess elite opinion.

But, as already discussed, even an expanded social science literature could only provide a tentative answer to the question of how and how much geoengineering will weaken the resolve to mitigate and adapt (and hence produce less mitigation and adaptation than otherwise to occur.) Consistent with the precautionary principle,[28] we almost certainly will need to proceed on the assumption that geoengineering can strongly weaken resolve to mitigate and adapt and thus we will have to ask, given that assumption, what can be done to mitigate the weakened resolve that geoengineering can produce?

As noted above, the Rami SM study suggests, consistent with Lin's arguments, that framing geoengineering as a partial, temporary, risky response to climate change is one path to limiting the potential of geoengineering to undermine climate change mitigation and adaptation. On first blush, it would seem easy to ensure that geoengineering in all relevant discourses is so framed: after all, the scientific and policy literature does not contain much in the way of argument that geoengineering is a total, easy, low-risk, permanent solution to climate change.[29] Rather, to the extent that commentators advocate for research into geoengineering or posit it is something worth exploring, especially with respect to solar radiation management, they generally depict it as a possible, temporary, risky measure to be used, if ever or at all, only while mitigation efforts are strengthened and the economy is decarbonized.[30]

[28] I use the term precautionary principle in a "thin" sense, simply to mean a principle that requires taking serious account of nonquantifiable, uncertain risks as part of decisionmaking. Getting the framing right for geoengineering also could be supported by a sophisticated cost-benefit analysis without explicit invocation of the precautionary principle. *See* Noah M. Sachs, *Rescuing the Strong Precautionary Principle From Its Critics* 2011 U ILL L REV. 1285 (discussing different versions of the principle).

[29] For example, David Keith, one of the most prominent scientists advocating for and engaging in geoengineering research, speaks of geoengineering in very measured, sober terms. *See, e.g.*, David Keith, Guardian, March 29, 2017, Fear of solar geoengineering is healthy – but don't distort our research ("Fear of solar geoengineering is entirely healthy. Its mere prospect might be hyped by fossil fuel interests to thwart emissions cuts. It could be used by one or a few nations in a way that's harmful to many. There might be some yet undiscovered risk making the technology much less effective in reality than the largely positive story told by computer models.").

[30] Jonas Anshelm and Anders Hansson, *Has the grand idea of geoengineering as Plan B run out of steam?*, ANTHROPOCENE REVIEW 64, 68 (2016).

"We. .. illustrat[e] the gradual but significant move towards more modest and critical descriptions of geoengineering as a climate control measure, away from emergency framings and notions of geoengineering as 'Plan B', and instead towards lowered expectations and ambiguity of the very notion of geoengineering.. .. Few now publicly question the position that geoengineering ought not be understood as a substitute for emissions reductions or an emergency option, and accordingly does not constitute 'Plan B'. Few actors seem to oppose the mass media's mainly negative interpretation of the Climate Intervention reports' treatment of CDR and SRM. Instead, Simon Nicholson of American University states that 'the idea advanced by the Royal Society that albedo modification is some kind of 'Plan B' has largely fallen out of favour'. . ."

However, it is easy to imagine how a more upbeat framing of geoengineering as a permanent, easy solution could gain traction in public discourse. Support for that framing could come from nations and business interests with an interest in continuing to promote fossil fuels (Russia, the Gulf states, developing countries that fear mitigation will slow their growth, the fossil fuel multinationals), political conservatives who see drastic mitigation as a threat to liberty as they perceive it, and individuals, corporations and even nations that think they might make a profit or gain advantage directly from geoengineering (for example, national leadership that believes geoengineering could be used to "improve" weather for agriculture and other uses, investors in geoengineering technology development). As the Union of Concerned Scientists has documented, there is an array of "think tanks" that have long worked to foster climate change skepticism,[31] and there is no reason they could not re-focus their efforts to promote geoengineering as a cheap, comprehensive way to avoid mitigation. Indeed, according to anecdotal reports, some conservative politicians and think tanks are already speaking of geoengineering in those terms.[32]

Once geoengineering is actively "sold" to the public, caveats regarding it may fade and the gleaming-success rhetoric of marketing may rule the day. As Clive Hamilton notes, modern culture is built on and has faith in "technological manipulation" of the natural world, and geoengineering can be pitched as another in a long line of technological solutions to the problem people face.[33]

[31] See Union of Concerned Scientists, Global Warming Skeptic Organizations (2013), August 2013, available at https://www.ucsusa.org/global-warming/solutions/fight-misinformation/global-warming-skeptic.html

[32] See, e.g. John Siciliano and Josh Siegel, Daily on Energy: Did Lamar Smith just outline a Republican climate plan?, Nov 8, 2017, http://www.washingtonexaminer.com/daily-on-energy-did-lamar-smith-just-outline-a-republican-climate-plan/article/2176978' (describing a House Hearing on Geoengineering convened by House Science, Space & Technology Committee Chairman Lamar Smit, a "noted skeptic of climate change"; S. Fred Singer, https://www.heartland.org/news-opinion/news/saving-humanity-from-catastrophic-global-cooling-a-task-for-geo-engineering?source=policybot ("While the science is certainly interesting and important, there is no need to delay the crucial and urgent tests of geo-engineering; they involve only minor costs and little risk to the atmospheric environment."); Eric Bickel & Lee Lane, *An Analysis of Climate Engineering as a Response to Climate Change,* https://www.heartland.org/_template-assets/documents/publications/Copenhagen%20Consensus%20geoengineering.pdf ("we believe it makes a strong case that the potential net benefits of SRM are large"); *Amy Goodman, Democracy Now, referencing reporting from Naomi Klein, Sept. 18, 2014 (*"I mean, you have the Heartland Institute describing geoengineering as, quote, "much less expensive than seeking to stem temperature rise solely through the reduction of greenhouse gas emissions"; Cato Institute arguing "geo-engineering is more cost-effective than emissions controls altogether"; Hudson Institute saying that geoengineering, quote, "could obviate the majority of the need for carbon cuts and enable us to avoid lifestyle changes." The very point you're making."), available at https://www.democracynow.org/2014/9/18/naomi_klein_on_motherhood_geoengineering_climate

[33] *See* Clive Hamilton, The Philosophy of Geoengineering ("the thinking that gives rise to geoengineering is the same thinking that first creates the world as an object suitable for technological manipulation. As a result, the only global warming escape routes that occur to us are technological ones, whether they be new forms of low-emission energy, carbon capture and storage or engineering the climate. So this view prompts the rhetorical question: How can we think our way

One question is whether more or less – and what kinds of – research into geoengineering will best help ensure that the more realistic, limited/temporary/risky framing remains dominant and is not supplanted or partially supplanted by an unnuanced, total/permanent/riskless framing. One could argue that the more research is done into geoengineering technology, the more plausible it will seem and the more temptation and ability there will be for certain interested parties to promote it as an easy, quick solution. In this view, once research programmes fully legitimize geoengineering and make it appear to be more readily deployable, it will be too late to set geoengineering aside as an unacceptable option; instead, Plan B may become Plan A.[34]

On the other hand, extensive research, if it is well-done and subject to peer review and focused not only on possible benefits of various geoengineering technologies, but also possible limits and risks,[35] could be needed to counter inflated claims on behalf of geoengineering. In the absence of high-quality research regarding geoengineering by a diverse range of scientists (including not just physicists but ecologists and biologists, as well as social scientists such as political scientists, economists and national security scholars), the public discourse may be too easily captured by those with an economic, ideological or other stake in promoting geoengineering as a total solution to climate change. Severely restricting funding of extensive research into geoengineering at mainstream venues (such as leading research universities and government agencies) may not prevent weakened resolve by keeping geoengineering an obscure and unexplored idea, as some might suppose; instead, research restrictions may make it easier for geoengineering to be developed with private or "solo actor" state research financing and sold to the public by those with an interest in doing such selling. And that, in turn, may result in less mitigation and adaptation that otherwise would have been achieved.[36]

We know that people engage in motivated reasoning in the selection of experts they credit as trustworthy. Political and cultural conservatives who object to mitigation policies, therefore, may well dismiss high-quality science illuminating the risks posed by a geoengineering.[37] But that does not mean all conservatives are beyond

out of a problem when the problem is the way we think?), available at http://clivehamilton.com/philosophy-of-geoengineering/

[34] *See* Clive Hamilton, *The Risks of Climate Engineering*, NY Times, Feb. 12, 2015 ("President Obama has been working assiduously to persuade the world that the United States is at last serious about Plan A — winding back its greenhouse gas emissions. The suspicions of much of the world would be reignited if the United States were the first major power to invest heavily in Plan B.").

[35] *See* Lin, supra note 3, at 709 (arguing that in addition, a portion of any funding for geoengineering research and development should be directed toward public outreach).

[36] David Morrow offers some helpful suggestions for scientists that may help mitigate the weakening of resolve, including an assessment of a broad range of technologies and scenarios;, messaging that highlights the limits of each technology; and active engagement with policy and policymakers. DR Morrow, *Ethical aspects of the mitigation obstruction argument against climate engineering research*, 372 Phil. Trans. R. Soc'y. 20,140,062 (2014), at 11-12.

[37] *See* Dan M. Kahan et al., *Cultural cognition of scientific consensus*, 14. J. Risk Research, Issue 2 (2011) ("But because the source of the enfeebled power of scientific opinion is different from

persuasion, and there are many people in the middle of the political or cultural spectrum. A well-developed body of science about geoengineering, directly addressing all its risks and limits, may persuade such people in the middle (and by extension, institutions) who are not strongly motivated to deny or minimize climate risks as an *apriori* matter. And persuading the persuadable may be enough to determine policy (at least if the persuadable includes those who are capable of influencing policy) .

The identity of the messengers also may be important as to whether geoengineering's risks and limits can be effectively communicated to polarized audiences. When we hear a messenger whom we regard as embodying our values, we are more likely to take the message to heart, even if the message is something we otherwise may not want to hear. Nixon could go to China and establish diplomatic relations there without public or military elite uproar because Nixon was identified as a foreign policy hawk.[38] Thus, a decorated general, a leading evangelical Christian pastor, a CEO of major public corporation or a self-made, billionaire technology mogul may be able to persuade political and cultural conservatives of the validity of framing geoengineering as a limited, necessarily temporary, and risky possible response to climate change. Finding proponents of this framing, especially for all the relevant national audiences, may well be a daunting task. But persuading key individual "thought leaders," presumably on a one-on one basis, using strong science as support, may be an important objective if, as it seems, the question of geoengineering is one that will not be obviated by decisively strong global mitigation efforts in the coming years.

5 Conclusion

This Chapter offers several arguments regarding geoengineering's relationship to mitigation and adaptation. First, compared to the terms moral hazard or risk compensation, weakened resolve better captures the concern regarding the possible adverse effect of geoengineering on mitigation and adaptation. Second, the inquiry into weakened resolve needs to be framed around who actually will determine

what is normally thought, the treatment must be something other than what is normally prescribed. It is not enough to assure that scientifically sound information – including evidence of what scientists them-selves believe – is widely disseminated: cultural cognition strongly motivates individuals – of all worldviews – to recognize such information as sound in a selective pattern that reinforces their cultural predispositions.")

[38] For examples of climate change framing that may transcend cultural divides, see https://www.huffingtonpost.com/entry/evangelical-climate-scientist-explains-why-christians-should-care-about-the-environment_us_586eadfee4b099cdb0fc3e5f; http://www.climatesciencewatch.org/2011/01/19/the-national-security-frame-a-path-forward-for-climate-change-communication/. For a discussion of the power of traditional actors adopting an innovative position, see Forrest Briscoe and Sean Safford, The Nixon-in-China Effect: Activism, Imitation, and the Institutionalization of Contentious Practices, 53 (3) Admin. Sci. Q. 460-491 (2008). https://doi.org/10.2189/asqu.53.3.460

mitigation and adaptation policy. Third, the current literature on weakened resolve would benefit from a broader geographical scope (to include major emitters such as China), the attitudes of elites as opposed to the general population, and the interaction between cultural and political polarization and weakened resolve. Fourth, because there is almost certainly some potential for weakened resolve even if we cannot *ex ante* establish how much, efforts should be made to mitigate any weakened resolve effect by framing geoengineering as a temporary, limited, risky response to climate change. Successfully maintaining that framing in public discourse may best be served by high-quality, peer-reviewed research regarding geoengineering technologies, with clear attention to their limits and risks. Restrictions on research at mainstream scientific venues itself may pose the risk that there will be a dearth of high-quality science to rebut efforts to rebut a framing of geoengineering as an easy, quick, low-risk solution to the problem of climate change.

References

1. Corry, O.: The international politics of geoengineering: The feasibility of Plan B for tackling climate change. Sec. Dialogue. **48**(4), 297–299 (2017)
2. Kahneman, D., Tversky, A.: Prospect Theory: An Analysis of Decision Under Risk. Econometrica. **47**, 263–265 (1979)
3. McLaren, D.: Mitigation deterrence and the "moral hazard" of solar radiation management. Earth's Future. **4**, 596–599 (2016)

Using Renewable Energy Policies to Develop Carbon Dioxide Removal

Anthony E. Chavez

Most analyses project that we must utilize carbon dioxide removal (CDR)) to avoid dangerous warming. Unfortunately, CDR is not ready for large-scale deployment. Two policies that successfully accelerated the development of renewable energy – Renewable Portfolio Standards (RPSs) and Feed-in Tariffs (FITs) – could achieve similar results with CDR. In fact, a combination of both policies might best incentivize the development and implementation of CDR.

1 The Need to Develop Carbon Dioxide Removal Technologies

Despite recent efforts to reduce carbon dioxide emissions, scientists still project that we will not avoid dangerous climate change. Models that calculate that we can avoid this result almost exclusively rely upon carbon dioxide removal options to stay below this level of warming. Although a number of CDR technologies are theoretically possible, they all have limitations. More germane here, they all remain far from the level of development and installation required.

The parties to the 2015 Paris Agreement agreed to aim to hold the rise in warming to "well below 2.0°C."[1] They further agreed to pursue efforts to hold warming to 1.5 °C.[2] The Paris Agreements and earlier global pacts have targeted a rise of

[1]Adoption of the Paris Agreement, UNFCC Conference of the Parties, 21st Sess., U.N. Doc. FCCC/CP/2015/10/Add.1 (Dec. 12, 2015), at art. 2(1)(a)http://unfccc.int/files/home/application/pdf/paris_agreement.pdf [hereinafter Paris Agreement].

[2]*Id.*

A. E. Chavez (✉)
Chase College of Law, Northern Kentucky University, Highland Heights, KY, USA
e-mail: chaveza1@nku.edu

© Springer Nature Switzerland AG 2021

W. Burns et al. (eds.), *Climate Geoengineering: Science, Law and Governance*, AESS Interdisciplinary Environmental Studies and Sciences Series,
https://doi.org/10.1007/978-3-030-72372-9_8

2.0 °C as the level to avoid because at that level "dangerous anthropogenic interference with the climate system" will be unavoidable.[3] Recent analyses indicate that even warming to the 1.5 °C level will cause serious regional consequences, such as extreme temperature warming, heavy precipitation, and droughts.[4]

Unfortunately, avoiding temperature rises of this magnitude are becoming increasingly unlikely. The Intergovernmental Panel on Climate Change (IPCC) concluded that we can emit only an additional 1000 Gt of CO_2 between 2011 and 2100 while retaining a 66% chance of keeping warming under 2 °C.[5] With annual emissions approximating 40Gt of CO_2 annually, society already emitted one-fifth of this amount in just 5 years.[6] Thus, under a business as usual scenario, we will use up the 2 °C carbon budget by as soon as 2038.

Consequently, integrated assessment models developed by the IPCC in its Fifth Assessment Report revealed that deployment of CDR technologies are likely a critical component for avoiding the 2 °C level at the end of the century. The IPCC noted that 166 of 900 integrated assessment models yielded a 66% chance of warming not exceeding the 2 °C level in 2100. 101 of these models required CDR to achieve this result.[7] In fact, they typically necessitated CDR on a "massive" scale.[8]

Although 2100 is still many decades away, efforts to develop, test, and deploy CDR – at scale – must commence shortly. The IPCC models indicate that keeping warming below 1.5 °C will require large-scale deployment of CDR within 10–20 years.[9] Even some projections to hold warming to 2.0 °C will necessitate CDR deployment to begin as soon as the 2020's.[10]

---

[3] Lena R. Boysen et al., *The Limits to Global-Warming Mitigation by Terrestrial Carbon Removal*, 5 EARTH'S FUTURE, MAY 17, 2017, 463, 463–474.

[4] V. Delmotte et al., GLOBAL WARMING OF 1.5 °C SPM 8 (2018).

[5] EUROPEAN ACADAMIES SCIENCE ADVISORY COMMITTEE, *Negative Emission Technologies: What Role in Meeting in Paris Agreement Targets?*, 35 EASAC POL'Y REP. 1, 4 (2018)

[6] *Id.* at 5.

[7] Christopher B. Field & Katharine J. Mach, *Rightsizing Carbon Dioxide Removal*, 356 SCIENCE, 706, 707 (May 19, 2017).

[8] Guy Lomax et al., *Investing in Negative Emissions*, 5 NATURE CLIMATE CHANGE, 498 (2015).

[9] R. Stuart Haszeldine et al., *Negative Emissions Technologies and Carbon Capture and Storage to Achieve the Paris Agreement Commitments*, 376 PHIL. TRANS. R. SOC. A 19–20 (Oct. 28, 2018).

[10] Matthew D. Eisaman, *Indirect Ocean Capture of Atmospheric CO2: Part II. Understanding the Cost of Negative Emissions*, 1 INTERNATIONAL JOURNAL OF GREENHOUSE GAS CONTROL (2018).

2 Many Technologies; None Ready

Carbon dioxide removal is an umbrella term that refers to an array of technologies that can effectuate removal of carbon dioxide from the air. A number of limitations, however, are likely to prevent any single technology from providing a "magic bullet" solution, necessitating that a suite of technologies be developed and deployed.

CDR technologies remove CO_2 from the atmosphere and sequester it underground permanently.[11] These technologies can be divided into two broad categories. The first involves methods that augment natural processes.[12] The second utilizes technological means to capture and bury the carbon dioxide.[13]

Although research on carbon dioxide removal technologies is constantly evolving, the most promising methods fall within the following eight categories:

- *Afforestation and reforestation* – afforestation involves the restoration of forests on lands deforested for at least 50 years; reforestation restores forests on lands more recently deforested.[14] The ability of forestation to remove CO_2 from the atmosphere depends upon a number of factors, including the availability of sufficient land, nutrients,[15] and water[16]; type and age of the trees[17]; and precipitation and CO_2 levels.[18] Possible sequestration from these activities could range from 1.5 to 14 $GtCO_2$ (billion tons of carbon dioxide) per year by 2030.[19] Although scientists calculate the annual costs of these activities to range from as low as \$7.50 per tCO_2 to as high as \$100 per tCO_2, most estimates do not exceed \$40 per tCO_2.[20]

[11] NATIONAL RESEARCH COUNCIL (NRC), CLIMATE INTERVENTION: CARBON DIOXIDE REMOVAL AND RELIABLE SEQUESTRATION 33 (2015). Carbon capture and utilization (CCU) systems, on the other hand, apply the captured CO_2 to a number of processes, including enhanced oil recovery, mineral carbonation, food and beverage carbonation, polymer processing, microalgae production, and enhanced coal bed methane recovery. Jennifer Wilcox, Peter C Psarras & Simona Liguori, Assessment of Reasonable Opportunities for Direct Air Capture, 12 ENVT'L. RES. LETTERS 1, 2 (2017).

[12] *Id.*

[13] *Id.*

[14] UN ENVT'L PROGRAMME (UNEP), *The Emissions Gap Report 2017: A UN Environment Synthesis Report*, at 60 (2017). These processes are necessitated by deforestation, which causes approximately 10% of anthropogenic greenhouse gas emissions. NRC, *supra* note 15 at 39.

[15] EASAC, *supra* note 6 at 17.

[16] Duncan McLaren, *Negatonnes—An Initial Assessment of the Potential For Negative Emission Techniques to Contribute Safely and Fairly to Meeting Carbon Budgets in the 21st Century*, 1 FRIENDS OF THE EARTH 1, 20 (2011).

[17] In general, net CO_2 removal peaks within 30–40 years, and then it declines to zero as the forest matures. NRC, *supra* note 15 at 40.

[18] *Id.*

[19] *Id.*

[20] McLaren, *supra* note 19 at 20; NRC, *supra* note 15 at 41–42.

- *Biochar* stores stable biomass in soil. Pyrolysis – heating vegetation slowly without using oxygen – combusts lumber waste, crop residues and other biomass to form biochar.[21] Because biochar resists decomposition, it stabilizes biomass buried in soil.[22] Biochar constitutes a negative emissions technology because it fixes atmospheric CO_2 in a stable form that can be easily sequestered.[23] Additionally, biochar can provide several co-benefits. These include increasing soil fertility and improving water and nutrient retention.[24] Scientists project that biochar can sequester as much as 1 $GtCO_2$ per year by 2030, and possibly up to 9.5 $GtCO_2$, by 2100.[25] Annual costs of biochar range from $18 to $166 per tCO_2.[26]
- *Land management* – soils serve as a carbon sink, yet cultivated soils have lost 50–70% of their carbon.[27] Soil tends to lose carbon through oxidation, such as when it is plowed.[28] All told, agricultural practices alone have released 10–12% of anthropogenic greenhouse gases.[29] Appropriate land management practices can increase soil carbon inputs and reduce soil carbon losses.[30] These practices include growing cover crops, leaving crop residues to decay, applying manure or compost, and using low- and no-till methods.[31] Possible sequestration from agricultural land management practices may be as high as 5.2 $GtCO_2$ per year.[32] Some of these practices (such as no-till) may already be cost-competitive with traditional practices. Anticipated annual costs range from $20 to $100 per tCO_2.[33]
- *Enhanced weathering* – as part of the natural carbon cycle, the weathering of certain silicate minerals reacting with CO_2 traps the CO_2 into carbonates.[34] The natural weathering process will remove atmospheric carbon, but it will require 100,000 years to return the climate to its preindustrial level.[35] Accelerated weathering augments the natural weathering process. It involves mining and grinding

[21] EASAC, *supra* note 6 at 18.

[22] UNEP, *supra* note 17 at 62.

[23] Niall McGlashan et al., *High-Level Techno-Economic Assessment of Negative Emissions Technologies*, 90 PROCESS SAFETY & ENVT'L. PROTECTION 501–10, 503 (2012).

[24] UNEP, *supra* note 17 at 62.

[25] McGlashan, *supra* note 26 at 503.

[26] UNEP, *supra* note 17 at 62.

[27] EASAC, *supra* note 6 at 18.

[28] McLaren, *supra* note 19 at 21.

[29] Stefan Frank et al., *Reducing Greenhouse Gas Emissions in Agriculture Without Compromising Food Security?*, 12 ENVT'L. RES. LETTERS 1, 2 (2017).

[30] UNEP, *supra* note 17 at 61.

[31] EASAC, *supra* note 6 at 18.

[32] *Id.* at 44.

[33] Peter Psarras et al., *Slicing the Pie: How Big Could Carbon Dioxide Removal Be?*, 6 WIRES ENERGY ENV'T 1, 1 (2017).

[34] EASAC, *supra* note 6 at 23.

[35] Jeremy Deaton, *Earth's "Weathering Thermostat" Keeps Climate in Check Over Very Long Periods of Time*, CLEANTECHNICA (Sept. 18, 2017), https://cleantechnica.com/2017/09/18/earths-weathering-thermostat-keeps-climate-check-long-periods-time/

particular minerals to small grain sizes to increase their surface area exposed for weathering.[36] While this method might be relatively inexpensive, ranging from $50 to $100 per tCO_2 annually,[37] it likely can sequester only 0.7–3.7 $GtCO_2$ per year.[38]

- *Ocean alkalinity enhancement* adds alkaline materials to the ocean to increase the amount of carbon the ocean absorbs.[39] Ocean alkalinity enhancement accelerates ocean carbon uptake and at the same time reverses ocean acidification.[40] Annual costs range from $30 to $60 per tCO_2.[41] If operated at the appropriate scale, this method could sequester sufficient carbon to return the atmosphere to its pre-industrial state.[42]
- *Ocean fertilization* deposits nutrients, such as iron, nitrogen or phosphorous, into the ocean to stimulate the growth of phytoplankton, which consume CO_2.[43] Scientists project that ocean fertilization could remove up to 3.7 $GtCO_2$ per year.[44] The costs for fertilization range broadly. Annual costs may be as low as $10–35 per tCO_2; however, if remineralization through respiration occurs, thereby reducing the efficiency of the carbon sequestration, the costs could rise to $450 tCO_2.[45]
- *Bioenergy and carbon capture with sequestration* (*BECCS*) combines carbon capture and sequestration technology with biomass burning in power plants. Since biomass burning is in theory carbon neutral, and in practice low carbon, the capture and sequestration of the system's emissions results in net negative emissions.[46] A critical advantage of BECCS as a carbon dioxide removal technology is that it also produces a salable product, electricity.[47] BECCS could

[36] Jessica Strefler et al., *Potential and Costs of Carbon Dioxide Removal by Enhanced Weathering of Rocks*, 13 ENV'T. RES. LETTERS 1, 1–2 (2018).

[37] EASAC, *supra* note 6 at 23.

[38] UNEP, *supra* note 17 at 64.

[39] *Id.*

[40] Andrew Lenton, *Assessing Carbon Dioxide Removal through Global and Regional Ocean Alkalinization under High and Low Emission Pathways*, 9 EARTH SYS. DYNAMICS 339–357, 340 (2018).

[41] McLaren, *supra* note 19 at 18.

[42] T. Kruger, *Increasing the Alkalinity of the Ocean to Enhance its Capacity to Act as a Carbon Sink and to Counteract the Effect of Ocean Acidification*, in GeoConvention, 4 (2010), http://www.searchanddiscovery.com/abstracts/pdf/2014/90172cspg/abstracts/ndx_krug.pdf [http://perma.cc/7CPX-CRMN]

[43] EASAC, *supra* note 6 at 27.

[44] NRC, *supra* note 14 at 61.

[45] EASAC, *supra* note 6 at 27.

[46] McLaren, *supra* note 19 at 17.

[47] McGlashan, *supra* note 26 at 504.

sequester between 2 and 18 $GtCO_2$ per year.[48] Most studies project BECCS annually to cost $50–100 per tCO_2.[49]

- *Direct air capture and carbon sequestration* (*DACCS*) separates CO_2 from the ambient air, processes it, and then buries it.[50] Because of the low concentration of carbon in the ambient air when compared to its presence in the emissions of a fossil fuel- or biomass-burning plant, DACCS systems require 2–10 times more energy to capture CO_2 than do power plants using CCS.[51] For this reason, DACCS plants will need to operate with renewable energy to assure that they produce net negative emissions.[52] DACCS has the technical potential to sequester as much as 20 $GtCO_2$ annually, but actual sequestration is most likely to range from 2 to 5 $GtCO_2$ per year.[53] An extensive review of DACCS projections concludes that the annual costs for sequestration may range from $200 per tCO_2 captured to as high as $1000 tCO_2.[54]

Several considerations regarding these technologies are important. First, no single method, except possibly ocean alkalinization, is likely to be able to sequester all of the carbon dioxide necessary to restore the climate to preindustrial levels. Currently, global CO_2 emissions approximate 39 $GtCO_2$ per year.[55]

Second, every method has constraints that limit the amount of CO_2 that it can sequester. For instance, BECCS would compete for limited resources with food production and could have serious impacts on biodiversity.[56] BECCS operated at scale could require 50% of global fertilizer and more than double the current global water withdrawals for irrigation.[57] In fact, several CDR approaches may compete with one another. BECCS, afforestation, reforestation, DACCS, and enhanced weathering all may draw upon the same land and water resources.[58] Methods that rely upon reactions with minerals – such as weathering and alkalinization – may

---

[48] UNEP, *supra* note 14 at 62. *See also* Elmar Kriegler et al., *Is Atmospheric Carbon Dioxide Removal a Gamer Changer for Climate Change Mitigation?*, 118 CLIMATIC CHANGE 45–57, 55 (May, 2013) (projecting BECCS deployment limited to a removal of 14–15 $GtCO_2$ per year).

[49] Matthias Honegger & David Reiner, *The Political Economy of Negative Emissions Technologies: Consequences for International Policy Design*, 18 CLIMATE POL'Y 306, 308–309 (2017).

[50] S. Fuss et al., *Research Priorities for Negative Emissions*, 11 ENV'T. RES. LETTERS 1, 3 (2016).

[51] NRC, *supra* note 14 at 68.

[52] UNEP, *supra* note 17 at 63–64.

[53] *Id.* at 64.

[54] EASAC, *supra* note 6 at 26.

[55] Secretary of Energy Advisory Board CO_2 Utilization Task Force (SEAB), *Task Force on RD&D Strategy for CO_2 Utilization and/or Negative Emissions at the Gigatonne Scale*, LETTER REPORT 37 (Dec. 12, 2016).

[56] Elmar Kriegler et al., *Is Atmospheric Carbon Dioxide Removal a Gamer Changer for Climate Change Mitigation?*, 118 CLIMATIC CHANGE 45–57, 55 (May, 2013).

[57] NRC, *supra* note 14 at 65.

[58] EASAC, *supra* note 6 at 12–13.

confront limitations deriving from the quantity of minerals that must be extracted, processed, and transported.[59]

Third, several of the CDR technologies will require decades of development before they will be ready for large-scale deployment. Many CDR technologies are little more than concepts and operate only as pilot projects.[60] Putting aside the technological aspects of CDR, the land preparations alone for sequestration under BECCS and DACCS will require decades. For instance, 5–10 years may be necessary to identify appropriate subsurface sites, to fully evaluate them, and then to complete permitting for sequestration operations.[61]

Fourth, few CDR methods, if any, are ready to be deployed at scale. BECCS, for example, is the most advanced of the CDR technologies. Nevertheless, current BECCS operations consist of only 15 pilot plants and one commercial plant.[62] Based upon present deployment rates for CCS projects, they will be able to capture only 700 $MtCO_2$ per year by mid-century. IPCC 2 °C models, however, require that they sequester 6000 $MtCO_2$ by then.[63] Similarly, emissions scenarios anticipate that several thousand DACCS plants will be operating by 2030; planned construction, however, only numbers in the tens.[64] Regarding biochar, while the theory and implementation is well established, deploying biochar at the necessary scale would require an increase of over 63 times the current charcoal production capacity.[65]

Fifth, timely development of CDR technologies is also important to develop the legal and accounting rules necessary to foster and regulate them. Liability issues especially need to be addressed for methods that utilize mechanical sequestration of carbon. Liability concerns can arise during operations and also after injection of the carbon dioxide.[66] There are special concerns regarding the effects that injection might have on the water supply.[67] Even development of forestation and land

---

[59] McLaren, *supra* note 19 at 17.

[60] Haszeldine, *et al*, *supra* note 12 at 11.

[61] *Id.* at 19.

[62] Wil Burns & Simon Nicholson, *Bioenergy and Carbon Capture with Storage (BECCS): the Prospects and Challenges of an Emerging Climate Policy* Response, 7 J. ENVT'L. STUD. & SCI. 527, 529 (2017). Even though the BECCS technology is relatively advanced, questions remain concerning the extent that its carbon sequestration will sufficiently offset emissions from direct and indirect land use changes. Naomi E. Vaughan & Clair Gough, *Expert Assessment Concludes Negative Emissions Scenarios May Not Deliver*, 11 ENVT'L. RES. LETT. 5 (2016).

[63] Haszeldine, *et al*, *supra* note 12 at 2.

[64] Glen P. Peters et al., *Key Indicators to Track Current Progress and Future Ambition of the Paris Agreement*, 121 NATURE CLIMATE CHANGE 1, 4 (2017).

[65] Niall R. McGashan et al., *Negative Emissions Technologies*, 8 Grantham Institute for Climate Change Briefing Paper, October, 2012, at 15.

[66] Mark de Figueiredo, et al., *The Liability of Carbon Dioxide Storage* 1 (undated).

[67] Ian Havercroft & Richard Macrory, *Legal Liability and Carbon Capture and Storage: A Comparative Perspective* 11 (Global Carbon Capture and Storage Institute/UCL Laws 2014), hub.globalccsinstitute.com/sites/default/files/publications/179798/legal-liability-carbon-capture-storage-comparative-perspective.pdf [http://perma.cc/NH74-H3XB]

management methods will require establishment of uniform standards for measuring carbon sequestration.[68]

Finally, the environmental impact of CDR technologies remains uncertain. A number of impacts may occur. Afforestation and reforestation may disrupt hydrological cycles, ecosystems, and biodiversity.[69] Energy crops used for BECCS will compete with food crops for land and water and necessitate conversion of habitat critical for biodiversity.[70] Carbon injection may increase seismicity.[71] Ocean fertilization may disrupt the ecology of the oceans,[72] and ocean alkalinity enhancement may have localized effects and detrimental impacts on ocean ecosystems.[73] Thus, whether deployed separately or in combination, CDR may entail significant impacts on land, water, or ecosystems.[74]

Thus, we can anticipate that we will need to utilize CDR technologies, yet they are both substantially underdeveloped and not fully understood. We need to institute policies that will encourage CDR's development and deployment.

3 RPSs and FITs Can Stimulate New Technologies

To stimulate the development and installation of CDR technologies, we should consider utilizing policies that led to the growth of renewable energy. Two policies have been instrumental to renewable energy development: Feed-in Tariffs (FITs) and Renewable Portfolio Standards (RPSs).

68 CTR. FOR CARBON REMOVAL, CARBON REMOVAL POLICY: OPPORTUNITIES FOR FEDERAL ACTION 8 (2017). https://static1.squarespace.com/static/5b9362d89d5abb8c51d474f8/t/5b 9427cd8a922dd0d7451136/1536436200923/Carbon%2BRemoval%2BPolicy%2BOpportunities%2Bfor%2BAction + %281%29.pdf [http://perma.cc/92YX-UQYN]

69 David P. Keller et al., *The Carbon Dioxide Removal Model Intercomparison Project (CDR-MIP): Rationale and Experimental Design*, 11 GEOSCIENTIFIC MODEL DEV. 1133, 1133–34 (2018).

70 EASAC, *supra* note 6 at 22.

71 NRC, *supra* note 14 at 111.

72 *Id.* at 61.

73 Phil Renforth & Gideon Henderson, *Assessing Ocean Alkalinity for Carbon Sequestration*, REVIEWS OF GEOPHYSICS 666 (2017).

74 Pete Smith, *Biophysical and Economic Limits to Negative CO_2 Emissions*, 6 NATURE CLIMATE CHANGE 42, 49 (2016).

3.1 RPSs Successfully Promoted Development of Renewable Energy

RPSs have been instrumental in fostering renewable energy growth, particularly in the United States. These policies set minimum requirements for technology adoption, which assure that these jurisdictions follow a pattern of continuous growth of the utilization of these technologies. Furthermore, RPSs utilize market mechanisms to satisfy their mandates, which favors reliance upon the lowest-cost solutions. This not only keeps compliance costs down, it also incentivizes innovation.

Iowa established the first RPS in 1983.[75] Starting in the mid-1990's, other states began to adopt their own RPSs.[76] Currently, 29 states and the District of Columbia have enacted RPSs.[77] A number of nations also use RPSs, including Australia, Canada, China, Japan, and the United Kingdom.[78]

RPSs mandate that electricity providers include minimum levels, or quotas, of renewable energy in their electricity mix.[79] These policies often do not favor particular energy sources, though they can include provisions to support specific technologies.[80]

RPSs impose multiple criteria with which electricity suppliers must comply. Most importantly, they set a minimum percent of electricity required to be generated from renewable sources and a timeline for compliance.[81] These requirements are usually set at low levels initially, increasing steadily over a period ranging from 10 to 20 years, and then remain at that level, unless the RPS is subsequently amended.[82] RPSs also specify the electricity sources that satisfy the mandate. Finally, they

[75] Michael T. Ferguson, *Green America: Renewable Standards, Tax Credits, and What's Next*, S&P GLOBAL 1 (Oct. 10, 2017), https://www.spglobal.com/en/research-insights/articles/green-america-renewable-standards-tax-credits-and-whats-next

[76] Galen Barbose, U.S. RENEWABLES PORTFOLIO STANDARDS: 2017 ANNUAL STATUS REPORT 6 (2017).

[77] *Id.* at 8.

[78] Ryan Wiser, Galen Barbose, & Edward Holt, *Supporting Solar Power in Renewables Portfolio Standards: Experience from the United States*, 39 ENERGY POLICY (2011) 3894–3905, 3894; Greg Buckman, *The Effectiveness of Renewable Portfolio Standard Banding and Carve-Outs in Supporting High-Cost Types of Renewable Electricity*, 39 ENERGY POLICY (2011)4105–4114, 4106.

[79] Qi Zhang, *Substitution Effect of Renewable Portfolio Standards and Renewable Energy Certificate Trading for Feed-In Tariff*, 227 APPLIED ENERGY 426, 426–427 (2017).

[80] Felix Mormann, *Constitutional Challenges and Regulatory Opportunities for State Climate Policy Innovation*, 41 HARV. ENVT'L. L. REV. 189, 198 (2017).

[81] Corey N. Allen, *Untapped Renewable Energy Potential: Lessons For Reforming Virginia's Renewable Energy Portfolio Standard From Texas And California*, 35 VA. ENVT'L. L. J. 117, 120 (2016).

[82] Environmental Protection Agency (EPA), Energy and Environment Guide to Action 5–10 (2015).

specify whether a regulated entity can satisfy its obligation through the purchase of renewable energy credits (RECs).[83]

This range of criteria included in a state's RPS results in each RPS being uniquely tailored to its jurisdiction.[84] States have tailored their RPS policies to fit each state's objectives, energy resources potential, and electricity market characteristics.[85] Besides the criteria described above, RPSs may also vary regarding specifics of qualifying electricity resources and technologies (vintage, location and deliverability), mechanisms used to favor particular resources, and specifics of RECs systems.[86]

Because RPSs have numerous design elements,[87] they are very flexible policy tools.[88] Legislatures typically tailor RPS elements to the circumstances and needs of each particular state.[89] This flexibility has allowed states with diverse renewable energy potentials to adopt successful RPSs.[90] Indeed, states have crafted RPS programs that are distinct over the range of criteria incorporated in RPSs.[91] Not surprisingly, no two states have identical RPSs.[92]

The flexibility of RPSs enables jurisdictions to revise and strengthen them regularly.[93] For instance, from 2015 to 2017, the legislatures in the 29 states with RPSs passed more than 200 RPS-related bills.[94] RPS flexibility has enabled states to learn from experience and modify their policies accordingly. They have updated RPS timetables, percentages, technologies, incentives, durations, and other provisions.[95] These changes have responded to the achievement of program goals, approaching target dates, changing market conditions, and other considerations.[96] Importantly, because of these modifications, the average RPS obligation rose from 7.6% of total

---

[83] Allen, *supra* note 84 at 120. RPSs also identify the entities required to comply, designate an administrator – usually a government agency – and specify their enforcement mechanisms. *Id.*

[84] GOVERNORS' WIND ENERGY COAL., RENEWABLE ELECTRICITY STANDARDS: STATE SUCCESS STORIES 9 (2013).

[85] EPA, *supra* note 85 at 5–2.

[86] Barbose, *supra* note 79 at 7.

[87] Governors' Wind Energy Coal., *supra* note 87 at 9.

[88] Luke J.L. Eastin, *An Assessment of the Effectiveness of Renewable Portfolio Standards in the United States*, THE ELECTRICITY J. 127 (2014).

[89] Warren Leon, THE STATE OF STATE RENEWABLE PORTFOLIO STANDARDS 10 (June 2013).

[90] Vicki Arroyo, *et al*, *State Innovation on Climate Change: Reducing Emissions from Key Sectors While Preparing for a "New Normal,"* 10 HARV. L. & POL'Y REV. 385, 398 (2016).

[91] Allyson Browne, *RPS Evolving: States Take on U.S. Climate Goals*, 31 NAT. RESOURCES & ENV'T 50, 50 (2017).

[92] *Id.*

[93] *Id.* at 51.

[94] *Id.*

[95] Leon, *supra* note 92 at 10.

[96] Browne, *supra* note 94 at 51.

electricity to 17.5%.[97] In fact, as of July 2017, six jurisdictions (California, Hawaii, New York, Oregon, Vermont, and the District of Columbia) now have renewable energy requirements of at least 50% by 2045 or sooner.[98]

By allowing obligated parties to buy and sell RECs, which certify renewable energy production, RPSs utilize market systems to achieve their quotas. Reliance upon market mechanisms incentivizes price competition among different technologies.[99] RPSs set percentages for renewable energy, but they permit electricity suppliers and generators to determine their methods of compliance.[100] Typically, these parties satisfy the RPS quotas by choosing lower-cost and lower-risk technologies.[101] This market pressure to utilize low-cost methods of generation drives innovation. Thus, RPSs encourage development and adoption of low-cost methods.[102] Over time, economies of scale and efficiencies gained through experience further drive down costs as markets expand.[103] Such results have occurred in recent years with wind power, solar photovoltaics, and batteries.[104]

RPSs provide other benefits, as well. For instance, they usually allow regulated entities 10–20 years to meet their quotas.[105] This is important since government policies are most effective when they provide long-term investment certainty.[106] In addition, graduated series of quotas support relatively smooth and continuous growth of the targeted technologies.[107]

While RPSs typically do not favor one technology over another, RPSs can include provisions targeted to incentivize specific technologies. These include caps

---

[97] Miriam Fischlein & Timothy M. Smith, *Revisiting Renewable Portfolio Standard Effectiveness: Policy Design and Outcome Specification Matter*, 46 POL'Y SCI. 277, 288 (2013).

[98] Barbose, *supra* note 79 at 6.

[99] Shahrouz Abolhosseini & Almas Heshmati, *The Main Support Mechanisms to Finance Renewable Energy Development*, RENEW. & SUST. ENERGY REV. (2014) 40:876–885, 881.

[100] Leon, *supra* note 92 at 8.

[101] Wiser, Barbose, & Holt, *supra* note 81 at 3896. The incentive to provide electricity at the lowest cost also incentivizes improving technologies to become more cost competitive. EPA, *supra* note 85 at 5–3.

[102] Some commentators suggest that a weakness of RPSs is that they are so market driven that they do not sufficiently encourage investment in less mature technologies. Buckman, *supra* note 81 at 4106–07. As discussed in the next section, RPSs can utilize carve outs or multipliers to stimulate development of these resources. *Id.* at 4107.

[103] John A. Mathews & Hao Tan, *Manufacture Renewables to Build Energy Security*, 513 NATURE 166, 167 (Sept. 11, 2014).

[104] Bobby Magill, Pioneers of Carbon Dioxide Removal See Boon for Renewables, BLOOMBERG BNA ENVIRONMENT & ENERGY REPORT (April 24, 2018).

[105] Paul Dvorak & Nathaniel Horner, *RPS Policies Are Driving Wind Turbine Innovation*, WINDPOWER (February 28, 2014), http://www.windpowerengineering.com/design/rps-policies-driving-wind-turbine-innovation/ [http://perma.cc/2TK9-RZHP]

[106] Leon, *supra* note 92 at 9.

[107] Herman K. Trabish, *Why Mandates Still Matter in the Age of Cheap Renewables*, UTILITY DIVE (January 3, 2018), https://www.utilitydive.com/news/why-mandates-still-matter-in-the-age-of-cheap-renewables/513797/

on reliance on particular technologies,[108] technology carve-outs, and credit multipliers.[109] A jurisdiction might impose a cap to prevent overreliance upon a particular technology and thereby assure technology diversification. Carve-outs (or set-asides) identify specific levels of electricity to be produced from a particular type of source. The jurisdiction "carves out" these quotas from the overall renewable energy percentage mandate.[110] Multipliers, on the other hand, provide that the generation of electricity by particular energy sources will earn multiples of RECs.[111] Jurisdictions utilize these approaches to encourage investment in particular technologies whose development is a policy objective of the state.[112] Often times these technologies are not currently competitive with other energy sources because of their higher cost, still-developing technology, or the presence of other market barriers.[113]

Analysts credit each approach with successes,[114] but they also recognize concerns can arise with each. Carve-outs, by requiring utilization of sources that are usually not the least-cost method, raise energy costs.[115] Multipliers, on the other hand, by allowing extra credit for generation from specified sources[116] (for instance, solar power), result in less overall production of the supported good (renewable energy).[117]

Jurisdictions with RPSs have experienced substantial growth in their renewable energy generation. For instance, during the period from 1997 to 2011, renewable energy production grew by 128.5% in non-RPS states, while in RPS states, it increased by 666.6%.[118] Fig. 1 presents the effects of RPSs on the growth of renewable energy in the United States from 2000 to 2016.

Figure 1 illustrates several key points about RPSs. First, renewable energy growth pursuant to the RPS "Minimum Growth Required" was, indeed, "slow and steady." For better and worse, this contrasts with the pattern typically presented by feed-in tariffs, discussed below. Second, as economies of scale take effect, the technologies become less expensive, and they are no longer deployed primarily to

---

[108] International Energy Agency, DEPLOYING RENEWABLES 2011 132 (2011).

[109] Wiser, Barbose, & Holt, *supra* note 81 at 3897.

[110] EPA, *supra* note 85 at 5–10.

[111] Buckman, *supra* note 81 at 4105. Multipliers are also identified as banding. *Id.*

[112] EPA, *supra* note 85 at 5–10.

[113] *Id.*

[114] Analysis has found that the use of set asides in RPSs has "heavily influenced" the deployment of solar energy in those states. Andrea Sarzynski, Jeremy Larrieu, & Gireesh Shrimali, *The Impact of State Financial Incentives on Market Deployment of Solar Technology*, ENERGY POLICY 46 (2012) 550–557, 551. Similarly, multipliers are credited with successfully supporting high-cost offshore wind development in the United Kingdom. Buckman, *supra* note 81 at 4114.

[115] Joshua Novacheck & Jeremiah X. Johnson, *The Environmental And Cost Implications of Solar Energy Preferences in Renewable Portfolio Standards*, 86 ENERGY POLICY 250, 256 (2015).

[116] *Id.* at 251.

[117] *Id.* at 254.

[118] Eastin, *supra* note 91 at 132.

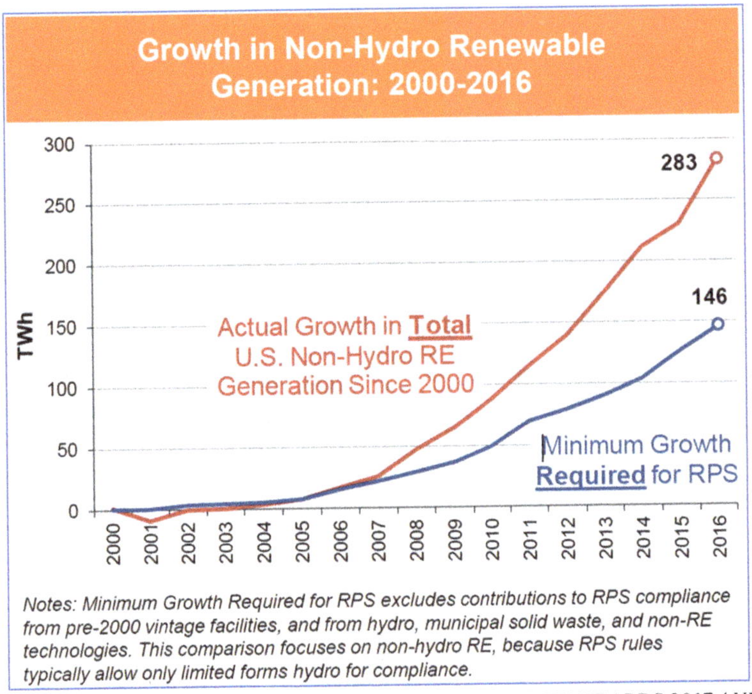

Source: Galen Barbose, U.S. RENEWABLES PORTFOLIO STANDARDS 2017 ANNUAL STATUS REPORT 12 (2017)

Fig. 1 Growth in Non-Hydro Renewable Generation: 2000–2016. (Source: Galen Barbose, U.S. RENEWABLES PORTFOLIO STANDARDS 2017 ANNUAL STATUS REPORT 12 (2017))

satisfy RPS mandates. Of the renewable energy deployed up to 2012, the proportion installed in RPS states was 67%.[119] By 2016, however, this percentage fell to 44%.[120]

3.2 *FITs Accelerated Development of Renewables*

Feed-in tariffs (FITs) are policies structured to accelerate investment in new technologies.[121] As discussed below, they have arguably been more successful in incentivizing renewable energy production than have been RPSs. However, as experience with FITs has continued, problems with their true costs have become apparent.

---

[119] Leon, *supra* note 92 at 4.

[120] Barbose, *supra* note 79 at 3.

[121] Yuzhuo Zhang, et al., *The Development of the Renewable Energy Power Industry under Feed-In Tariff and Renewable Portfolio Standard: A Case Study of China's Photovoltaic Power Industry*, 9 SUSTAINABILITY (2017) 532, 2; https://doi.org/10.3390/su9040532

FITs assure a guaranteed payment for the provision of electricity from an approved usually renewable – source.[122] FITs consist of two parts. The first, the feed-in prong, assures electricity producers access to the local power grid; the second, the tariff prong, requires that utilities purchase the output of these generators consistent with the FIT terms.[123] FIT contracts guarantee payment for the full output from a supplier at a pre-determined amount for a guaranteed period of time. Typically, the period of the contract lasts at least 15 years, though commonly they extend as long as 20 years.[124]

The tariffs are determined through one of three approaches. The first method starts with the actual, levelized cost to generate the electricity and then adds an amount for profit.[125] Thus, this approach assures a reasonable return on investment.[126] The second approach sets the tariff equivalent to the value of the good (electricity) provided. Regulators can set this rate as equivalent to a utility's avoided costs of acquiring the electricity through other, conventional sources.[127] Alternatively, they can calculate a "value" for the provided energy. Typically, regulators set this price with reference to external factors. For renewable energy, this includes considerations such as the value of climate mitigation, air quality, health impacts, and energy security.[128] The third method sets a fixed-price incentive without reference to actual or avoided costs.[129]

The theoretical basis for a FIT policy differs from that underlying an RPS regime. Essentially, RPSs regulate quantities of renewable energy generation, while FITs control their prices.[130] As noted previously, RPSs establish markets for technologies by setting quotas for their products.[131] FITs, on the other hand, set a price for electricity from specified sources, thereby guaranteeing reliable revenue streams for that production with an assured profit.[132] This guarantees that investors in projects – whether small or large – will not only be able to sell their product, they will also receive a favorable return on their investments.[133] FITs thus remove most financial

---

[122] Toby Couture & Karlynn Cory, STATE CLEAN ENERGY POLICIES ANALYSIS (SCEPA) PROJECT: AN ANALYSIS OF RENEWABLE ENERGY FEED-IN TARIFFS IN THE UNITED STATES 2 (2009).

[123] Felix Mormann, *Clean Energy Federalism*, 67 FLA. L. REV. 1621, 1628 (2016).

[124] Karlynn Cory, Toby Couture, & Claire Kreycik, FEED-IN TARIFF POLICY: DESIGN, IMPLEMENTATION, AND RPS POLICY INTERACTIONS 2 (2009).

[125] *Id.*

[126] Couture & Cory, *supra* note 125 at 2.

[127] Cory, Couture, & Kreycik, *supra* note 127 at 2.

[128] *Id.*

[129] Couture & Cory, *supra* note 125 at 2.

[130] Tae-hyeong Kwon, *Rent and Rent-seeking in Renewable Energy Support Policies: Feed-in Tariff vs. Renewable Portfolio Standard*, RENEW. & SUST. ENERG. REV. (2015) 44:676–681, 676.

[131] Leon, *supra* note 92 at 8.

[132] Cory, Couture, & Kreycik, *supra* note 127 at 9.

[133] *Id.*

risks for investors.[134] Some analysts consider the guarantee of reliable revenue streams to be one of the most important elements of FITs.[135] On top of this, lengthy contracts provide long-term stability for investors.[136] Thus, FITs assure stable markets and profitable investments.[137]

FITs provide another benefit. In view of the lengthy contract and guaranteed, profitable rate that they assure, FITs facilitate access to financing for projects.[138] This reduces the high up-front cost of financing by enabling developers to finance a larger portion of the project with debt financing.[139] This is significant since greater deployment of CCS technologies has been hindered not by excessive costs but by a lack of financing.[140]

This is an important distinction between FITs and RPSs. Jurisdictions utilizing RPSs typically require project developers to submit competitive bids in the hope of winning contracts.[141] Experience in Europe indicates that the guaranteed and profitable contracts assured under FITs foster quicker development and financing of renewable energy than occurs pursuant to RPSs.[142] Consequently, industry tends to favor FITs because of the stable and profitable markets that they establish.[143]

Similar to RPSs, FITs also can specifically target multiple technologies for development. Since FITs set rates for electricity generation, policymakers can set different rates according to a number of different considerations. Indeed, FITs commonly establish separate rates based upon technologies, project sizes, and other factors.[144] For example, some European countries have established between 20 and 30 separate FITs rates.[145]

FITs have been extremely successful at encouraging renewable energy development. Overwhelmingly, they are the most popular renewable energy policy in the world, and they account for a larger share of renewable energy than do RPSs.

---

[134] Richard Schmalensee, *Evaluating Policies to Increase Electricity Generation from Renewable Energy*, REV. OF ENVT'L. ECON. AND POL'Y, (2012) 6;45–64, 60 [doi:https://doi.org/10.1093/reep/rer020]

[135] Cory, Couture, & Kreycik, *supra* note 127 at 9.

[136] Zhang, *supra* note 82 at 426.

[137] Peng Sun & Pu-yan Nie, *A Comparative Study of Feed-In Tariff and Renewable Portfolio Standard*, 74 RENEWABLE ENERGY (2015) 255–262, 255.

[138] Couture & Cory, *supra* note 125 at 17–18.

[139] *Id.* at 4.

[140] James Temple, *The Daunting Math of Climate Change Means We'll Need Carbon Capture*, MIT TECHNOLOGY REVIEW (April 24, 2018), https://www.technologyreview.com/s/610927/the-daunting-math-of-climate-change-means-well-need-carbon-capture/

[141] *Id.* at 22.

[142] Cory, Couture, & Kreycik, *supra* note 127 at 9.

[143] Q. Zhang, *supra* note 82 at 427.

[144] Mormann, *supra* note 126 at 1662.

[145] UNEP, *supra* note 17 at 38.

Globally, FITs account for 75% of solar energy and 45% of wind power.[146] In Europe, FITs have engendered quick and substantial renewable energy expansion.[147] FITs have also supported massive growth of renewable energy in Asia.[148]

Consistent with these results, analyses suggest that FITs have been more effective than RPSs at fostering renewable energy development. Although RPSs create markets, FITs have more successfully encouraged development of the technologies necessary to populate these markets.[149] The European experience suggests that renewable energy development and financing occurs more quickly with FITs than RPSs.[150] One study, based upon International Energy Agency data, concluded that FITs encourage four times more clean energy deployment than do RPSs.[151]

Despite these robust results, FITs typically become victims of their own success. In a recent pattern, FITs effectively promote the installation of renewable energy in the short term, yet they generate overwhelming costs in the long term.[152] As commentators have noted, "the paradox inherent in feed-in tariffs is that they are designed to gradually self-destruct."[153] Ironically, this results from two aspects critical to their success: their premium tariff rates and the long duration of their contracts. Essentially, FITs lock in high rates for decades. As discussed, these provisions have the desired effect of inspiring rapid deployment. And, the resulting growth has another desired and expected result – the costs of production fall. However, jurisdictions remain locked in – for years, possibly decades – to rates that were, by definition, excessive when first established. As a result, the FITs premiums impose substantial, long-term burdens.[154] These burdens fall either directly on the utility customers through higher surcharges or indirectly on the taxpayers if the government covers the premium.[155]

Jurisdictions have applied certain fixes to minimize this problem. Some have adjusted rates periodically to reflect new realities regarding the cost of electricity production.[156] For instance, Germany applied a tariff degression, an annual reduction in the tariff by a pre-established percentage. Spain, on the other hand, annually

---

[146] Q.Y. Yan, et al., *Overall Review of Feed-In Tariff and Renewable Portfolio Standard Policy: A Perspective of China*, 75 EARTH ENVIRON. SCI. 2 (2016).

[147] Cory, Couture, & Kreycik, *supra* note 127 at 2.

[148] Chris Lo, *Renewable Energy: Are Feed-In Tariffs Going Out of Style*?, POWER TECHNOLOGY, (January 18, 2017), https://www.power-technology.com/features/featurerenewable-energy-are-feed-in-tariffs-going-out-of-style-5718419/

[149] Mormann, *supra* note 126 at 1658.

[150] Cory, Couture, & Kreycik, *supra* note 127 at 9.

[151] Mormann, *supra* note 126 at 1660.

[152] Yan, *supra* note 149 at 2.

[153] Lincoln L. Davies & Kirsten Allen, *Feed-in Tariffs in Turmoil*, 116 W. VA. L. REV. 937, 997 n.18 (2014),

[154] Q. Zhang, *supra* note 82 at 433.

[155] UNEP, *supra* note 17 at 81.

[156] Cory, Couture, & Kreycik, *supra* note 127 at 5.

adjusted its FITs rates.[157] Of course, in both instances, previously-contracted rates still remained applicable and still caused the FITs to self-destruct.

Because of this tendency, some of the most prominent FITs success stories have become cautionary tales. The FIT in Spain helped double the percent of electricity sourced from renewables from 18.4% in 2006 to 37.4% in 2015.[158] However, because Spain decided to cap retail electricity prices, the system accumulated a "tariff debt" of €26 billion.[159] Initially, utilities shouldered this debt, but in 2009 the Spanish government needed to assume it. Within 3 years, Spain abolished its FIT.[160] Another success story that lacks a happy ending comes from Germany. Germany's FIT helped increase the percent of its electricity produced from renewable energy from 4% in 1990 to more than 30% in 2017.[161] Its consumers, however, paid for this renewable energy – in 2016 as much as €25 billion in surcharges.[162] Consequently, after a series of cuts in the FIT subsidy over several years,[163] in 2016 Germany decided to end its FIT.[164] A number of other countries have also recently abandoned or severely limited their FITs, including Finland,[165] Greece,[166] Portugal,[167] and South Korea.[168]

---

[157] *Id.*

[158] Lo, *supra* note 151.

[159] Davies & Allen, *supra* note 156 at 977.

[160] *Id.* Spain exacerbated its problems when it subsequently cancelled contracted FIT payments retroactively and replaced them with a complex payment program. Investors brought dozens of lawsuits over these cuts. Spain lost the first decision in one of these cases, and the court ordered it to pay €128 million. Blanca Díaz López, *Spain Loses Its First Renewable Energy Case in International Courts*, PV MAGAZINE, (May 5, 2017), https://www.pv-magazine.com/2017/05/05/spain-loses-its-first-renewable-energy-case-in-international-courts/

[161] Lo, *supra* note 151.

[162] Jeffrey Ball, *Germany's High-Priced Energy Revolution*, FORTUNE, (March 14, 2017), https://finance.yahoo.com/news/germany-high-priced-energy-revolution-103034269.html

[163] *Id.*

[164] Joshua S Hill, *Germany Confirms End To Renewable Energy Feed-in Tariffs*, CLEANTECHNICA, (July 12, 2016), https://cleantechnica.com/2016/07/12/germany-confirms-end-renewable-energy-feed-tariffs/

[165] Dittmar & Indrenius, D&I ALERT – ENERGY, INFRASTRUCTURE & NATURAL RESOURCES 2 (November 25, 2016) https://www.dittmar.fi/service/energy-infra/

[166] RES Legal, *Greece: Overall Summary*, RES LEGAL EUROPE (undated), (last visited July 28, 2018), http://www.res-legal.eu/search-by-country/greece/

[167] RES Legal, *Feed-in tariff (Tarifas feed-in)*, RES LEGAL EUROPE (December 12, 2017), http://www.res-legal.eu/search-by-country/portugal/single/s/res-e/t/promotion/aid/feed-in-tariff-tarifas-feed-in/lastp/179/

[168] Davies & Allen, *supra* note 156 at 995.

4 Combining FITs and RPSs Can Facilitate The Development OF CDR

RPSs and FITs both have demonstrated track records of spurring investment into new technologies. With few exceptions, jurisdictions have treated FITs and RPSs as mutually exclusive policies.[169] While each separately could substantially increase CDR's development, a combination of the two likely would be more successful. The two policies could work together synergistically by reinforcing each other's methods and helping to compensate for their respective weaknesses.[170]

FITs, as demonstrated above, are especially effective at fostering initial investment and deployment of new technologies. Because FITs provide a guaranteed return on investment, they encourage investment and initial deployment of technologies.[171] However, as discussed previously, FITs eventually "self-destruct." The long-term contracts at premium rates that are the strength of FITs become their downfall. Although these rates do stimulate investment into these technologies, as these technologies become more abundant, their premium rates prohibitively burden ratepayers and taxpayers.

Consequently, rather than implement FITs unilaterally, jurisdictions should enact them as part of RPS programs. By their nature, RPSs can provide a framework into which other mechanisms could be incorporated.[172] For instance, FITs could provide complementary support to achieve RPS quotas.[173] FITs can help spur initial investments to achieve early RPS targets. Then, after the early phases of investment help to establish the technologies, the FITs subsidy should be ratcheted down. At the same time, the RPS quota should continue to rise to ensure that new deployments continue.[174] Policies that guarantee prices can facilitate early technological development, while quantity-based requirements are more effective for mature technologies.[175] Thus, FITs and RPSs are natural complements. Indeed, Dong, et al., found that jurisdictions that combined these two policies experienced the greatest level of

[169] Mormann, *supra* note 126 at 1628.

[170] Couture & Cory, *supra* note 125 at 22.

[171] Cory, Couture, & Kreycik, *supra* note 127 at 13.

[172] Zhao Xin-gang, et al., *The Policy Effects of Feed-In Tariff and Renewable Portfolio Standard: A Case Study of China's Waste Incineration Power Industry*, WASTE MANAG. 68 (2017) 711–723, 711–12.

[173] Sun & Nie, *supra* note 140 at 256.

[174] Xin-gang, et al., *supra* note 175 at 721.

[175] Francesco Nicolli & Francesco Vona, *Heterogeneous Policies, Heterogeneous Technologies: The Case of Renewable Energy*, ENERGY ECON. 56 (2016) 190–204, 190.

installed capacity.[176] For instance, China's FIT played an integral role in enabling the country to achieve its RPS targets.[177]

States that have already adopted RPSs could add carve-outs requiring utilization of BECCS. As discussed previously, RPSs could incorporate FITs, carve- outs, and multipliers to target BECCS as means for compliance. To the extent that the states adopted their RPSs to limit atmospheric CO_2 (through the reduction of carbon emissions),[178] the broader inclusion of other CDR technologies would further this goal. Alternatively, states could establish a separate program, say a CDR Portfolio Standard, solely dedicated to the reduction of CO_2 emissions and the development of CDR.

There are other reasons to embrace the use of FITs and RPS to incentivize CDR technologies. First, both policies have developed multiple technologies at the same time. RPSs typically recognize between one and two dozen technologies as eligible to satisfy their mandates. For example, Wisconsin's RPS identifies 26 eligible technologies.[179] FITs, for their part, can provide differentiated rates based upon specific technology types,[180] as demonstrated by countries in Europe using 20 to 30 different rates.[181] This ability to develop multiple technologies will be critical, in view of the limited potential of all individual CDR method if used at scale.[182] As a result, multiple CDR technologies will almost certainly be required to compensate for anthropogenic greenhouse gas emissions.[183]

Using RPSs and FITs jointly will also help CDR to leverage the advantages of market systems. As noted previously, RPSs foster markets for new technologies.[184] This is important, since the development of new technologies requires circumstances which facilitate their testing and improvement while being supported by actual markets.[185] Moreover, FITs, as discussed above, attract the financing necessary to populate these markets.[186] Investment in new technologies requires strong

---

[176] C.G. Dong, *Feed-in Tariff vs. Renewable Portfolio Standard: An Empirical Test of Their Relative Effectiveness in Promoting Wind Capacity Development*, ENERGY POLICY 42 (2012) 476–485, 484.

[177] Yan, *supra* note 149 at 8.

[178] Ottmar Edenhofer, Ramón Pichs-Madruga, & Youba Sokona (eds.), SPECIAL REPORT ON RENEWABLE ENERGY SOURCES AND CLIMATE CHANGE MITIGATION 18 (2012).

[179] DSIRE, *Program Overview: Wisconsin*, (last updated November 18, 2015), http://programs.dsireusa.org/system/program/detail/190

[180] Cory, Couture, & Kreycik, *supra* note 127 at 7.

[181] UNEP, *supra* note 17 at 38.

[182] Fuss, *supra* note 53 at 3.

[183] Jan C. Minx, et al., *Fast Growing Research on Negative Emissions*, ENVIRON. RES. LETT. 12, 2 (2017).

[184] Leon, *supra* note 92 at 8.

[185] Yuki Ishimoto et al., *Putting Costs of Direct Air Capture in Context* 12 (Inst. of Applied Energy, Working Paper No. 002, Jun., 2017).

[186] Mormann, *supra* note 126 at 1658.

and certain policy and price signals,[187] These are precisely what FITs provide.[188] Furthermore, once FIT's costs become burdensome, states can wind down them down, while relying upon RPS mandates to ensure that technology deployments continue.[189]

RPS policies can benefit CDR development because of their ability to incentivize innovation and reduce costs. Deploying CDR will be expensive. For instance, the NRC projected the costs of carbon capture as ranging from $50 to more than $1000 per tCO_2.[190] The costs of sequestration, it estimated, would range from $6 to hundreds of dollars per tCO_2.[191] Estimated costs of specific technologies cover similarly broad ranges, as indicated previously.[192] Furthermore, the scale of CDR will be substantial. The overall system to capture and bury carbon will likely need to be as extensive as that which extracted it.[193] Thus, utilization of RPSs to drive innovation and lower costs will be critical.

Another benefit of using the RPS framework to develop CDR technologies involves the accounting and RECs systems RPSs have already established. Carbon dioxide removal will present significant issues of tracking and accounting.[194] Measuring the amount of greenhouse gases that are captured and permanently sequestered is more complicated than are other forms of environmental accounting, such as tracking emissions.[195] For instance, different ecosystems, soils, and biomass complicate the calculation of carbon dioxide removed from the atmosphere.[196] Nevertheless, consistent accounting rules also have not been developed for CDR processes.[197] CDR also will likely involve multi-jurisdictional transactions necessitating uniform measurement, reporting, and verification of activities.[198] RPSs already require tracking systems for RECs transactions,[199] compliance with annual

---

[187] Niall Mac Dowell et al., *The Role of CO_2 Capture and Utilization in Mitigating Climate Change*, 7 NATURE CLIMATE CHANGE 243, 243 (2017).

[188] Mormann, *supra* note 126 at 1660.

[189] Xin-gang, et al., *supra* note 175 at 721 (proposing to raise the RPS quota after a subsidy is reduced or eliminated).

[190] NRC *supra* note 14 at 106. Part of this range derives from the method of carbon capture. Capture from an emissions source is dramatically less expensive than from the ambient air. Accordingly, the latter may cost up to ten times more than capture directly from an emissions source. Psarras, *supra* note 36 at 4.

[191] NRC *supra* note 14 at 106.

[192] *Supra*, pages 3–5.

[193] NRC *supra* note 14 at 105.

[194] Lomax, *supra* note 11 at 499.

[195] *Id.*

[196] *Id.*

[197] Fuss, *supra* note 53 at 7.

[198] Glen P. Peters & Oliver Geden, *Catalysing a Political Shift from Low to Negative Carbon*, 7 NATURE CLIMATE CHANGE 619–621, 621 (2017).

[199] Arroyo, *supra* note 93 at 399.

reporting requirements,[200] and establishment of regional tracking systems. These regional systems are sufficiently flexible to enable future interconnection and expansion.[201]

Incorporating CDR into RPSs will require jurisdictions to address several considerations. First, states will need to decide how quickly to require compliance with a CDR mandate. A phase-in period will be both necessary and helpful. It will be necessary since most CDR methods currently are not ready for implementation.[202] The one CDR technology that is well understood and also produces electricity, BECCS, has not been deployed to scale and will require years before it can be.[203] Furthermore, time will be required to develop the transportation systems, storage facilities, and plants necessary to operate BECCS at scale.[204] Extended implementation will also allow jurisdictions to modify their accounting systems to measure and track the capturing and burying of carbon dioxide. Jurisdictions will need to develop methods to measure the carbon captured, the amount successfully sequestered, the permanence of sequestration. Moreover, they will need to develop comparable measurements across different environments and technologies.[205] Furthermore, they will need to determine the tariff rates for each covered technology. This may require calculation of the cost of each technology or of its value.[206]

Phasing-in implementation will also enable states to expand coverage of RPSs to sectors beyond the energy sector. Currently, RPS mandates apply only to parties involved in the provision of electricity – investor-owned utilities, municipalities, and electric cooperatives.[207] For CDR to yield truly negative emissions, however, they must also compensate for the emissions from additional sectors.[208] For instance, current estimates project that, even though emissions in the U.S. electricity sector will decline over the next decade, emissions will continue to rise in the industrial

---

[200] EPA, *supra* note 85 at 5–11.

[201] Leon, *supra* note 92 at 6.

[202] Keller, *supra* note 72 at 4.

[203] Burns & Nicholson, *supra* note 65 at 3.

[204] Henrik Karlsson & Lennart Byström, GLOBAL STATUS OF BECCS PROJECTS 2010 40 (2011).

[205] CTR. FOR CARBON REMOVAL, *supra* note 71 at 8. *See also* Feifei Shen, *China's Prep for Carbon-Market Trading May Take Up to Two Years*, BLOOMBERG BNA ENVIRONMENT & ENERGY REPORT (Dec. 21, 2017) (noting that China may spend up to two years preparing data reporting and other systems before starting trading in its new carbon market). China's program is discussed *infra*.

[206] *See supra*, notes 126–30 and accompanying text.

[207] National Conference of State Legislatures, STATE RENEWABLE PORTFOLIO STANDARDS AND GOALS (July 20, 2018), http://www.ncsl.org/research/energy/renewable-portfolio-standards.aspx

[208] Cory, Couture, & Kreycik, *supra* note 127 at 2.

and agricultural sectors.[209] Furthermore, the transportation sector, now the largest single source of domestic emissions, continues to experience increasing emissions.[210]

Thus, states need to extend their RPSs' coverage to include non-electricity sectors. The inclusion of these new sectors will require a phased-in transition. China currently is planning to institute such a program. In 2018, China initiated an emissions trading system, which will utilize a phase-in process. After developing rules for the system and testing it through simulated trading, China will then require compliance by its electricity sector. In a final phase targeted to begin after 2020, China will extend its trading program to non-ferrous metal and cement sectors.[211]

Enacting RPSs and FITs at the level of individual states or nations can be beneficial. As the Supreme Court has noted, states can serve "as laboratories for experimentation to devise various solutions where the best solution is far from clear."[212] Indeed, in the federalism system of the United States, states function as "innovation centers."[213] Not only do they experiment with policies, they also "compete" with other states to "develop the most effective and efficient regulatory program."[214] Indeed, the popularity of RPSs suggests to some that this decentralized policymaking engendered a race to the top.[215] Furthermore, instituting these policies at the local level is consistent with the bottom-up approach favored by the Paris Agreement.[216]

5 Conclusion

Carbon dioxide removal technologies will become essential to avoid dangerous climate change. Unfortunately, these technologies are not ready to serve this role. Feed-in tariffs and renewable portfolio standards were instrumental in incentivizing

[209] Kate Larsen et al., *Taking Stock 2017: Adjusting Expectations For US GHG Emissions* 4–5 (2017), https://rhg.com/qp-content/uploads/2017/05/RHG_ENR_Taking_Stock_24May2017.pdf [http://perma.cc/EZ4U-3AB6]

[210] Rhodium Group, *Preliminary US Emissions Estimates for 2018* (January 8, 2019), https://rhg.com/research/preliminary-us-emissions-estimates-for-2018/

[211] *See* Dean Scott, *China's Trimmed Carbon Trading Will Still Boost Worldwide Action*, BLOOMBERG BNA ENVIRONMENT & ENERGY REPORT (Dec. 21, 2017) (noting that China's carbon trading system will initially cover power generators but subsequently expand to encompass metals, chemicals, and building materials).

[212] *United States v. Lopez*, 514 U.S. 549, 581, (1995) (KENNEDY, J., concurring).

[213] Allison C.C. Hoppe, *State-Level Regulation as The Ideal Foundation For Action on Climate Change: a Localized Beginning to the Solution of a Global Problem*, 101 CORNELL L. REV. 1627, 1650 (2016).

[214] *Id.* at 1650–51.

[215] Thomas P. Lyon & Haitao Yin, *Why Do States Adopt Renewable Portfolio Standards?: An Empirical Investigation*, 31 ENERGY J. 153 (2010).

[216] Rob Bellamy, *Incentivize Negative Emissions Responsibly*, 3 NATURE ENERGY 532–534, 532 (2018).

the deployment of renewable energy. They can similarly foster the development and installation of CDR.

References

1. Allison, C.C.: Hoppe, state-level regulation as the ideal foundation for action on climate change: A localized beginning to the solution of a global problem. Cornell L. Rev. **101**, 1627–1650 (2016)
2. Bellamy, R.: Incentivize negative emissions responsibly. Nature Energy. **3**, 532–534–532–532 (2018)
3. Boysen, L.R., et al.: The limits to global-warming mitigation by terrestrial carbon removal. Earth's Future. **5**, 463–463–474 (2017)
4. Dowell, N.M., et al.: The role of CO_2 capture and utilization in mitigating climate change. Nat. Climate Change. **7**, 243 (2017)
5. Eisaman, M.D.: Indirect ocean capture of atmospheric CO2: Part II. Understanding the cost of negative emissions. Int. J. Greenhouse Gas Control. **1** (2018)
6. Fischlein, M., Smith, T.M.: Revisiting renewable portfolio standard effectiveness: policy design and outcome specification matter. Pol'y Sci. **46**, 277–288 (2013)
7. Frank, S., et al.: Reducing greenhouse gas emissions in agriculture without compromising food security? Envt'l. Res. Lett. **12**, 1–2 (2017)
8. Fuss, S., et al.: Research Priorities for Negative Emissions. Env't. Res. Lett. **11**, 1–3 (2016)
9. Honegger, M., Reiner, D.: The political economy of negative emissions technologies: Consequences for international policy design. Climate Pol'y. **18**, 306–308–309 (2017)
10. Keller, D.P., et al.: The carbon dioxide removal model intercomparison project (CDR-MIP): Rationale and experimental design. Geosci. Model Dev. **11**, 1133–1133–34 (2018)
11. Lenton, A.: Assessing carbon dioxide removal through global and regional ocean Alkalinization under high and low emission pathways. Earth Sys. Dynamics. **9**, 339–357–339–340 (2018)
12. Lomax, G., et al.: Investing in negative emissions. Nat. Climate Change. **5**, 498 (2015)
13. Lyon, T.P., Yin, H.: Why do states adopt renewable portfolio standards?: An empirical investigation. Energy J. **31**, 153 (2010)
14. McGlashan, N., et al.: High-Level Techno-Economic Assessment of Negative Emissions Technologies. Process Safety Envt'l. Protect. **90**, 501–10–501–501503 (2012)
15. Mormann, F.: Constitutional challenges and regulatory opportunities for state climate policy innovation. Harv. Envt'l. L. Rev. **41**, 189–198 (2017)
16. Novacheck, J., Johnson, J.X.: The environmental and cost implications of solar energy preferences in renewable portfolio standards. Energy Pol. **86**, 250–256 (2015)
17. Peters, G.P., Geden, O.: Catalysing a political shift from low to negative carbon. Nat. Climate Change. **7**, 619–621–619–621 (2017)
18. Psarras, P., et al.: Slicing the pie: how big could carbon dioxide removal be? Wires Energy Env't. **6**, 1–1 (2017)
19. Smith, P.: Biophysical and economic limits to negative CO_2 emissions. Nat. Climate Change. **6**, 42–49 (2016)
20. Xin-gang, Z., et al.: The policy effects of feed-in tariff and renewable portfolio standard: A case study of China's waste incineration power industry. Waste Manag. **68**, 711–723 (2017) 711–12
21. Zhang, Q.: Substitution effect of renewable portfolio standards and renewable energy certificate trading for feed-in tariff. Appl. Energy. **227**, 426–426–427 (2017)

Associated and Incremental Storage: Opportunities for Increased CO_2 Removal with Enhanced Oil Recovery

Tara Righetti

The vast majority of technologies and processes to address climate change contemplate a road to decarbonization, eliminating society's dependence on fossil fuel use and dependency. However, it has become increasingly clear that carbon dioxide removal techniques are also necessary to achieve emissions reductions of a magnitude necessary to limit global temperature increases to 1.5 °C above pre-industrial levels. Some of these carbon dioxide removal technologies, including Direct Air Capture and bioenergy with carbon capture and storage (BECCS), rely on carbon capture, and geologic storage or utilization to remove CO_2 from the atmosphere.[1] Delayed deployment of carbon dioxide removal techniques will require greater storage capacities to meet established targets.[2] Although there are numerous technical options for geologic storage, including offshore storage,[3] saline storage, and basalt carbon mineralization represent technical option for carbon management, these technologies face significant barriers including regulatory uncertainty, the lack of a comprehensive legal framework, and the absence of necessary infrastructure.

Tara Righetti is a professor at the University of Wyoming College of Law and the UW School of Energy Resources. Professor Righetti's research is supported in party by the National Science Foundation EPSCoR program (NSF EPSCoR grant OIA-1632899). Professor Righetti acknowledges Madeleine Lewis (JD/MA'19) for her research assistance.

[1] Global CCS Inst., The Global Status of CCS: 2017, at 7, 9 (2017).

[2] Christopher Zahasky and Samuel Krevor, Global Geologic Carbon Storage Requirements of Climate Change Mitigation Scenarios, Energy and Environmental Science (2020).

[3] Romany Webb & Michael Gerrard, *Overcoming Impediments to Offshore CO_2 Storage: Legal Issues in the United States and Canada*, 49 Envt'l L. Rep. 10,634 (2019).

T. Righetti (✉)
University of Wyoming College of Law, Laramie, WY, USA
e-mail: Tara.Righetti@uwyo.edu

© Springer Nature Switzerland AG 2021

W. Burns et al. (eds.), *Climate Geoengineering: Science, Law and Governance*, AESS Interdisciplinary Environmental Studies and Sciences Series,
https://doi.org/10.1007/978-3-030-72372-9_9

In contrast, carbon storage associated with enhanced oil recovery (EOR) benefits from income streams from CO_2 utilization and established cooperative federalism regulatory frameworks and international technical standards. Not only does CO_2-EOR represent an opportunity for carbon neutral crude, but it also represents one of the most accessible options for immediate CO_2 sequestration.[4]

Expanding utilization of carbon capture in enhanced oil recovery (EOR) and encouraging the transition of CO_2-EOR projects to projects for additional, incremental storage, could provide significant reductions in global carbon.[5] Unlike many other climate engineering opportunities,[6] carbon storage in the United States has localized impacts and is already subject to a comprehensive legal and governance framework. Although a transition towards renewable sources is already underway, fossil fuels remain an abundant and cost-effective source of energy.[7] Due to existing investments in the energy status quo, fossil fuels will likely remain prevalent for at least the next few decades, with the climatic impacts of burning those fuels lasting much longer.[8] This chapter explores the role of carbon capture, utilization and storage (CCUS) for mitigating carbon emissions within a comprehensive climate engineering framework that is driven by the momentum of new technology, but which also welcomes innovations from the existing energy network to meet long-term climate change goals. It analyzes opportunities to increase the carbon dioxide reduction potential of enhanced oil recovery and to encourage transition of depleted CO_2-EOR assets into projects for incremental storage.

1 Geologic Storage

Geologic storage of carbon dioxide is an important component of mitigation pathways and is inextricable from several intensive – or deep – negative emissions technologies.[9] Carbon, Capture, and Sequestration (CCS) and Carbon, Capture,

---

[4] *See* The White House, United States Mid-Century Strategy for Deep Decarbonization (2016), *available at* https://unfccc.int/files/focus/long-term_strategies/application/pdf/mid_century_strategy_report-final_red.pdf (designating to CCUS technology a significant role in reducing carbon emissions by 2050).

[5] *See infra* notes 30 - 47 and accompanying text.

[6] Anthony Chavez, *A Napoleonic Approach to Climate Change: The Geoengineering Branch,* 5 Wash. Lee J. Energy, Climate Env't 111,124-125 (2013).

[7] *See generally* Nat'l Ass'n of Regulatory Utility Comm'n's, *Carbon Capture, Utilization, and Sequestration: Technology and Policy Status and Opportunities* (Nov. 5, 2018), available at https://pubs.naruc.org/pub/8C07B393-A9A0-3F04-4832-D43790E10B91 (analyzing future of CCUS technology in the context of declining coal and rising natural gas usage).

[8] *Id.* Although the vitality of traditional dominant energy sources like coal are declining, natural gas remains abundant and affordable to consumers of electricity. *Id.*

[9] *See* James Hansen, *Young people's burden: Requirement of negative emissions*. 8 Earth System Dynamics 577 (2017); *International Energy Agency, Carbon Capture and Storage: The Solution of Deep Emissions Reductions*, OECD/IEA (2015), *available at* https://www.iea.org/

Utilization, and Sequestration involve processes through which CO_2 is captured and injected underground, either for enhanced oil recovery or for long term or permanent storage.[10] The sequestered CO_2 may be captured from anthropogenic sources such as coal and natural gas fired power plants, natural gas separation facilities, or net-negative bioenergy facilities (BECCS)[11] or captured through "Direct Air Capture technologies" that extract carbon dioxide from ambient air[12] or in closed-loop industrial processes.[13] These processes may make it possible to decarbonize fossil and bioenergy generation[14] and, through negative emissions, reduce atmospheric CO_2 below current levels.[15] The permanent underground sequestration of CO_2 has been promoted as one of the "stabilization wedges" critical to an economically sustainable approach to achieving global climate reduction goals.[16] The Intergovernmental Panel on Climate Change (IPCC) has concluded that large scale implementation of CCUS technologies are necessary to achieve both a 1.5 °C and 2 °C degree scenarios "that do not radically reduce energy demand or do not offer carbon-neutral alternatives to liquids and gases that do not rely on bioenergy."[17] As

---

publications/freepublications/publication/CarbonCaptureandStorageThesolutionfor deepemissionsreductions.pdf.

[10] Academic literature refers to both CCUS and CCS, often using the terms interchangeably. However, there are differences between projects where CO_2 is exclusively stored and projects where CO_2 is utilized for EOR, or the production of chemicals or other industrial products, *see*, Rosa M. Cuellar-Franca & Adisa Azapagic, *Carbon Capture, Storage, and Utilization Technologies: A Critical Analysis and Comparison of Their Life Cycle Environmental Impacts*, 9 J. CO_2 UTILIZATION 82, 83 (2015).

[11] Joris Kornneeff, et al., *Global Potential for Biomass and Carbon Dioxide Capture, Transport and Storage up to 2050,* 11 INT'L J. GREENHOUSE GAS CONTROL 117, 119 (2012).

[12] David Keith, *Why capture CO_2 from the atmosphere*, 325 SCI. 1654, 1654 (2009).

[13] *See* Marco Mazzotti, et al., *Direct air capture of CO_2 with Chemicals: Optimization of a two-loop hydroxide carbonate system using a countercurrent air-liquid contactor*, 1 CLIMATIC CHANGE 118, 120 (2013) ("Direct Air Capture involves a system with an 'air contactor' where ambient air flows over a chemical sorbent that selectively removes the CO_2 …[which is] then released as a concentrated stream for disposal or reuse.")

[14] R. Stuart Haszeldine, *Can CCS and NETs Enable the Continued Use of Fossil Carbon Fuels after CoP21?,* 32(2) OXFORD REV. ECON. POL'Y 304, 310 (2016).

[15] Global CCS Inst., *supra* note 2 at 18, 20.

[16] *See* Stephen Pacala & Robert Socolow, *Stabilization Wedges: Solving the Climate Program for the Next 50 Years With Current Technologies*, 305 SCI. 968 (Aug. 13, 2004), *available at* http://science.sciencemag.org/content/305/5686/968.full.

[17] INTERGOVERNMENTAL PANEL ON CLIMATE CHANGE, GLOBAL WARMING OF 1.5 °C AN IPCC SPECIAL REPORT ON THE IMPACTS OF GLOBAL WARMING OF 1.5 °C ABOVE PRE-INDUSTRIAL LEVELS AND RELATED GLOBAL GREENHOUSE GAS EMISSION PATHWAYS, IN THE CONTEXT OF STRENGTHENING THE GLOBAL RESPONSE TO THE THREAT OF CLIMATE CHANGE, SUSTAINABLE DEVELOPMENT, AND EFFORTS TO ERADICATE POVERTY 134–35 (V. Masson-Delmotte, et al., eds., 2018); Some scientists have questioned whether bioenergy with carbon capture and sequestration are viable at the scale forecasted by the majority of 2 °C models, *see,* Jeff Tollefson, *Is The 2 Degree C World a Fantasy?* NATURE (November 24, 2015), *available at* http://www.nature.com/news/is-the-2-c-world-a-fantasy-1.18868

such, geologic carbon sequestration is among the mitigation technologies available to combat climate change.[18]

Even without including saline storage, geologic storage capacity is sufficient to accommodate the demand of mitigation pathways used within the majority of climate models.[19] The U.S. Department of Energy (DOE) estimates that in the U.S. alone there is adequate sequestration capacity for geologic storage to contain more than 3300 billion metric tons of CO_2.[20] Sequestration requires rock formations with impervious layers and free of faulting to prevent the injected CO_2 from migrating or escaping into other formations, such as fresh water aquifers, or to the surface.[21] The underground reservoirs where CO_2 can be sequestered may be depleted oil or gas fields or newly discovered non-hydrocarbon storage sites such as deep saline aquifers or coal seams.[22] For example, the Jackson Dome in the Pisgah Anticline in Mississippi is thought to have securely stored more than 200 metric tons of naturally occurring CO_2 for 65 million years.[23]

Despite the abundance of geologic storage capacity, geologic storage has recently been de-emphasized as a mechanism for emissions reduction under the Clean Air Act. The principal rule and regulations that would have required or encouraged the use of CCS for climate reduction in the United States have been revised or repealed. CCS was identified as the "best system of emissions reduction" under the Clean Air Act[24] for fossil fuel boilers and integrated gasification combined cycle units under President Obama's Clean Power Plan.[25] The Affordable Clean Energy Rule, which replaced and repealed the Clean Power Plan, replaced CCS with heat rate

---

[18] U.S. Dep't of Energy, Carbon Capture Utilization and Storage: Climate Change, Economic Competitiveness, and Energy Security (August 2016), *available at* https://energy.gov/sites/prod/files/2016/09/f33/DOE%20%20Carbon%20Capture%20Utilization%20and%20Storage_2016-09-07.pdf

[19] *Id.*

[20] U.S. Department of Energy, Carbon Sequestration Atlas of the United States and Canada, 15, (2007). In 2017, U.S. energy-related carbon dioxide (CO_2) emissions in 2017 were approximately 5.14 billion metric tons, *see U.S.* Energy Related CO_2 Emissions Fell Slightly in 2017, U.S. Energy Info. Admin. (Sept. 5, 2018), *available at* https://www.eia.gov/environment/emissions/carbon/

[21] Michael. J. Nasi & Jacob Arechiga, *Greenhouse Gas Reduction Technologies for Power Generation*, RMMLF Special Institute, Climate Change Law and Regulations: Planning for a Carbon-Constrained Regulatory Environment ch. 9B (2015).

[22] Stefan Bachu, *Identification of Oil Reservoirs Suitable for CO_2-EOR and CO_2 Storage (CCUS) Using Reserves Databases, with Application to Alberta, Canada*, 44 Int'l J. Greenhouse Gas Control 152, 153 (2016); Stephanie M. Haggerty, Note, *Legal Requirements for Widespread Implementation of CO_2 Sequestration in Depleted Oil Reserves*, 21 Pace Envt'l L. Rev. 197, 200–01 (2003).

[23] Sally Benson et al., *Underground Geological Storage*, in IPCC Special Report on Carbon Dioxide Capture and Storage 210 (Bert Metz et al. eds., 2005), *available at* http://www.ipcc.ch/pdf/special-reports/srccs/srccs wholereport.pdf

[24] 42 U.S.C. § 7411(a)(1) (West 2018).

[25] *Carbon Pollution Emission Guidelines for Existing Stationary Sources: Electric Utility Generating Units*, 80 Fed. Reg. 64,662 (Oct. 23, 2015).

improvement.[26] Similarly, EPA rules previously promoted Carbon Capture, Sequestration, and Utilization as the "best system of emissions reduction" for new coal-fired generating units under section 111(b) of the Clean Air Act.[27] The EPA's current proposed rule reverses this finding due to the high cost and limited geographic availability of CCUS, instead encouraging steam cycle efficiency and best operating practices.[28]

Separately, state laws and policies and federal tax credits continue to encourage carbon storage and utilization, leading to an increase in interest in onshore carbon storage projects. Many of these potential projects are located within, or beneath, existing oilfields. As such, they benefit from existing infrastructure and technical knowledge. Policies that encourage to CO_2-EOR projects to maximize associated storage provide a bridge towards decarbonization of transportation fuels while providing commercial incentives for carbon dioxide reduction projects including Direct Air Capture and BECCS. These projects may also provide needed economic opportunity to fossil-dependent energy communities facing devastating workforce and economic impacts from the energy transition. Policies and laws should further encourage transition of depleted CO_2-EOR assets into incremental geologic storage projects.

2 Associated Storage with Enhanced Oil Recovery

Of the CO_2 that is captured, and transported and sequestered today nearly all of it is used in tertiary recovery of oil from hydrocarbon reservoirs, often referred to as enhanced oil recovery (EOR).[29] Conventional oil production may only produce as much as 80%, or as little as 10%, of the initial oil and gas in place.[30] As pressure within the reservoir diminishes, oil remains trapped within the pore space.[31] Injection of CO_2 mobilizes some of the oil within the pore spaces so that it can flow towards a production well.[32] For example, injection of CO_2 has been used to increase

[26] Repeal of the Clean Power Plan; Emission Guidelines for Greenhouse Gas Emissions from Existing Electric Utility Generating Units; Revisions to Emission Guidelines Implementing Regulations, 84 FR 32520 (July 8, 2019). The EPA is currently finalizing implementing regulations for the Affordable Clean Energy Rule and future emission guidelines promulgated under CAA section 111(d).

[27] *See* 42 U.S.C. § 7411(b) (West 2018).

[28] *See* Review of Standards of Performance for Greenhouse Gas Emissions From New, Modified, and Reconstructed Stationary Sources: Electric Utility Generating Units, 83 FR 65617 (Dec. 21, 2018).

[29] Ian J. Duncan, *CO_2 -EOR 101: An Overview of CO_2 Enhanced Oil Recovery, Enhanced Oil Recovery: Legal Framework for Sustainable Management of Mature Oil Fields*, 2015-4 Rocky Mt. Min. L. Fdn. 4-1 (2015).

[30] *Id.*

[31] *Id.* at 4-2.

[32] *Id.*

hydrocarbon production in the Permian Oil Field in Texas since at least 1972, resulting in the injection of more than 12 trillion cubic feet (Tcf), or 900 million short tons, of CO_2 as of December 2018.[33] These improved recovery operations are estimated to have increased production by approximately 1.86 billion barrels. EOR operations also result in permanent underground storage of some of the injected CO_2. As long as tertiary recovery operations continue, CO_2 is recycled and reinjected with only minimal losses throughout the process.[34] At the conclusion of operations approximately 90% of the total CO_2 injected will remain within the depleted hydrocarbon reservoir, a process that is referred to as "associated storage."[35] Associated storage accounts for the majority of anthropogenic CO_2 that has been sequestered to date.[36]

Historic processes for CO_2-EOR have not themselves resulted in production that is net-negative.[37] The eventual combustion of the produced hydrocarbons has still resulted in more CO_2 than was captured and sequestered via associated storage. First, although CO_2-EOR has resulted in the injection and storage of substantial volumes of CO_2, almost all of the CO_2 used in EOR operations is natural CO_2 - meaning it is drilled and produced from naturally occurring subsurface deposits.[38] Thus it does not displace anthropogenic sources or result in carbon dioxide removal. As of 2016, only eight of more than 114 commercial CO_2 injection projects used anthropogenic CO_2, injecting an estimated total of 21 metric tons annually.[39] Even where anthropogenic sources are used, the net emissions benefits may vary based on consideration of a number of factors, including the source of CO_2, the method of

[33] Figure provided via personal correspondence with Meltzer Consulting.

[34] Robert C. Ferguson, et al., *Storing CO_2 with Enhanced Oil Recovery*, ENERGY PROCEDIA 1 (2009) 1989-1996; J. Greg Schnacke et al., *Carbon Dioxide Infrastructure: Pipeline Transport Issues and Regulatory Concerns – Past, Present, and Future, Enhanced Oil Recovery: Legal Framework for Sustainable Management of Mature Oil Fields*, 2015 RMMLF SPECIAL INSTITUTE 10, 10-8 (2015).

[35] Stephen L. Melzer, *Carbon Dioxide Enhanced Oil Recovery (CO_2 EOR): Factors Involved in Adding Carbon Capture, Utilization and Storage (CCUS) to Enhanced Oil Recovery* 11 (February 2012) (report prepared for the National Enhanced Oil Recovery Initiative, Center for Climate and Energy Solutions) (hereafter "*Melzer EOR Report*")

[36] Philip M. Marston & Patricia A. Moore, *From EOR to CCS: The Evolving Legal and Regulatory Framework for Carbon Capture and Storage*, 29 ENERGY L. J. 421, 424–25 (2008) ("[t]he amount of CO_2 that has been incidentally stored [as residual unrecoverable CO_2 injected for EOR] over the last several decades dwarfs the volumes injected by CCS pilot projects around the world.").

[37] Life cycle analyses of the net environmental impacts of associated storage from CO_2-EOR are unclear, see, Michael Godec et al., *Evaluation of Technology and Policy Issues Associated with the Storage of Carbon Dioxide via Enhanced Oil Recovery in Determining the Potential for Carbon Negative Oil*, 114 ENERGY PROCEDIA 6563, 6573-74 (2017).

[38] Bob Berwyn, *Wait, They're drilling for CO_2 in Colorado?* THE COLORADO INDEPENDENT (March 10, 2015), *available at* https://www.coloradoindependent.com/2015/03/10/wait-theyre-drilling-for-co2-in-colorado/

[39] Massachusetts Institute of Technology, Commercial EOR Projects using Anthropogenic Carbon Dioxide, https://sequestration.mit.edu/tools/projects/index_eor.html (Sept. 2016).

capture, and the distance from the source to the enhanced recovery asset.[40] For example, the life cycle emissions of production from CO_2-EOR using certain types of anthropogenic CO_2 could be as much as 60% lower than CO_2-EOR using natural source CO_2 and average domestic oil production.[41] The carbon dioxide removal benefits of CO_2-EOR thus vary significantly between projects.

New technologies have the potential to increase the carbon dioxide removal potential of CO_2-EOR. The majority of life-cycle emissions analyses consider only historic volumes of associated storage.[42] For the past 40 years, the exclusive goal of CO_2-EOR has been increased hydrocarbon production. In this context, CO_2 is not viewed as a waste material, but rather as a valuable and expensive commodity that must be purchased and the loss of which should be minimized.[43] Without a financial incentive for associated storage of CO_2, project operators have sought to purchase the minimal amount of CO_2 necessary, and deliberately limiting the storage achieved by CO_2-EOR.[44] Thus life-cycle analyses based on historic storage do not reflect the GHG reduction potential of associated storage.[45] The use of "next-gen" technologies that seek to maximize storage as well as hydrocarbon production could vastly increase the amounts of CO_2 sequestered during CO_2-EOR operations, possibly even achieving net negative hydrocarbon production.[46]

The track record of CO_2-EOR operations demonstrate the ability to successfully, and commercially, inject $CO_{2,}$ resulting in sequestration at small scales in a manner that has not to date resulted in catastrophic environmental harms.[47] However, despite the lack of significant incidents to date, the capture, injection, and use of CO_2 for EOR may pose significant environmental risks.[48] The risks include potential unintended releases of CO_2 into the environment during transportation, injection, or as a

---

[40] Michael L. Godec, et al., *Potential Issues and Costs Associated with Verifying CO_2 Storage During and After CO_2-EOR*, 114 ENERGY PROCEDIA 7399, 7402 (2017).

[41] Hussain, et al., *Comparative life-cycle inventory (LCI) of greenhouse gas (GHG) emissions of enhanced oil recovery (EOR) methods using different CO_2 sources*, 16 INT'L J. GREENHOUSE GAS CONTROL 129–144, (2013)

[42] Michael Godec et al., *supra* note 38 at 6565.

[43] *See* Godec, *Potential Issues*, *supra* note 41 at 7402.

[44] Godec, et al., *supra* note 38 at 6566 ("[s]ince the purchased cost of injected CO_2 was often the largest cost component of a CO_2-EOR project, CO_2-EOR operators attempted to optimize incremental oil production in individual CO_2-EOR projects by minimizing the amount of CO_2 injected per incremental barrel of oil produced.").

[45] Godec, et al., *supra* note 38, at 6567.

[46] U.S. Dep't of Energy/National Energy Technology Laboratory, DOE/NETL-2011/1504, *Improving Domestic Energy Security and Lowering CO_2 Emissions with "Next Generation" CO_2-Enhanced Oil Recovery (CO_2-EOR)*, (report prepared by Advanced Resources International, Jun. 20, 2011), *available at* http://www.netl.doe.gov/energy-analyses/pubs/NextGen_CO2_EOR_06142011.pdf

[47] Data regarding incidents in CO_2 wells may be difficult to study. *See* Porse, S.L., Wade, S., & Hovorka, S.D., *Can We Treat CO_2 Well Blowouts Like Routine Plumbing Problems? A Study of the Incidence, Impact, and Perception of Loss of Well Control*, 63 ENERGY PROCEDIA 7149 (2014).

[48] Duncan, *supra* note 30, at 7-8.

result of improper storage or monitoring.[49] For example, migration through leaking or improperly plugged wells could result in damage to overlying fresh water aquifers.[50] Similarly, a catastrophic breach or release of CO_2 from the storage reservoir to the atmosphere, such as at Aliso Canyon, could have severely adverse local impacts on human health and the physical environment.[51] Should some sequestered CO_2 leak back into the atmosphere, it would negate some of the climate benefits, or could even result in additional net climate harm after considering the opportunity cost and energy inputs required for sequestration projects.[52] Although the net impacts may be difficult to measure, these projects also pose environmental impacts associated with expanded energy production including hazardous wastes products resulting from CO_2 recovery techniques,[53] induced seismicity,[54] and land use impacts associated with expanded infrastructure.[55]

Carbon capture technologies, particularly those involving EOR or other carbon utilization, are also criticized for presenting a potential moral hazard that could lock society "into a high-temperature pathway."[56] The term "moral hazard," borrowed from insurance law, refers to a scenario in which costly behavior is unmitigated or incentivized as a result of inequitable distribution of risk.[57] In the context of carbon dioxide removal approaches, the moral hazard concern is that carbon sequestration

---

[49] Alexandra B. Klass & Elizabeth J. Wilson, *Carbon Capture and Sequestration: Identifying and Managing Risks*, 8 Issues L. Scholarship 1, 1 (2009).

[50] U.S. Gen. Accounting Office, RCED-89-97, Report to the Chairman Environment, Energy and Natural Resources Subcommittee, Committee on Government Operations, House of Representatives, Drinking Water Safeguards Are Not Preventing Contamination from Injected Oil and Gas Wastes 19 (1989).

[51] *See* S. Conley, et al., *Methane emissions from the 2015 Aliso Canyon blowout in Los Angeles, CA*, 351 Science 1317 (25 Feb 2016), available at http://science.sciencemag.org/content/351/6279/1317.full. From October 2015 to February 201 a major blowout of natural gas from a storage reservoir beneath Aliso Canyon, near Los Angeles, caused leakage of gas and toxins harmful to human health. *Id.*; Wilson, E.J., Friedmann, S.J., & Pollak, M.F., *Research for Deployment: Incorporating Risk, Regulation, and Liability for Carbon Capture and Sequestration*, 41 Envt'l Science & Tech. 5945, 5946 (2007).

[52] Klass and Wilson, *supra* note 50; Klaus Keller et al., *Carbon Dioxide Sequestration: How Much and When?*, 88 Climatic Change 267, 268 (2008).

[53] Khoo, H.H. & Tan, R.B.H., *Life cycle evaluation of CO_2 recovery and mineral sequestration alternatives*, 25 Envt'l Science & Tech. 208, 212 (2006).

[54] Trae Gray, *A 2015 Analysis and Update on U.S. Pore Space Law—The Necessity of Proceeding Cautiously With Respect to the "Stick" Known as Pore Space*, 1 Oil & Gas, Nat. Resources & Energy J. 227, 326 (2015).

[55] Kevin Anderson & Glen Peters, *The Trouble With Negative Emissions*, 354 Sci. 6309 (14 Oct 2014); Karsten Pruess et al., *Numerical Modeling of Aquifer Disposal of CO_2*, 8 Soc'y Petroleum Engineers J. 49, 52–53 (2003).

[56] Dominic Lenzi, *The ethics of negative emissions*, Global Sustainability 2 (18 July, 2018); Albert C. Lin, *Does Geoengineering Present a Moral Hazard?*, 40 Ecology L.Q. 673, 676–77 (2013).

[57] William Burns, *Geoengineering the Climate: An Overview of Solar Radiation Management Options*, 46 Tulsa L. Rev. 283, 297 (2010).

may prolong the viability of fossil fuels by reducing the political will to adopt stringent mitigation policies that would reduce GHG emissions.[58] Investment in CO_2-EOR could also have energy production ramifications. Some groups fear that continued reliance on fossil fuels and access to credits for CO_2-EOR projects under federal and state programs could displace funding and resources from the development of more sustainable energy alternatives.[59] Opponents also argue that policies that support continued hydrocarbon production defer the fundamental realignment of consumer habits and a transformation of economic structures necessary to facilitate a transition away from fossil fuels.[60] Aversion towards more expensive or inconvenient technologies may lead to over-optimism among scientists[61] regarding the potential for cost reductions and increased carbon dioxide reduction potential of CCUS. Cumulatively, these factors may drive policy makers away from approaches that force consumption changes by the imposition of pollution taxes or emissions restrictions.[62] Combined with potential impacts to the physical environment, these concerns have led certain environmental advocacy groups and policy analysts to question whether carbon capture technologies, particularly those associated with EOR, should even be considered as part of a climate solution.[63]

At the same time, however, the decision not to further pursue carbon utilization techniques on the grounds of moral hazard could bear equal or greater consequences.[64] The orchestration of society's transition from fossil fuels to more sustainable energy sources is already underway. Completing that transition, however, could take many decades, demanding radical changes from both consumers and industry.[65] In the time it would require for society to collectively reduce emissions and harness

---

[58] Paul Baer, *An Issue of Scenarios: Carbon Sequestration as Investment and the Distribution of Risk*, 59 Climatic Change 283, 287 (2003); Intergovernmental Panel on Climate Change, Global Warming of 1.5 ° C 4-21 (2018).

[59] Alexandra B. Klass & Elizabeth J. Wilson, *Climate Change Carbon Sequestration and Property Rights*, 2010 U. Ill. L. Rev. 363, 371–72 (2010), *but see*, Jesse Reynolds, *A Critical Examination of the Climate Engineering Moral Hazard and Risk Compensation Concern*, 2 Anthropocene Rev. 174, 185 (2015) (for the argument that evidence of moral hazard is "inconclusive").

[60] Jay Michaelson, *Geoengineering: A Climate Change Manhattan Project*, 17 Stan. Envt'l L.J. 73, 132–34 (1998) (discussing deep environmentalist approaches to climate change, which anticipate a structural shift in consumer habits and technology).

[61] Christine Merk, et al., *Do climate engineering experts display moral hazard behaviour?*, 19 Climate Pol'y 231, 232 (2018).

[62] Troy H. Campbell and Aaron C. Kay, *Solution Aversion: On the Relation Between Ideology and Motivated Disbelief*, 107 J. Personality & Social Psychology 809, 811 (2014); Klass and Wilson, *supra* note 60, at 364–65.

[63] David Biello, *Can Carbon Capture Technology Be Part of the Climate Solution*, Yale Env't 360 (September 8, 2014).

[64] Anthony E. Chavez, *Using Legal Principles to Guide Geoengineering Deployment*, 24 N. Y. U. Envt'l L. J. 59, 70 (2016); Reynolds, *supra* note 60, at 183.

[65] Anthony E. Chavez, *A Napoleonic Approach to Climate Change: The Geoengineering Branch*, 5 Wash. & Lee J. Energy, Climate, & Env't 93, 105–06 (2013) ("[c]onversion to new energy technologies occurs [slowly] … On average, energy technologies have required 30 years to advance from being technically available to reaching materiality. This pattern was consistent

more sustainable energy sources, we may pass critical climatic thresholds, manifesting itself in serious ramifications for human institutions and ecosystems.[66] CCUS, conversely, could be implemented on smaller scales, sequestering greenhouse gas emissions at least until the time that such a widespread transition or other geoengineering solutions become feasible.[67] In aid of the eventual goal of decarbonization, rather than in contravention of it, CCUS could decelerate the effects of climate change by capturing a portion of carbon emissions from power plant or other industrial point sources.[68]

3 Standardization, Regulation, and Subsidization

The study of CCS and CCUS is pertinent to conversations regarding climate engineering for the road map it provides for development of internationally accepted technical standards, application of a cooperative federalism regulatory framework, and provision of federal incentives.[69] The International Organization for Standardization (ISO) has published seven standards for carbon dioxide capture, transportation, and storage.[70] Standards are developed over a several year period by an international ISO technical committee comprised of international technical experts, including representatives from developing countries. In addition to its seven previously published standards, the technical committee also has four other standards under development for carbon dioxide capture, lifecycle risk management, CO_2 stream composition, and quantification and verification.[71] For example, the committee for Carbon Dioxide, transportation, and geologic storage (ISO/TC 265) recently approved the final draft of an international standard for the carbon dioxide capture, transportation, and geologic storage for CO_2-EOR.[72] This standard covers topics including demonstration of containment, well standards, quantification of

---

across all technologies, including nuclear power, natural gas, biofuels, wind, and solar photovoltaic.").

[66] Michaelson, *supra* note 61, at 102–05.

[67] Alexandra B. Klass & Elizabeth Wilson, *supra* note 60, at 423.

[68] Jeremy David & Howard Herzog, *The Cost of Carbon Capture*, CARBON CAPTURE & SEQUESTRATION TECHNOLOGIES AT MIT 2 (Sept. 30, 2016), *available at* http://sequestration.mit.edu/pdf/David_and_Herzog.pdf

[69] Jonas J. Monast, et al., *A Cooperative Federalism Framework for CCS Regulation*, 7 ENVT'L & ENERGY L. & POL'Y J. 1, 6 (2012).

[70] *See* Int'l Org. for Standardization, *Standards Catalogue, ISO/TC 265 Carbon dioxide capture, transportation, and geological storage*, *available at* https://www.iso.org/committee/648607/x/catalogue/p/1/u/1/w/0/d/0 (last visited July 8, 2019).

[71] Int'l Org. for Standardization, *ISO/DIS 27916: Carbon Dioxide Capture, Transportation and Geologic Storage—Carbon Dioxide Storage Using Enhanced Oil Recovery*, *available at* https://www.iso.org/standard/65937.html (last visited July 7, 2019).

[72] *Id.*

stored CO_2, monitoring, record keeping, and project termination and closure.[73] These consensus standards not only provide technical guidance and encourage international conformity to generally accepted best practices, but may also be incorporated by reference into law.[74]

CCUS may also provide a model for cooperative federalism frameworks for climate engineering governance. CO_2-EOR operations are subject to a robust array of state and federal regulations in the United States designed to protect against potential environmental harms, including leakage of CO_2 into existing aquifers or the atmosphere. The subsurface injection of CO_2 for hydrocarbon recovery is managed pursuant to the Underground Injection Control program (UIC program) under the Safe Drinking Water Act.[75] The UIC program classifies underground injection activities into six separate classes.[76] The majority of CO_2 injection wells are permitted as a Class II hydrocarbon associated injection well and managed under a state or federal UIC program.[77] Class II well regulations prohibit the use of injection pressure that could "initiate new fractures or propagate existing fractures in the confining zone"adjacent to underground drinking water sources, require a demonstration of mechanical integrity every 5 years, and require monthly monitoring of injection pressures and other factors.[78] In states that have primacy over administration of the UIC program, regulation of Class II injection wells are delegated to state oil and gas conservation agencies.[79] In addition to requirements of the UIC program, many states with robust injection programs have also instituted additional regulations to address risks of induced seismicity such as the traffic-light system for UIC well permitting which has been instituted in Oklahoma and Illinois.[80]

CO_2-EOR operations in the United States, and the capture and transport of CO_2 associated with those operations, are also potentially subject to a number of other federal and state environmental laws requiring analysis of environmental impacts or providing regulation and control of hazardous waste. This network of federal

[73] *Id.*

[74] National Technology Transfer and Advancement Act of 1995, Pub. L. No. 104-113, § 12(d), 110 Stat. 775 (1996); Office of Mgmt. & Budget, Circular A-119, Federal Participation in the Development and Use of Voluntary Consensus Standards and in Conformity Assessment Activities, 63 Fed. Reg. 8546, 8549 (Feb. 19, 1998) (available at http://www.whitehouse.gov/omb/circulars_a119); Emily S. Bremer, *On the Cost of Public Standards in Private Law*, 63 U. Kan. L. Rev. 279, 296 (2015

[75] 42 U.S.C. §§ 300f –300j-26 (West 2018); Elizabeth Wilson, et al., *Regulating the Ultimate Sink: Managing the Risks for Geologic CO_2 Storage*, 37 Envt'l Sci. & Technology 3476, 3478 (2003).

[76] Arnold W. Reitze Jr., *Federal Control of Carbon Capture*, 41 Envt'l L. Rep. 10,796, 10,803 (2011).

[77] *See* Marston & Moore, *supra* note 37, at 467; Susan Zachos, *Overview of Class II Underground Injection Control Program*, 35A RMMLF-INST 4, 4-1 (1994).

[78] 40 C.F.R. § 146.23 (West 2018).

[79] *See, e.g.*,N.M. Stat. Ann. § 70-2-12 (West 2018); Wyo. Stat. Ann. § 30-5-104(d)(v) (West 2018).

[80] 62 Ill. Adm. Code 240.796 (West 2018); Okla. Corp. Comm'n, *Statement on Proactive Approach to Seismic Activity*, (Dec. 1, 2014), *available at* http://www.occeweb.com/SeismicStatementB.pdf

environmental laws encompasses the National Environmental Policy Act (NEPA)[81] the Comprehensive Environmental Response, Liability, and Compensation Act (CERCLA),[82] and the Resource Conservation and Recovery Act.[83] These acts do not prescribe regulatory rules specific to CO_2-EOR operations, but often apply broadly to carbon-emitting and industrial activities. NEPA, for example, requires the preparation of an Environmental Impact Statement (EIS) when a proposed federal action will "significantly [affect] the quality of the human environment."[84] In practice, NEPA affects the timing of authorization of specific projects, and is invoked where federal funding is a component of project funding, for the development of resource management plans authorizing EOR activities, for the creation of enhanced recovery units including more than 20% federal minerals, or where CO_2 pipelines cross federal land.[85] Similar state statutes in California, New York may require state agencies such as the oil and gas conservation agencies responsible for permitting Class II wells to consider whether carbon-emitting activities pose a significant environmental impact, and, if so, to prescribe mitigation measures.[86] Both CERCLA and RCRA address the management and clean-up of hazardous wastes dangerous to human health or the environment.[87] While CO_2 has not been defined by EPA as a hazardous substance under either CERCLA or RCRA,[88] other constituents of a CO_2 stream, or interactions between those components and the injection environment, could subject CO_2 Pipeline Operators or injection projects to regulation under those acts.[89]

CO_2-EOR operations are also subject to state laws and regulations governing drilling, surface operations, and the creation of injection units. The state oil and gas conservation agency may be responsible for all of the following: permitting; securing operating bonds; determining setbacks from property lines, wells, or occupied structures; setting well construction requirements; regulating the activities for protection of human health and the environment; and verifying compliance with state

[81] 42 U.S.C. §§4321-4370 h (West 2018).

[82] 42 U.S.C. §§ 9601-9675 (2012).

[83] 42 U.S.C. § 6901 (West 2018).

[84] 42 U.S.C. § 4332(c) (West 2018).

[85] Lin, *supra* note 57, at 2555; Tara Righetti, *Siting Carbon Dioxide Pipelines*, 3 Oil & Gas, Nat. Resources & Energy J. 907, 931-933 (2017).

[86] Cal. Pub. Res. Code, §§ 21,000–21189.57 (West 2018); N.Y. Envtl. Cons.. Law Ann. § 8-0109 (McKinney 2006); Cal. Code Regs. tit. 14, § 15,064 (West 2018); California Carbon Capture and Storage Review Panel, *California Draft Report A Review of Carbon Capture and Storage In California* 14 (2010).

[87] *See* 42 U.S.C. §§ 9601-9675 (2012).

[88] 40 C.F.R. 9, 260, 261.4(h) (2014).

[89] Jeffrey W. Moore, *The Potential Law Of On-Shore Geologic Sequestration Of CO_2 Captured From Coal-Fired Power Plants*, 28 Energy L.J. 443, 445 (2007); Marston & Moore, *supra* note 37 at 471.

statutes enacted for the protection of split-estate surface interests.[90] In addition to granting the drilling permit, the majority of state agencies also have the authority to combine— involuntarily, if necessary—subsurface interests into an enhanced recovery unit to accomplish coordinated development and to allocate production and costs among owners within the unit.[91] Where lands within an injection unit are fragmented with both federal and fee ownership, a small fraction of federal lands may be included in units formed pursuant to state law. Many states further require recording of completion records and regular mechanical integrity testing, monitoring or reporting.[92]

Federal rules have further encouraged development of comprehensive programs for monitoring and reporting of GHG emissions associated with CO_2-EOR. In 2010, the EPA established reporting requirements pursuant to the GHG Reporting Program (GHGRP) Subpart PP for suppliers of carbon dioxide, Subpart W for oil and gas equipment, and Subpart UU for facilities that inject CO_2 underground to enhance hydrocarbon recovery.[93] These regulations require reporting of data related to emissions and CO_2 management during CO_2-EOR. Combined with commercial concerns, these rules have encouraged project operators to develop comprehensive CO_2 "reservoir surveillance" programs to monitor pressures, identify changing risk profiles during injection, and to optimize production.[94]

Congress, states, and the Department of Energy (DOE) have provided incentives related to technological development and storage of CO_2 associated with EOR. For instance, the DOE's Office of Fossil Energy and the National Energy Technology Laboratory have provided direct support by way of grants to expand research and commercial applications of carbon storage. This support is expected to continue and may increase associated with President Biden's plans to revitalize energy communities and address climate change. Recently, the Interagency Working Group on Coal and Power Plant Communities and Economic Revitalization identified retrofitting traditional energy generation and industrial facilities with carbon capture technologies as a pathway to create good-paying jobs in energy communities.[95] In addition to DOEs, that report identified that congress had allocated \$8.5 billion to DOE's Loan Program Office to support new and innovative technologies to decarbonize

[90] *See* Tara Righetti, *Environmental Considerations in Oil and Gas Conservation and Permitting*, 64 Rocky Mtn. Min. L. Inst. 5-1 (2018).

[91] Marie Durrant, *Preparing for the Flood: CO_2 enhanced oil recovery*, 59 RMMLF-INST 11-1 (2013).

[92] *See, e.g.*, Wyo. Oil and Gas Conservation Comm'n Rules, ch. 4, § 10 (2018).

[93] 40 C.F.R. 98.232 (West 2019); 40 C.F.R. §98.422 (West 2019); 40 C.F.R. §98.470 (West 2019).

[94] Steven Melzer, Carbon Dioxide Enhanced Oil Recovery (CO_2 EOR): Factors Involved in Adding Carbon Capture, Utilization and Storage (CCUS) to Enhanced Oil Recovery, Report to National Enhanced Oil Recovery Initiative, Center for Climate and Energy Solutions, 8-9 (2012).

[95] Interagency Working Group on Coal and Power Plant Communities and Economic Revitalization, Initial Report to the President on Empowering Workers Through Revitalizing Energy Communities (April 2021).

traditional industrial and energy sources through technologies such as CCUS.[96] Combined with tax credits for subsurface storage, federal programs for carbon capture retrofits provide a significant incentive for commercial decarbonization projects.

Geologic storage is further encouraged through Section 45Q of the U.S. Federal Tax Code. Section 45Q provides owners of carbon capture equipment with tax credits for each metric ton of CO_2 disposed of in "secure geological storage."[97] Previously, demonstration of secure geological storage required an operator to comply with the more stringent Section RR of the EPA's GHG Reporting Program, and thus the credit was rarely claimed. Section 45Q was modified and reauthorized with passage of the 2018 Bipartisan Budget Act.[98] The 2018 modifications increase the credit amount, make credits transferable, expand eligible uses of qualified CO_2, make Direct Air Capture facilities eligible for the credit, and direct the Secretary of the Treasury to establish regulations to define secure geologic storage.[99] The Energy Act of 2020, passed as Division Z of the omnibus appropriations bill on December 27th, again included extensions of the 45Q credit as well as providing additional funding to DOE for carbon capture demonstration projects, directed EPA to support direct air capture and carbon capture research, and directing the Council on Environmental Quality to establish guidance for carbon capture projects.[100]

4 Incremental Storage: Challenges and Opportunities

Following completion of EOR operations, depleted oil fields are excellent candidates for additional, incremental storage.[101] The holding potential of the depleted reservoir, and its capacity for pressurization, have been demonstrated over geologic time: without the natural trap and seal, oil and gas could not have been stored for millions of years.[102] Further, many of these depleted oil fields, particularly those where there have been tertiary recovery operations, already have much of the necessary infrastructure to transport, pressurize, and inject CO_2, including injection wells, CO_2 pipelines, and compressors. As such, these assets are well-suited for utilizing captured CO_2 and may be a "bird in the hand" for immediate CO_2 removal. Despite these natural geologic and infrastructure attributes, incremental storage projects are subject to many of the same financial constraints presented by geologic storage projects. Further, certain legal and regulatory challenges pose significant obstacles to transitioning EOR facilities into operations for incremental, permanent storage.

96 *Id.*

97 26 U.S.C. 45Q (West 2018).

98 The Bipartisan Budget Act of 2018, Pub. L. No. 115-123, 132 Stat. 232

99 *Id.*

100 Energy Act of 2020, div. Z, Pub. L. No. 116-260, 134 Stat. 1182.

101 Marston & Moore, *supra* note 37 at 437.

102 *Id.*

Incremental and geologic storage operations in the United States require an entirely different set of property entitlements compared with CO_2 injection for tertiary recovery. The United States is one of a few countries in the world with private ownership of mineral rights.[103] In many cases, mineral rights may be owned by a different person or entity than the surface property interests.[104] On private land, the pore space necessary for geologic storage and other non-mineral components of the subsurface are part of the surface estate.[105] Conversely, the mineral owner has the right to explore for and produce the hydrocarbons contained within that pore space, including the right to use the surface (which is inclusive of the pore space) as is reasonably necessary to hydrocarbon production.[106] Thus, when the mineral owner leases its right to the oil and gas operator it conveys the right to inject CO_2 or other substances into the pore space in order to increase the recovery of hydrocarbons. That right ends, however, when the economic recovery of hydrocarbons ends, but may be restored at such point as additional hydrocarbons become recoverable. As a result, transitioning assets from CO_2-EOR would likely require identifying the owners of the pore space and securing additional injection rights from both surface owners and potentially, non-development rights from mineral owners for any residually producible hydrocarbons.[107] Even where mineral and surface ownership is unified, the oil and gas lease rights end when hydrocarbons cease to be produced in paying quantities.[108] Pore space owners may be hesitant to grant injection leases due to fear about liability for leakage or post-closure issues including contamination of groundwater. Obtaining pore space injection rights on federal land is further complicated by regulatory uncertainty regarding agency authority and the lack of a federal program for granting pore space leases or injection easements.[109]

---

[103] Sylvia L. Harrison, *Disposition of the Mineral Estate on United States Public Lands: A Historical Perspective*, 10 Pub. Land L. Rev. 131, 134 (1989).

[104] Chartiers Block Coal v. Mellon, 25 A. 597, 598 (1893).

[105] Jean Feriancek, *Resolving Ownership of Pore Space*, 26 Nat. Resources & Env'T 49, 49–50 (2012); Troy A. Rule, *Property Rights and Modern Energy*, 20 Geo. Mason L. Rev. 803, 810 (2013); A precise determination of ownership would require an analysis of the conveyance that created the split estate. *See* Bruce M. Kramer, *Horizontal Drilling and Trespass: A Challenge to the Norms of Property and Tort Law*, 25 Colo. Nat. Resources, Energy & Envt'l L. Rev. 291, 296–97 (2014).

[106] Feland v. Placid Oil Co., 171 N.W.2d 829 (N.D.1969); Fischer v. Continental Res., Inc., 49 F. Supp. 3d 637, 646 (D.N.D.2014); Duncan, *supra* note 30, at 7-2; Howard R. Williams, *Williams & Meyers on Oil and Gas Law* § 202.1 (2015).

[107] Wendy B. Jacobs, Global Climate Change and U.S. Law 581 (Michael Gerrard & Jody Freeman eds., 2d ed., 2014) ("it will be no simple logistical matter to determine when precisely the pore space within the mineral estate has been fully mined and has reverted to the surface owner.").

[108] 8 Patrick H. Martin & Bruce M. Kramer, Williams Meyers Manual of Oil and Gas Terms §§ 1125-26 (2015).

[109] Agencies may be able to grant injection easements pursuant to Title V of the Federal Land Management Policy Act. *See*, 43 U.S.C. §1761(a) (West 2018), *and*, Bureau Land Mgmt., Colo. State Office, Colorado State Office, Instructional Memorandum No. CO-2016, Class II Injection Facilities and Wells (Mar. 28, 2016). For an analysis of these issues, *see*, *Tara*

Injection of CO_2 for incremental storage, rather than hydrocarbon purposes, may also subject the project operator to liability for trespass. Once an enhanced recovery unit is created, a mineral owner engaging in injection operations within the unit is generally protected against claims of trespass by a common law doctrine known as the negative rule of capture.[110] That rule provides that a landowner "may inject into a formation substances which may migrate through the structure to the land of others, even if it thus results in the displacement under such land of more valuable with less valuable substances."[111] Thus, the transboundary injection of water or gas for enhanced oil recovery operations has generally been found to result in neither trespass nor conversion.[112] The same, however, cannot be said of operations for incremental injection. Like operations for CO_2-EOR, injection of CO_2 for incremental storage is likely to result in transboundary migration of CO_2 as well as displacement of brine or changes in pressurization within the storage complex.[113] Although the majority of cases asserting subsurface trespass resulting from transboundary migration of wastewater have failed to result in damage awards under both statutory and common law tort arguments,[114] no common law rule such as the rule of capture exists to insulate storage operators from claims of trespass.[115] Evaluating the issue of subsurface trespass and the standing of adjacent surface and mineral owners to sue, have reached different results.[116]

The creation of carbon sequestration units provides administrative remedies related to migrating substances but does not dispatch the related tort issue of trespass. The Interstate Oil and Gas Compact Commission (IOGCC) has developed a Model Statute and Model Rules and Regulations for Carbon Storage.[117] A number of states have adopted statutes for the creation of units for carbon sequestration.[118]

Righetti, Kris Koski, Jesse Richardson, and Sam Taylor, The Carbon Storage Future of Public Lands __ PACE ENVT'L. L. REV. __(2021).

[110] Owen Anderson, *Lord Coke, the Restatement, and Modern Subsurface Trespass Law*, 6 TEX. J. OIL GAS & ENERGY L. 203, 233–234 (2010–2011).

[111] R.R. Comm'n v. Manziel, 361 S.W.2d 560, 568 (Tex. 1962).

[112] Anderson, *supra* note 111, at 233–234.

[113] *See* Edward Rubin et al., *Technical Summary, in Sally Benson* et al., *Underground Geological Storage*, in IPCC SPECIAL REPORT ON CARBON DIOXIDE CAPTURE AND STORAGE (Bert Metz et al. eds., 2005).

[114] *See* Chance v. BP Chemicals, 670 N.E.2d 985 (Ohio 1996); Burlington Res. Oil & Gas Co., LP v. Land & Sons Inc., 259 P.3d 766 (Mont. 2011).

[115] Tara Righetti, *Correlative Rights and Limited Common Property in the Pore Space: A Response to the Challenge of Subsurface Trespass in Carbon Capture and Sequestration*, 47 ENVT'L L. REP. NEWS & ANALYSIS 10,420, 10,429–30 (2017).

[116] Kris Koski, Jesse Richardson, Tara Righetti, & Sam Taylor, UNITED STATES ENERGY ASSOCIATION, *Study on State's Policies & Regulations per CO_2-EOR Storage Conventional, ROZ and EOR in Shale: Permitting, Infrastructure, Incentives, Royalty Owners, Eminent Domain, Mineral-Pore Space, and Storage Lease Issues* (2020).

[117] INTERSTATE OIL & GAS COMPACT COMMISSION TASK FORCE ON CARBON CAPTURE & GEOLOGIC STORAGE, A LEGAL AND REGULATORY GUIDE FOR STATES & PROVINCES 15, 22 (2007).

[118] KY. REV. STAT. ANN. § 353.808 (West 2011); WYO. STAT. ANN. § 34-1-153 (2011).

These rules provide for the creation of injection units within which one party can conduct injection operations upon the approval of some percentage of landowners within the unit. Rules for carbon storage acknowledge the possibility of migration, both within the storage complex and to surrounding formations or the surface. For example, to guard against any such migration, the Model Rules recommend that the operator in its injection plan address how "the mechanisms of confinement" will "prevent migration of CO_2 beyond the proposed storage reservoir."[119] Wyoming permits modification of units to include areas to which injected substances have migrated based on "the fair and equitable determination of pore space storage capacity."[120] Inclusion within a unit, however, does not *per se* bar claims of trespass. In the context of wastewater injection, it has been consistently acknowledged by courts that the receipt of an administrative permit alone is not a defense against tort.[121] State laws permitting unitization and allowing state regulatory agencies to adjust the size of units to account for subsurface migration, however, may give state agencies primary jurisdiction over subsurface trespass issues thus requiring potential plaintiffs to first exhaust administrative remedies. There is federal process to create carbon storage units on federal lands, nor is it clear whether and to what extend federal lands can be included in carbon storage units created pursuant to state law.

Further, incremental storage operations may be subject to more stringent EPA regulations. Injection wells for geologic storage *without* EOR are permitted and regulated under Class VI of the U.S. EPA's UIC Program.[122] The Class VI Wells contain comprehensive requirements for site characterization; area of review and corrective action; well construction; mechanical integrity, monitoring, recording-keeping and reporting; well plugging, post-injection site care and closure; financial responsibility; and emergency and remedial response.[123] Although many of these requirements also apply to Class II operations, or are incorporated within existing best practices for CO_2-EOR,[124] the Class VI rules arguably provide more complete and rigorous oversight of CO_2 injection for geologic storage than is required by Class II programs.[125] Transitioning a CO_2-EOR facility to a project for incremental CO_2 storage where the primary purpose is long term storage of CO_2 may thus

[119] IOGCC, *supra* note 118, at 26.

[120] Wyo. Stat. Ann. §§ 35-11-313 —316 (2011).

[121] *See* FPL Farming Ltd. v. Environmental Processing Sys., L.C., 351 S.W.3d 306 (Tex. 2011); Snyder Ranches, Inc. v. Oil Conservation Comm'n of N.M., 798 P.2d 587 (N.M. 1990).

[122] 40 C.F.R. §§ 144.11– 144.19, 144.51 (West 2018).

[123] 40 C.F.R. § 144.51.

[124] Godec, *supra* note 41, at 7407.

[125] Nat. Res. Def. Council, *Strengthening the regulation of enhanced oil recovery to align it with the objectives of geologic carbon dioxide sequestration* (2017), *available at* https://www.nrdc.org/sites/default/files/regulation-eor-carbon-dioxide-sequestration-report.pdf. Further, unlike the Class II program, only North Dakota has primacy over the Class VI injection program, *See, State of North Dakota Underground Injection Control Program; Class VI Primacy Approval*, 83 Fed. Reg. 17,758 (Apr. 24, 2018).

require additional permitting and regulatory oversight. EPA regulations provide that, "where there is an increased risk to USDWs compared to Class II operations," the project operator must apply for and obtain a Class VI permit.[126] Accordingly, a Class VI permit may not be required in all scenarios.[127] In Wyoming, which has primacy over both the Class VI and Class II programs of the SDWA, the Wyoming Department of Environmental Quality, which has authority for administering the Class VI program, is currently in the midst of negotiating a memorandum of understanding with the Wyoming Oil and Gas Conservation Commission, which has authority over the State's Class II program, to implement discrete aspects of the program, including, for example, the potential conversion of Class II wells to Class VI wells. In most states however, the Class II program is administered by state agencies and the Class VI program is administered by the EPA.[128] As a result, coordination between state and federal agencies, as well as development of new rules, will be necessary to convert Class II wells.

EPA also has separate, more stringent, requirements for reporting, storage certification and monitoring for geologic sequestration projects. Subpart RR of the EPA's Greenhouse Gas reporting program applies to any well or group of wells that inject CO_2 for long-term containment in subsurface geologic formations, including all wells permitted as Class VI under the EPA UIC program.[129] In order to qualify under Subpart RR, the operator must submit a Monitoring Reporting and Verification (MRV) Plan and have it approved by EPA.[130] These rules require that sources of CO_2,[131] as well as leakage[132] be monitored, estimated, and reported to EPA. Approval of a plan by the EPA could be subject to litigation. The costs and benefits of Subpart RR relative to the reporting requirements of Subpart UU are controversial. Some environmental advocates argue that the GHGRP requirements of Subpart RR are not stringent enough to assure proper sequestration or to prevent and mitigate leakage.[133] Others argue that the differences between subpart UU and RR introduce unnecessary obstacles to more widespread use of anthropogenic sources of CO_2 in CO_2-EOR.[134] For example, a 2017 workshop report of the Department of Energy identified concerns that CO_2 pipeline operators will exclude certain upstream sources to avoid potentially subjecting the entire stream to GHG reporting requirements, or that downstream users may avoid accepting anthropogenic or comingled

126 40 CFR § 144.19 (West 2018).

127 Memorandum from Peter C. Grevatt, Director Office of Ground Water and Drinking Water, EPA 1 (Apr. 23, 2018), *available at* https://www.epa.gov/sites/production/files/2015-07/documents/class2eorclass6memo_0.pdf

128 Koski, et al., *supra* note 117.

129 40 CFR § 98.441, subpart RR (2010).

130 40 CFR § 98.448 (2010).

131 40 CFR § 998.446(d) (2010).

132 40 CFR § 998.446(f)(3) (2010).

133 Nat. Res. Def. Council, *supra* note 126, at 7–8.

134 J. Greg Schnacke et al., *supra* note 35 at 10-30–32.

CO_2 to avoid plan approval or reporting requirements.[135] The availability of higher credits for injection of anthropogenic CO_2 under the 45Q program, however, may provide commercial incentives for operators and carriers to use captured CO_2.

Transition of CO_2-EOR assets to geologic storage assets also requires addressing issues of long term environmental and tort liability and transition of ongoing monitoring responsibilities.[136] Operators of geologic storage projects may be liable for future releases of CO_2 under a number of statutes and tort theories.[137] Given the firm life of most of CO_2-EOR operators, one may presume that neither operators nor landowners will want to assume the additional risk presented by geologic storage operations and thus that transfer of stewardship from a private operator to a public entity must occur.[138] Presently, there is no long-term federal program to transfer liability and monitoring and verification requirements for geologic storage operations.[139] In order to attract geologic storage projects, however, a few states have enacted legislation to transfer liability to a federal program, should one exist, or to the state, and to absolve the storage operator from future tort liability resulting from carbon storage.[140] Other states have provided for transfer of ongoing monitoring responsibilities, without limiting the liability of the injection operator.[141] While these programs may be sufficient to attract demonstration scale projects, in order to realize the full potential of carbon storage as a climate engineering tool, a uniform and comprehensive system providing for transfers of custody, monitoring, and liability to public management is needed.[142] For example, the 2010 Report of the Interagency Task Force on Carbon Capture and Storage suggested that liability concerns could be addressed through limitations on claims, legislation facilitating private insurance, government ownership, liability, or indemnification, post closure transfers of liability, or through creation of a liability fund.[143] These or similar programs should be crafted to discourage moral hazards regarding site selection and to assure that public and environmental health are managed over the long term.[144]

---

135 U.S. Dep't of Energy, Siting and Regulating Carbon Capture, Utilization, and Storage Infrastructure, Workshop Report (2017).

136 Alexandra B. Klass & Elizabeth J. Wilson, *Climate Change and Carbon Sequestration: Assessing a Liability Regime for Long-Term Storage of Carbon Dioxide*, 58 Emory L.J. 103 (2008).

137 See *Id.*

138 *Id* at 172.

139 *Id.* at 149.

140 20 Ill. Comp. Stat. 1107/25 (West 2018); Ky. Rev. Stat. § 353.810(3) (West 2018); Mont. Code Ann. § 82-11-181 (West 2018); Tex. Nat. Res. Code Ann. § 119.004 (West 2018).

141 Wyo. Stat. Ann. § 35-11-318 (West 2018).

142 Klass & Wilson, *supra* note 137, at 176.

143 Dep't of Energy, Report of the Interagency Task Force on Carbon Capture and Storage, 109-123, (2010).

144 *Id.* at 172–73.

5 Conclusion

Changes to law and policy may encourage more widespread adoption of anthropogenic CO_2-EOR and the transition of EOR projects into geologic storage projects for incremental storage.[145] Policies to encourage more widespread use of CO_2-EOR and transition of assets for incremental geologic storage should be balanced with health, safety, and environmental concerns.

The associated storage achieved through CO_2-EOR operations can be maximized by encouraging use of next-gen technologies. Assuming stable demand for fossil energy sources, these anthropogenic CO_2-EOR projects have potential to displace higher emitting hydrocarbon resources.[146] Legal and policy changes that encourage use of next-generation technologies by providing methods to certify storage volumes and making available financial incentives may significantly increase the carbon dioxide reduction potential of commercially-driven associated storage.[147] The GHG reduction potential of associated storage can be further achieved through implementation of programs that discourage or prohibit the use of natural CO_2, and which create incentives for the transportation and use of CO_2 from anthropogenic and direct air capture sources.

Greater deployment of CCUS can further be encouraged through the development of rules and protocols to reduce uncertainty regarding the eligibility of carbon storage projects for state and federal tax credits and participation in state carbon reduction programs. These efforts are already underway. Currently, congress is considering a number of proposals that would provide a direct pay option for 45Q and increase the amount of the credit, thus increasing the subsidy for carbon removal projects and decreasing operators dependence on tax equity investors. In early 2019 California finalized its CCS Protocol and Air Resources Board regulations regarding the state's Low Carbon Fuel Standard. The protocol clarifies how life cycle analyses will be applied to quantify the amount of utilized CO_2 eligible for credits pursuant to the state's Low Carbon Fuel Standard.[148] Likewise, the Department of the Treasury recently finalized guidance regarding demonstration of Secure Geologic Containment pursuant to 45Q.[149] Similar clarity is needed for other state carbon reduction programs including the Regional Greenhouse Gas Initiative and Texas's severance tax rate reduction for EOR projects using anthropogenic CO_2. Development of these rules will reduce uncertainty for operators while providing

---

[145] *See*, Klass and Wilson, *supra* note 60, at 423–29; Marston & Moore, *supra* note 37, at 487–90.

[146] Godec, *supra* note 38 at 6567.

[147] *Id.*

[148] California Air Resources Board, *Carbon Capture and Sequestration Protocol under the Low Carbon Fuel Standard* (August 13, 2018), https://ww2.arb.ca.gov/sites/default/files/2019-03/CCS_Protocol_Under_LCFS_8-13-18.pdf

[149] Department of the Treasury, Credit for Carbon Oxide Sequestration, 26 C.F.R. Part 1 (Jan. 6, 2021).

greater transparency and consistency regarding administration of tax and carbon credit incentive programs.

Finally, the transition of CO_2-EOR assets for incremental storage can be encouraged by streamlining permitting and reporting requirements and addressing issues with public management and post-closure liability. The storage potential represented by depleted hydrocarbon reservoirs will remain unutilized or underutilized if operators are required to assume additional liability or risk that are inconsistent with the industry's business models. Like EPA's requirements for Class VI wells, these enhanced monitoring or reporting requirements should only be imposed where the operations pose an increased risk compared to associated storage with CO_2-EOR. For instance, requirements of subpart RR could be harmonized with the existing requirements of subpart UU, the ISO standards, state regulatory programs, and the rigorous reservoir surveillance programs already implemented by the majority of operators. Further, federal legislation providing for uniform transfer of projects to public management and to address post-closure funding and responsibility issues is needed.

Carbon storage is not a comprehensive solution to climate change. Improvements in efficiency and mitigation and decarbonization solutions are needed in order to achieve a 1.5–2 °C scenario. However, of the geoengineering solutions available, those that contemplate a role for geologic storage associated with EOR have numerous comparative benefits: EOR is already subject to a robust regulatory regime, is commercially driven, has unrealized potential for GHG reduction, and has been in use with few environmental incidents over the past 50 years. Policies to encourage further use of geologic storage through associated and incremental storage may increase the efficiency of carbon utilization and storage, facilitate decarbonization, and help to abate greenhouse gas emissions during the transition to carbon negative sources.

Regulating Geoengineering: International Competition and Cooperation

Soheil Shayegh, Garth Heutel, and Juan Moreno-Cruz

1 Introduction

Solar geoengineering (SGE) consists of increasing the reflectivity of the Earth's atmosphere with the intention of reducing the impacts of climate change. Solar geoengineering offers, in terms of direct costs, a relatively inexpensive means to limit warming. In addition to its low cost, modeling and natural analogues suggest that a main advantage of solar geoengineering would be how quickly the climate system would respond to it. The largest risks posed by deployment of these options are the possible side effects could cause and the fact that the distribution of the benefits and damages would not be uniform across the globe.[1] These characteristics

[1] *See, e.g.*, John Latham et al., *Climate engineering: exploring nuances and consequences of deliberately altering the Earth's energy budget*, PHILOSOPHICAL TRANSACTIONS. SERIES A, MATHEMATICAL, PHYSICAL, AND ENGINEERING SCIENCES 372, 2031 (2014) and CLIMATE INTERVENTION: REFLECTING SUNLIGHT TO COOL EARTH (2015) for reviews of the science behind solar geoengineering, and see Gernot Klepper, and Wilfried Rickels, *Climate engineering: Economic considerations and research challenges*, 8(2) REV. ENVT'L ECON. & POLICY 270, 289 (2014);Garth Heutel, Juan B. Moreno-Cruz, and Katharine Ricke, *Climate engineering economics*, 8 ANNUAL REV. RESOURCE ECON. 99, 118 (2016) for reviews of the economics of solar geoengineering.

S. Shayegh (✉)
RFF-CMCC European Institute on Economics and the Environment (EIEE), Milan, Italy
e-mail: soheil.shayegh@eiee.org

G. Heutel
Department of Economics, Georgia State University, Atlanta, GA, USA

NBER, Cambridge, MA, USA

J. Moreno-Cruz
School of Environment, Enterprise and Development, University of Waterloo, Waterloo, ON, Canada

© Springer Nature Switzerland AG 2021

W. Burns et al. (eds.), *Climate Geoengineering: Science, Law and Governance*, AESS Interdisciplinary Environmental Studies and Sciences Series,
https://doi.org/10.1007/978-3-030-72372-9_10

make solar geoengineering one of the most difficult climate approaches to regulate from an international perspective. First of all, the possibility of deployment of solar geoengineering options could decrease the incentives for countries to reduce their greenhouse gas emissions, thus creating a need to escalate the use of geoengineering. Alternatively, if perceived damages from solar geoengineering are too large, abatement could be used as a disincentive for the deployment of solar geoengineering. Second, because the approach is inexpensive, it could be implemented by a single country, or a small coalition of countries. Thus, there is the threat that this country or coalition could impose its will on the rest of the planet.[2]

We study the issue of governance for solar geoengineering using both a static analytical model and a dynamic numerical model. In both models, we solve for abatement and solar geoengineering strategies under three different cooperation scenarios. First, we consider the centralized case, or the case of full cooperation, in which a single decision-making agent (the social planner) chooses all regions' outcomes to maximize net utility. Second, we consider the other extreme case of no cooperation whatsoever; with each region acting independently, choosing only its own abatement and geoengineering level to maximize its own utility, and taking other regions' actions as fixed. Third, we consider the case of limited cooperation, or coalitions, in which just a subset of regions act cooperatively and the rest act independently. The analytical model demonstrates that total social welfare decreases as the extent of cooperation decreases, and the resulting abatement and geoengineering strategies becomes less stringent. These findings confirm the existence of the classic "free-rider" problem in this setting.

Next, we modify a well-known integrated assessment model (IAM) of climate change policy, the DICE model,[3] in two ways. First, we include SGE as a policy tool alongside abatement. Second, we allow for two homogeneous players that can cooperate or not, depending on the simulation. One important difference between the analytical model and the numerical model is the inclusion of damages from SGE deployment in the numerical model. We model damages from SGE in the numerical model in two different ways - either local or global. The results depend on this assumption about SGE damages. When damages are local, there is a free-rider problem with both SGE and abatement, as predicted by the analytical model. Less coordination leads to less abatement and less SGE. However, when SGE damages are global, there is still a free-rider problem for abatement, but now there is a "free-driver" problem for SGE. Less coordination leads to less abatement but *more* SGE.

---

[2] *See, e.g.*, Juan B. Moreno-Cruz, *Mitigation and the geoengineering threat*, 41 RESOURCE AND ENERGY ECONOMICS} 248, 263 (2015) and Katharine L. Ricke, Juan B. Moreno-Cruz, and Ken Caldeira, *Strategic incentives for climate geoengineering coalitions to exclude broad participation*, 8(1) ENVIRONMENTAL RESEARCH LETTERS 014021 (2013) and Martin L. Weitzman, *A Voting Architecture for the Governance of Free-Driver Externalities, with Application to Geoengineering*, 117(4) *The Scandinavian Journal of Economics* 1049, 1068 (2015).

[3] William Nordhaus, *Estimates of the social cost of carbon: concepts and results from the DICE-2013R model and alternative approaches*, 1(1/2) JOURNAL OF THE ASSOCIATION OF ENVIRONMENTAL AND RESOURCE ECONOMISTS 273, 312 (2014).

Our work is closely related to a recent study[4] that also uses an IAM with SGE to study the free-driving effect of geoengineering. While we use DICE, researchers in the other study use WITCH, a regional IAM with a detailed energy sector. In their theoretical model, the free-driving effect depends on the SGE implementation costs and impacts. They assume that SGE damages are the result of global SGE deployment. However, in our model, we have separated the damages of SGE deployment depending on its origin. Damages in each region can be a function of the local level of SGE deployed by that region or of the total level of SGE deployed by all regions.

The incentives to over-provide SGE are also found in a theoretical model.[5] The free-driving effects come from the benefits that one country receives from unilateral deployment of SGE over other countries. Weitzman has shown that the combination of low SGE cost and private benefits from its deployment will result in over-provision of geoengineering or free driving.[6]

In the following section, we present our base-case analytical model. Section 3 refines the analytical model by adding damages from SGE. Section 4 presents the details of our numerical simulation model, and Sect. 5 presents our simulation results.

2 Analytical Model

Abatement policies are aimed at reducing the amount of emissions from economic activities. Solar geoengineering policies, on the other hand, are designed to reduce the impacts of greenhouse gas (GHG) emissions, namely, the rise in atmospheric temperature. In a simple climate model, we present here, unabated emissions will add to the already existing amount of GHG in the atmosphere and will eventually raise the global mean temperature through an increase in radiative forcing. Solar geoengineering reduces radiative forcing, directly reducing temperature. The temperature rise will reduce the economic output through sea level rise, extreme weather events, or disruptions in agricultural practices. The loss of economic output creates an incentive for present abatement efforts to reduce GHG emissions, and geoengineering efforts to directly reduce temperature. Both strategies are costly, and as a result, the optimal level of each can be found through balancing its short-term costs against long-term benefits.

---

[4] Johannes Emmerling, and Massimo Tavoni, *Quantifying non-cooperative climate engineering* (2017).

[5] *See, e.g.*, Juan B. Moreno-Cruz, *Mitigation and the geoengineering threat*, 41 RESOURCE AND ENERGY ECONOMICS} 248, 263 (2015) and Juan B. Moreno-Cruz, and Sjak Smulders *Revisiting the economics of climate change: the role of geoengineering* 71(2) RESEARCH IN ECONOMICS 212, 224 (2017).

[6] Martin L. Weitzman, *A Voting Architecture for the Governance of Free-Driver Externalities, with Application to Geoengineering*}, 117(4) THE SCANDINAVIAN JOURNAL OF ECONOMICS 1049, 1068 (2015).

We construct a simple model of economic output in order to capture the interactions between the climate system and the economic system. There are $\boldsymbol{N}$ players, which we will refer to as countries (alternatively, these could be regions), and which for now are assumed to be homogeneous. Each country $\boldsymbol{i}$ has two control variables: the level of emissions, $\boldsymbol{E_i}$, which indicates net emissions after abatement, and the level of geoengineering, $\boldsymbol{G_i}$. Both emissions and geoengineering affect radiative forcing in a linear relationship, and radiative forcing affects temperature through a linear function. Both assumptions will be relaxed later in the numerical model. Emissions and geoengineering are chosen at the country level, while radiative forcing and temperature are global. We denote by $\Delta \boldsymbol{R}$, the change in radiative forcing, which is a function of global emissions $\boldsymbol{E} = \sum_{i=1}^{N} \boldsymbol{E_i}$ and global geoengineering $\boldsymbol{G} = \sum_{i=1}^{N} \boldsymbol{G_i}$:

$$\Delta R = \alpha E - \beta G, \tag{1}$$

where $\boldsymbol{\alpha}$ is the scaling parameter and $\boldsymbol{\beta}$ is a parameter controlling the effectiveness of geoengineering. In the extreme case when $\boldsymbol{\beta = 0}$ geoengineering is ineffective and therefore the only option to reduce climate damages will be through controlling the level of emissions.

Atmospheric temperature increase due to change in radiative forcing is:

$$\Delta T = \theta \Delta R \tag{2}$$

where $\boldsymbol{\theta}$ is a parameter representing climate sensitivity. Each country has a utility that is a function of its emissions, the amount of solar geoengineering, and global temperature:

$$U_i\left(E_i, G_i, \Delta T\right) = E_i - \frac{1}{2}\eta\left(E_i\right)^2 - \frac{1}{2}\gamma\left(G_i\right)^2 - \frac{1}{2}\delta \Delta T^2 \tag{3}$$

where $\boldsymbol{\eta}$, $\boldsymbol{\gamma}$, and $\boldsymbol{\delta}$ are the parameters of emissions cost function, solar geoengineering cost function, and climate change damage cost function, respectively. Both solar geoengineering and emissions reduction costs are local – accrued only by region $\boldsymbol{i}$. In Sect. 3 we will also consider the damages from SGE in local and global cases. We next consider three specifications for equilibrium behavior, depending on the level of coordination across countries.

2.1 Full Cooperation (First Best)

First, we consider the case of full international cooperation of all $\boldsymbol{N}$ countries. This is equivalent to a central planner choosing the optimal levels of emissions and solar geoengineering for each country, taking into account all countries' actions simultaneously. This will yield the first-best outcome. The planner's problem is:

$$\max_{\substack{E_1,\cdots,E_N \\ G_i,\cdots,G_N}} \sum_{i=1}^{N} U_i\left(E_i, G_i, \Delta T\right) \tag{4}$$

Since we assume the ***N*** countries are identical, we can assume that the solutions are identical for each country and solve. Define the solutions to this first-best problem as $\boldsymbol{E_i^{fb}}$ and $\boldsymbol{G_i^{fb}}$. These solutions are:

$$E_i^{fb} = \frac{\gamma + N^2\delta\theta^2\beta^2}{\eta\gamma + N^2\delta\theta^2\left(\eta\beta^2 + \gamma\alpha^2\right)}$$

$$G_i^{fb} = \frac{N^2\delta\theta^2\alpha\beta}{\eta\gamma + N^2\delta\theta^2\left(\eta\beta^2 + \gamma\alpha^2\right)} \tag{5}$$

These individual levels of emissions and geoengineering can be summed to the global levels of emissions $\boldsymbol{E^{fb}}$ and geoengineering $\boldsymbol{G^{fb}}$ by adding all ***N*** identical countries' actions:

$$E^{fb} = \frac{N\gamma + N^3\delta\theta^2\beta^2}{\eta\gamma + N^2\delta\theta^2\left(\eta\beta^2 + \gamma\alpha^2\right)} \tag{6}$$

$$G^{fb} = \frac{N^3\delta\theta^2\alpha\beta}{\eta\gamma + N^2\delta\theta^2\left(\eta\beta^2 + \gamma\alpha^2\right)}$$

Atmospheric temperature change can be calculated from plugging in these optimal values into Eqs. (1 and 2):

$$\Delta T^{fb} = \theta\left(\alpha E^{fb} - \beta G^{fb}\right) = \frac{N\gamma\theta\alpha}{\eta\gamma + N^2\delta\theta^2\left(\eta\beta^2 + \gamma\alpha^2\right)} \tag{7}$$

When $\boldsymbol{\beta = 0}$ (i.e. geoengineering is ineffective) or when $\boldsymbol{\gamma \to \infty}$ (i.e. geoengineering is too costly), the optimal level of geoengineering is $\boldsymbol{G^{fb} = 0}$, and the optimal level of emissions is $E^{fb} = N(\eta + N^2\delta\theta^2\alpha^2)^{-1}$.

2.2 *Competition (Independent Action)*

Now we assume that each of the ***N*** countries acts completely independently, choosing their respective privately optimal levels of abatement and geoengineering without cooperation with other countries and assuming that other countries' actions are fixed. Thus we solve for a (symmetric) Nash equilibrium. Country i's problem is:

$$\max_{E_i, G_i} U_i\left(E_i, G_i, \Delta T\right) \tag{8}$$

As in the previous subsection, we can solve for resulting levels of emissions E_i^{comp} and geoengineering G_i^{comp} using the first-order conditions, taking into account the homogeneity of the solutions.

$$E_i^{comp} = \frac{\gamma + N\delta\theta^2\beta^2}{\eta\gamma + N\delta\theta^2\left(\eta\beta^2 + \gamma\alpha^2\right)} \tag{9}$$

$$G_i^{comp} = \frac{N\delta\theta^2\alpha\beta}{\eta\gamma + N\delta\theta^2\left(\eta\beta^2 + \gamma\alpha^2\right)}$$

The total level of emissions E_i^{comp} and the total level of geoengineering G_i^{comp} are calculated as the sum of the all N countries' actions:

$$E^{comp} = \frac{N\gamma + N^2\delta\theta^2\beta^2}{\eta\gamma + N\delta\theta^2\left(\eta\beta^2 + \gamma\alpha^2\right)} \tag{10}$$

$$G^{comp} = \frac{N^2\delta\theta^2\alpha\beta}{\eta\gamma + N\delta\theta^2\left(\eta\beta^2 + \gamma\alpha^2\right)}$$

The change in atmospheric temperature is:

$$\Delta T^{comp} = \theta\left(\alpha E^{comp} - \beta G^{comp}\right) = \frac{N\gamma\theta\alpha}{\eta\gamma + N\delta\theta^2\left(\eta\beta^2 + \gamma\alpha^2\right)} \tag{11}$$

We can compare these results with those from the case of full cooperation. Equation (9) shows that the individual level of emissions is higher in the competition case compared to the full cooperation case (Eq. 5), and that the individual level of geoengineering is lower in the competition case compared to the full cooperation case. In other words, both levels of abatement and geoengineering decrease in the competition case compared to the full cooperation case.

Consequently, comparing Eqs. (11 and 7), the temperature change is larger in the competition case than in the full cooperation case. This confirms our hypothesis that in the competition case, due to the problem of free-riding, countries have less incentive to lower their emissions or to use geoengineering. As the number of countries N increases, both the level of emissions E^{comp} and the level of geoengineering G^{comp} increase.

$$\frac{\partial E^{comp}}{\partial N} = \frac{\eta\gamma\left(\gamma + 2N\delta\theta^2\beta^2\right) + N^2\delta^2\theta^4\beta^2\left(\eta\beta^2 + \gamma\alpha^2\right)}{\left(\eta\gamma + N\delta\theta^2\left(\eta\beta^2 + \gamma\alpha^2\right)\right)^2} > 0 \tag{12}$$

$$\frac{\partial G^{comp}}{\partial N} = \frac{N\delta\theta^2\alpha\beta\left(2\eta\gamma + N\delta\theta^2\left(\eta\beta^2 + \gamma\alpha^2\right)\right)}{\left(\eta\gamma + N\delta\theta^2\left(\eta\beta^2 + \gamma\alpha^2\right)\right)^2} > 0.$$

When $\beta = 0$ (i.e. geoengineering is ineffective) or when $\gamma \rightarrow \infty$ (i.e. geoengineering is too costly), the optimal level of geoengineering is $G^{comp} = 0$ and the equilibrium level of emissions will be $E^{comp} = N(\eta + N\delta\theta^2\alpha^2)^{-1}$.

2.3 Coordination (Coalition/Partial Cooperation)

So far we have studied the two extreme cases of international climate policy regulations: full cooperation and competition. In reality, most countries are likely to formulate positions somewhere in between these two cases. While there is a level of global coordination that tries to bring all countries together in achieving a global climate target, countries are, for the most part, acting independently. A recent example of such coordinating efforts was the development of nationally determined contributions (NDCs) as part of the Paris Agreement. NDCs are a set of actions that each individual country is going to take in order to achieve a global goal (e.g. keeping the global mean temperature rise below 2 °C). The key elements of this new approach are decision-making in the national level and setting climate targets at the global level.

We investigate this by modeling the case of coordination or partial cooperation. We model this by assuming that there is a set of M countries that are part of a coalition. This set is determined exogenously; we do not model the incentives behind coalition formation. Within the coalition, the M countries act fully cooperatively, as if there is a central planner choosing each country's E_i and G_i to maximize the total utility of all M coalition countries, taking the actions of the remaining $N - M$ countries as exogenous. The non-coalition $N - M$ countries each act completely independently, each choosing just its own $E_i$and $G_i$to maximize just its own utility U_i. The result from these optimization problems is a set of $2M$ first-order conditions, for emissions and geoengineering of the coalition countries, and $2(N - M)$ first-order conditions for the non-coalition countries. We again assume that all countries are symmetric with respect to all features of their utility functions, but here there is asymmetry between the coalition and non-coalition countries. Thus, the set of first-order conditions is reduced to four equations for four unknowns: the emissions and geoengineering of each coalition country E_i^{coal} and G_i^{coal}, and the emissions and geoengineering of each non-coalition country $E_i^{noncoal}$and $G_i^{noncoal}$.

The equilibrium solutions are:

$$E_i^{cool} = \frac{\gamma + M\delta\theta^2\left(A\beta^2 - B\alpha^2\gamma\right)}{\eta\gamma + AM\delta\theta^2\left(\eta\beta^2 + \gamma\alpha^2\right)} \tag{13}$$

$$G_i^{cool} = \frac{MN\delta\theta^2\alpha\beta}{\eta\gamma + AM\delta\theta^2\left(\eta\beta^2 + \gamma\alpha^2\right)}$$

$$E_i^{noncoal} = \frac{\gamma + M\delta\theta^2\left(A\beta^2 + \left(\alpha^2\gamma/\eta\right)\left(M-1\right)\right)}{\eta\gamma + AM\delta\theta^2\left(\eta\beta^2 + \gamma\alpha^2\right)}$$

$$G_i^{noncoal} = \frac{N\delta\theta^2\alpha\beta}{\eta\gamma + AM\delta\theta^2\left(\eta\beta^2 + \gamma\alpha^2\right)}$$

where $\boldsymbol{A} \equiv \boldsymbol{M} + \frac{\boldsymbol{N}-\boldsymbol{M}}{\boldsymbol{M}}$ and $\boldsymbol{B} \equiv \frac{(\boldsymbol{M}-1)(\boldsymbol{N}-\boldsymbol{M})}{\boldsymbol{M}\eta}$.

Total emissions and total geoengineering are $E^{coord} = ME_i^{coal} + (N-M)E_i^{noncoal}$ and $G^{coord} = MG_i^{coal} + (N-M)G_i^{noncoal}$, which can be simplified as:

$$E^{coord} = \frac{N\gamma + AMN\delta\theta^2\beta^2}{\eta\gamma + MA\delta\theta^2\left(\eta\beta^2 + \gamma\alpha^2\right)} \tag{14}$$

$$G^{coord} = \frac{AMN\delta\theta^2\alpha\beta}{\eta\gamma + MA\delta\theta^2\left(\eta\beta^2 + \gamma\alpha^2\right)}$$

The resulting temperature increase is:

$$\Delta T^{coord} = \theta\left(\alpha E^{coord} - \beta G^{coord}\right) = \frac{N\gamma\theta\alpha}{\eta\gamma + MA\delta\theta^2\left(\eta\beta^2 + \gamma\alpha^2\right)} \tag{15}$$

The coordination case is an intermediate case between the two previous cases modeled. When $\boldsymbol{M} = \boldsymbol{N}$, the solutions here are identical to those in Sect. 2.1. When $\boldsymbol{M} = \mathbf{1}$, these solutions are identical to those in Sect. 2.2.

We can conduct comparative statics on these solutions to see how policy is affected by the degree of coordination, measured by the size of the coalition $\boldsymbol{M}$.

$$\frac{\partial \Delta T^{coord}}{\partial M} = \frac{-N\gamma\theta\alpha\left(2M-1\right)\left(\delta\theta^2\left(\eta\beta^2 + \gamma\alpha^2\right)\right)}{\left(\eta\gamma + MA\delta\theta^2\left(\eta\beta^2 + \gamma\alpha^2\right)\right)^2} < 0 \tag{16}$$

Temperature is lower when there is more coordination, since the free rider problem becomes smaller and smaller as there are more coalition members.

$$\frac{\partial E^{coord}}{\partial M} = \left(2M-1\right)N\delta\theta^2\frac{\beta^2\left(1 - \gamma\eta - \delta\theta^2 AM\left(\eta\beta^2 + \gamma\alpha^2\right)\right) - \gamma^2\alpha^2}{\left(\eta\gamma + MA\delta\theta^2\left(\eta\beta^2 + \gamma\alpha^2\right)\right)^2} \tag{17}$$

Every term in Eq. (17) is negative except for the first **1** in the numerator, so the right-hand side of the equation is negative unless the entire rest of the numerator is dominated by that **1**. That is, with more coordination (higher $\boldsymbol{M}$), there is lower emissions.

$$\frac{\partial G^{coord}}{\partial M} = \frac{(2M-1)N\delta\theta^2\alpha\beta\eta\gamma}{\left(\eta\gamma + MA\delta\theta^2\left(\eta\beta^2 + \gamma\alpha^2\right)\right)^2} \tag{18}$$

With more coordination (higher M), there is more geoengineering.

The analytical model provides intuitive results for how coordination affects policy outcomes and temperatures. But, it makes many crucial simplifications to arrive at these solutions. One crucial assumption that needs further investigation is to what extent the damages from deployment of SGE may affect optimal decisions. In the next section, we theoretically investigate optimal policies under two different assumptions about SGE damages: one in which they are local and another where they are global. Following that, we consider a numerical simulation model that allows for either local or global SGE damages.

3 SGE Damages

We modify our theoretical model to include a representation of SGE damages. We use a quadratic cost function similar to other costs in the model to account for SGE damages. We consider two cases with local and global SGE damages and investigate the optimal policies under each case.

3.1 Local SGE Damages

First we consider the case with local SGE damages. In this case we add an additional term to Eq. (3) to represent these damages:

$$U_i\left(E_i, G_i, \Delta T\right) = E_i - \frac{1}{2}\eta\left(E_i\right)^2 - \frac{1}{2}\gamma\left(G_i\right)^2 - \frac{1}{2}\delta\Delta T^2 - \frac{1}{2}\lambda\left(G_i\right)^2 \tag{19}$$

The last term captures the SGE damages, and λ is the parameter of these damages. Since only G_i enters country i's damage function, these damages are local, not global. Following the calculations for the full cooperation case presented in Sect. 2.1 we derive $E_i^{fb-local}$ and $G_i^{fb-local}$, the optimal emission and SGE levels, as the solutions to this first-best (full cooperation) problem:

$$E_i^{fb-local} = \frac{(\gamma+\lambda) + N^2\delta\theta^2\beta^2}{\eta(\gamma+\lambda) + N^2\delta\theta^2\left(\eta\beta^2 + (\gamma+\lambda)\alpha^2\right)} \tag{20}$$

$$G_i^{fb-local} = \frac{N^2\delta\theta^2\alpha\beta}{\eta(\gamma+\lambda) + N^2\delta\theta^2\left(\eta\beta^2 + (\gamma+\lambda)\alpha^2\right)}$$

As is obvious from these equations, the local SGE damages appear in the optimal solution as an additional SGE cost.

We can derive similar solutions for the competition case following the calculations presented in Sect. 2.2. As in the previous subsection, we can solve for resulting levels of emissions $\boldsymbol{E}_i^{comp-local}$ and geoengineering $\boldsymbol{G}_i^{comp-local}$ using the first-order conditions, taking into account the homogeneity of the solutions.

$$E_i^{comp-local} = \frac{(\gamma+\lambda)+N^2\delta\theta^2\beta^2}{\eta(\gamma+\lambda)+N\delta\theta^2\left(\eta\beta^2+(\gamma+\lambda)\alpha^2\right)} \tag{21}$$

$$G_i^{comp-local} = \frac{N\delta\theta^2\alpha\beta}{\eta(\gamma+\lambda)+N\delta\theta^2\left(\eta\beta^2+(\gamma+\lambda)\alpha^2\right)}$$

Comparing the levels of SGE in the equations above reveals that $\boldsymbol{G}_i^{fb-local} > \boldsymbol{G}_i^{comp-local}$, which means as we move from the full cooperation case to competition case, each region will take advantage of other regions' SGE deployment and will provide less SGE compared to the full cooperation case. This deviation from the first-best outcome is a standard free-riding problem. We will provide numerical evidence for this behavior in Sect. 4.

3.2 Global SGE Damages

In the case with global SGE damages, Eq. (3) is modified to:

$$U_i(E_i,G_i,\Delta T,G) = E_i - \frac{1}{2}\eta(E_i)^2 - \frac{1}{2}\gamma(G_i)^2 - \frac{1}{2}\delta\Delta T^2 - \frac{1}{2}\lambda(G)^2 \tag{22}$$

where $\boldsymbol{G}$ is the sum of SGE from all $\boldsymbol{N}$ regions and represents the global damages from SGE, with $\boldsymbol{\lambda}$ still capturing the magnitude of these damages. Similar to the case with local damages, we derive $\boldsymbol{E}_i^{fb-global}$ and $\boldsymbol{G}_i^{fb-global}$, the optimal emission and SGE levels, as the solutions to the first-best (full cooperation) problem:

$$E_i^{fb-global} = \frac{(\gamma+N^2\lambda)+N^2\delta\theta^2\beta^2}{\eta(\gamma+N^2\lambda)+N^2\delta\theta^2\left(\eta\beta^2+(\gamma+N^2\lambda)\alpha^2\right)} \tag{23}$$

$$G_i^{fb-global} = \frac{N^2\delta\theta^2\alpha\beta}{\eta(\gamma+N^2\lambda)+N^2\delta\theta^2\left(\eta\beta^2+(\gamma+N^2\lambda)\alpha^2\right)}$$

The levels of emissions $\boldsymbol{E}_i^{comp-global}$ and geoengineering $\boldsymbol{G}_i^{comp-global}$ in the competition case are:

$$E_i^{comp-global} = \frac{(\gamma + N\lambda) + N\delta\theta^2\beta^2}{\eta(\gamma + N\lambda) + N\delta\theta^2\left(\eta\beta^2 + (\gamma + N\lambda)\alpha^2\right)} \tag{24}$$

$$G_i^{comp-global} = \frac{N\delta\theta^2\alpha\beta}{\eta(\gamma + N\lambda) + N\delta\theta^2\left(\eta\beta^2 + (\gamma + N\lambda)\alpha^2\right)}$$

Comparing the SGE levels under the full cooperation case to the competition case, it is ambiguous as to which is larger. In fact, we find a condition under which the solution switches from free-riding behavior (i.e. providing less SGE in the competition case compared to the full cooperation case) to free-driving behavior (i.e. providing more SGE in the competition case compared to the full cooperation case). If the SGE damage parameter λ is greater than $\frac{\gamma\eta}{N^3\delta\theta^2\alpha^2}$, then the SGE level in the competition case is greater than the optimal SGE level in the full cooperation case: $G_i^{comp-global} > G_i^{fb-global}$ (free-driving). On the other hand, if the SGE damage parameter λ is less than $\frac{\gamma\eta}{N^3\delta\theta^2\alpha^2}$, the SGE level in the competition case is less than the optimal SGE level in the full cooperation case: $G_i^{comp-global} < G_i^{fb-global}$ (free-riding). In other words, when global SGE damages are relatively high, then competition results in a free-driving effect where countries actually conduct too much SGE, because they only account for its effect on their own utility and not on the damages that their SGE cause to other countries. This only occurs when global SGE damages λ are high enough so that its free-driving effect dominates the free-riding effect from the benefits of SGE (which are always global).

Next, we move on to our numerical model, where $N = 2$ and the SGE damages are assumed comparable with other costs and therefore the first condition (free-driving) holds.

4 Numerical Model

We base our numerical simulation on a well-known integrated assessment model that is widely used in academic research and policy making to find the optimal emission levels in the face of imminent damages from temperature change. The Dynamic Integrated Climate-Economy (DICE) model was designed and developed by William Nordhaus at Yale University. It is a centralized decision making tool with a representative-agent economic model. There is an endogenous capital stock and an exogenous technological and population growth dynamic inside the model. Carbon emissions are directly linked to economic production, but they can be reduced through two processes: first, the carbon intensity of output is decreasing over time through an exogenous procedure, and second, abatement action can reduce the emissions. The carbon cycle in the model consists of a three-layer model of the atmosphere and upper and lower oceans. The atmospheric carbon concentration affects the atmosphere's radiative forcing and the atmospheric temperature consequently. Finally, the climate and economy sections of the model are linked together through a damage function that indicates the loss in total economic output due to a change in atmospheric temperature. The objective of the model is to maximize the net present value of total social welfare by finding the optimal carbon abatement trajectories. The model has a 5-year time period and runs for 60 periods.

Details of the DICE model are available publicly and also at William Nordhaus's website.[7] We are using the version DICE-2013R.

4.1 Modifications to DICE

We modify the DICE model in the same way as in our previous study.[8] In this section, we present only a brief summary of how the DICE model has been modified. More details of the modifications that we make are available in our other papers.[9] Those papers and their appendices contain the full list of model equations and the calibration methodology. Here, we merely summarize our modifications to DICE.

There are five ways in which we modify DICE to incorporate solar geoengineering.

- In addition to a policy choice variable $\boldsymbol{a}_t$ for the intensity of emissions abatement, we add a second policy choice variable, $\boldsymbol{g}_t$, representing the intensity of solar geoengineering.
- There is a direct cost of geoengineering implementation, modeled analogously to the way that abatement cost is modeled in DICE. Based on prior literature, this cost is quite small, reflecting the fact that the direct costs of solar geoengineering are cheap relative to abatement. To completely offset the radiative forcing caused by greenhouse gases costs about 0.27% of global GDP.
- In addition to its implementation costs, we model damages from solar geoengineering. These damages are modeled analogously to the way that climate change damages are modeled in DICE. It should be emphasized that these damages are highly uncertain, and so in our parameterization we are very conservative about the value. Thus, we assume these damages are very high. The amount of

[7] William Nordhaus, *Estimates of the social cost of carbon: concepts and results from the DICE-2013R model and alternative approaches*, 1(1/2) JOURNAL OF THE ASSOCIATION OF ENVIRONMENTAL AND RESOURCE ECONOMISTS 273, 312 (2014) and also available at https://sites.google.com/site/williamdnordhaus/dice-rice

[8] Garth Heutel, and Juan Moreno-Cruz, and Soheil Shayegh, *Solar geoengineering, uncertainty, and the price of carbon*, 87 JOURNAL OF ENVIRONMENTAL ECONOMICS AND MANAGEMENT 24, 41 (2018)

[9] Garth Heutel, and Juan Moreno-Cruz, and Soheil Shayegh, *Solar geoengineering, uncertainty, and the price of carbon*, 87 JOURNAL OF ENVIRONMENTAL ECONOMICS AND MANAGEMENT 24, 41 (2018) and Garth Heutel, and Juan Moreno-Cruz, and Soheil Shayegh, *Climate tipping points and solar geoengineering*, 132 JOURNAL OF ECONOMIC BEHAVIOR & ORGANIZATION 19, 45 (2016). These papers model epistemic uncertainty over certain parameter values, though here we restrict analysis to the deterministic case.

geoengineering needed to offset the warming effects of $\boldsymbol{CO_2}$ leads to damages of 3% of gross global GDP, which is about equal to damages from climate change itself under moderate warming. As in Sect. 3 above, we model SGE damages in two ways: local and global. We numerically verify that that this modeling choice has a direct impact on the incentives for the regions and results in either a free-rider or a free-driver effect.

- The radiative forcing equation is modified to include the effect of geoengineering. The radiative forcing is a sum of the original specification of radiative forcing from DICE and the radiative forcing caused by solar geoengineering $\boldsymbol{g_t}$.
- Finally, we modify the climate change damage function to reflect the fact that damages are not only a function of temperature, but are also a function of atmospheric and ocean carbon concentrations. This is crucial when modeling solar geoengineering policy, because solar geoengineering reduces temperature but does not reduce atmospheric or ocean carbon. We set 80% of climate change damages from temperature increase, 10% from atmospheric carbon concentrations, and 10% from ocean carbon concentrations.

Furthermore, to study coordination among countries or regions, we must extend the model beyond a global, representative agent model. While DICE has been regionally disaggregated via the RICE model[10], here we take a much simpler approach. We assume that there are two homogeneous countries indexed by $\boldsymbol{i}$ and $\boldsymbol{j}$, and we calibrate each country simply by dividing all of the relevant stock variables by half. Costs of abatement and geoengineering are borne just by the individual country, but the damages from climate change and geoengineering and the radiative forcing effect of geoengineering are global and depend on the total amount from both countries.

4.2 International Coordination

As with the analytical model in Sect. 2, we consider three different frameworks for international governance of climate policy, including geoengineering deployment.

- **Cooperation** First is the case of full cooperation, analogous to the treatment in Sect. 2.1. Both countries are working together as one to maximize the sum of the two countries' utilities. This is equivalent to a social planner choosing abatement and geoengineering in all periods for both countries:

$$\max_{\{a_{i,t},g_{i,t},a_{j,t},g_{j,t}\}_{t=1}^{T}} U_i\left(\{a_{i,t},g_{i,t},a_{j,t},g_{j,t}\}_{t=1}^{T}\right)+U_j\left(\{a_{i,t},g_{i,t},a_{j,t},g_{j,t}\}_{t=1}^{T}\right) \tag{25}$$

[10] William D. Nordhaus, and Zili Yang, *A regional dynamic general-equilibrium model of alternative climate-change strategies*, 1996 THE AMERICAN ECONOMIC REVIEW 741, 765

where U_i and U_j represent the net present value of utility for each country over the entire T periods and are a function of all choice variables over each period from both countries.

- **Competition** The next scenario is the case of competition, or independent action, as in Sect. 2.2. Each country is trying to maximize its welfare independently, holding constant the action of the other country:

$$\max_{\{a_{i,t},g_{i,t}\}_{t=1}^{T}} U_i\left(\left\{a_{i,t},g_{i,t},a^*_{j,t},g^*_{j,t}\right\}_{t=1}^{T}\right) \tag{26}$$

$$\max_{\{a_{j,t},g_{j,t}\}_{t=1}^{T}} U_i\left(\left\{a^*_{i,t},g^*_{i,t},a_{j,t},g_{j,t}\right\}_{t=1}^{T}\right)$$

In country i's maximization problem, the actions of country j are taken as fixed - $a^*_{j,t}$ and $g^*_{j,t}$, and likewise for country j.

- **Coordination** The last scenario is the case of coordination, or partial cooperation. This is analogous to the treatment in Sect. 2.3, but here in the numerical model coordination is modeled somewhat differently than it was in the analytical model. The analytical model had a subset of M of the N total countries forming a coalition. Here, with just $N = 2$ countries, any strict subset is just 1 and identical to the competition case. Therefore, we assume that each country is acting independently, choosing just its own abatement and geoengineering levels, but is maximizing the sum of its own welfare and a portion of the other country's welfare. We call this portion ω the *coordination factor*, and it measures the degree of coordination, similar to how M, the size of the coalition, does in the analytical model. The coordination factor ω can be between $\mathbf{0}$ (corresponding to the competition case) and $\mathbf{1}$ (corresponding to the cooperation case). In a more formal way, it means simultaneously solving the welfare maximization problem of each agent by applying the coordination factor, ω, to obtain the partial sum of the two agents' welfare and then solving the first order conditions for both agents simultaneously:

$$\max_{\{a_{i,t},g_{i,t}\}_{t=1}^{T}} U_i\left(\left\{a_{i,t},g_{i,t},a^*_{j,t},g^*_{j,t}\right\}_{t=1}^{T}\right) + \omega U_j\left(\left\{a_{i,t},g_{i,t},a^*_{j,t},g^*_{j,t}\right\}_{t=1}^{T}\right) \tag{27}$$

$$\max_{\{a_{j,t},g_{j,t}\}_{t=1}^{T}} \omega U_i\left(\left\{a^*_{i,t},g^*_{i,t},a_{j,t},g_{j,t}\right\}_{t=1}^{T}\right) + U_j\left(\left\{a^*_{i,t},g^*_{i,t},a_{j,t},g_{j,t}\right\}_{t=1}^{T}\right)$$

When the coordination factor $\omega = \mathbf{0}$, this case becomes identical to the competition case. When $\omega = \mathbf{1}$, the solution is identical to the solution in the cooperation case. In the simulations, we consider two different values for ω: a low coordination value $\omega = \mathbf{1/3}$ and a high coordination value $\omega = \mathbf{2/3}$.

5 Results

We perform two sets of simulations, corresponding to the two assumptions about SGE damages described above. In the first set, we assume that the damages from deploying SGE in each region are only a function of the local deployment of SGE. In this case, each region is only affected by the SGE cost and SGE damage that are incurred due to the deployment of SGE in that region. In the second set of simulations, however, we assume that the SGE damages are global, meaning that each region's SGE damages are a function of the total amount of SGE deployed by all regions.

5.1 *Local SGE Damages*

The results under this assumption are shown in Fig. 1. Panel *A* shows optimal SGE under different levels of coordination. It verifies our analytical result in terms of the free-riding problem in the case of non-cooperative strategies. As we move from a cooperative world to the world with less coordination and more competition, the level of SGE decreases. In all cases, the SGE level starts out with a jump and gradually increases as the damages from climate change increase. It eventually peaks in

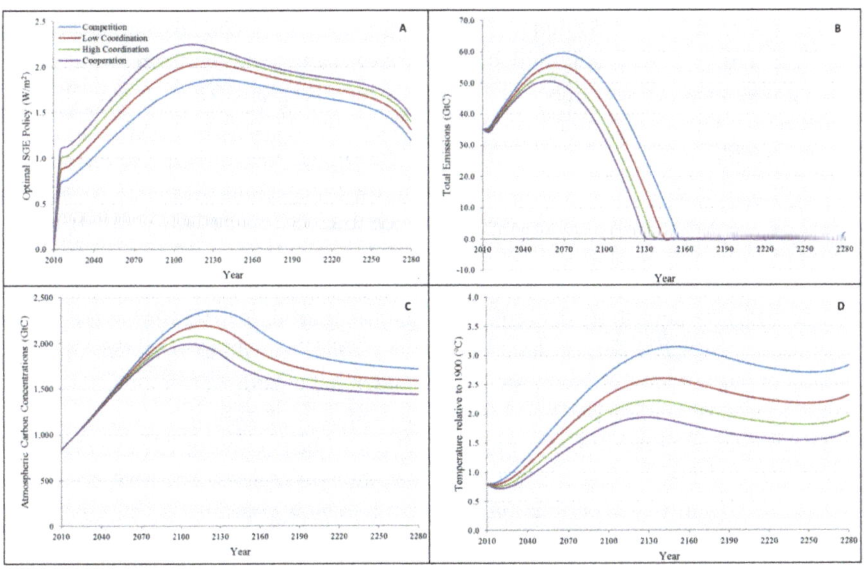

Fig. 1 Climate policies and outcomes for the model with local SGE damages. Each panel shows four scenarios: cooperation, high coordination, low coordination, and competition. Panel **A** shows the optimal SGE policy. Panel **B** represents the total emissions. Panel **C** shows the total carbon concentration in the atmosphere and panel **D** is the temperature change relative to 1900

around year 2110 and reaches its maximum value between 2.2 W/m^2 in the full cooperation case and 1.8 W/m^2 in the competition case. Since the results shown here are only for one of the two identical regions, this translates into 3.6 – 4.4 W/m^2 reduction in solar radiative forcing in the next 100 years.

Similar free riding can be observed for abatement. Panel ***B*** shows the level of emissions under the four different coordination assumptions. Cooperation yields the highest abatement and therefore the lowest level of emissions, while the emissions are highest under competition. Emissions over the next 100 years increase to up to 60 *GtC* in the competition case and 50 *GtC* in the full cooperation case. By 2130, all emissions are abated in the full cooperation case. In contrast, the competition case delays reaching the 100% abatement point to year 2160. When the 100% abatement point is reached, there will be less need for reducing the temperature through SGE and therefore the level of SGE gradually decreases.

The results from these two panels are in line with our theoretical model from Sect. 2, which assumes local SGE damages. Comparing Eqs. (5 and 9), it can be shown that for $\boldsymbol{N > 1}$:

$$E_i^{fb} < E_i^{comp} \tag{28}$$

$$G_i^{fb} > G_i^{comp}$$

These equations show the free-riding effect in the context of climate change policy. For both abatement and SGE actions, moving away from a cooperation regime to a competitive regime reduces the regional incentives for adopting a more stringent climate policy. While the cost and damages of climate actions (abatement and SGE) are locally incurred, the benefit of these actions in the form of reduction in the global mean temperature is felt globally by all regions. Therefore, each individual region has no incentive to commit to the optimal (cooperative case) policy.

As a result of the free-riding effect in abatement, atmospheric concentration increases as the level of cooperation between the two regions decreases (panel ***C***). While in the cooperation case, carbon concentration reaches only up to about 2000 *GtC* by 2110, it peaks 20 years later at about 2300 *GtC* in the competition case. After abatement efforts in each case reach the 100% abatement rate, the atmospheric concentration starts declining and stabilizes around 1450 *GtC* in the cooperation case and 1700 *GtC* in the competition case. Meanwhile, as shown in panel ***D***, temperature gradually increases to just under $2.0^{\circ}C$ above pre-industrial in the cooperation case while it reaches $3.0^{\circ}C$ in the competition case.

The middle two lines in all panels of Fig. 1 show the two intermediate cases with high and low degrees of coordination between the two regions. The high coordination case is closer to the cooperation case, while the low coordination case is closer to the competition case.

5.2 Global SGE Damages

The results for simulations under this assumption are shown in Fig. 2. Panel *A* shows SGE under different coordination levels. In contrast to the results under the assumption that SGE damages are local, we now observe a free-driving effect rather than a free-riding effect from non-cooperative cases. As we move from the cooperative case to the competition case, the level of SGE increases. In all cases, the SGE level starts out with a jump and gradually increases as the damages from climate change increase. It eventually peaks in around year 2120 and reaches its maximum value between 1.0 W/m^2in the full cooperation case and 1.2 W/m^2 in the competition case. Given that the results shown here are only for one of the two identical regions, this translates into 2.0 – 2.4 W/m^2 reduction in solar radiative forcing in the next 100 years.

The free-riding effect, however, still can be observed for abatement. Panel ***B*** shows the level of emissions under different strategies. In contrast to SGE, the cooperation case has the highest abatement and therefore the lowest emissions, while the competition case has the lowest abatement and highest emissions. Emissions over the next 100 years increase to 57 *GtC* in the competition case and 46 *GtC* in the full cooperation case. By 2120, all emissions are abated in the full cooperation case. In contrast, competition delays reaching the 100% abatement point to 2150. As in

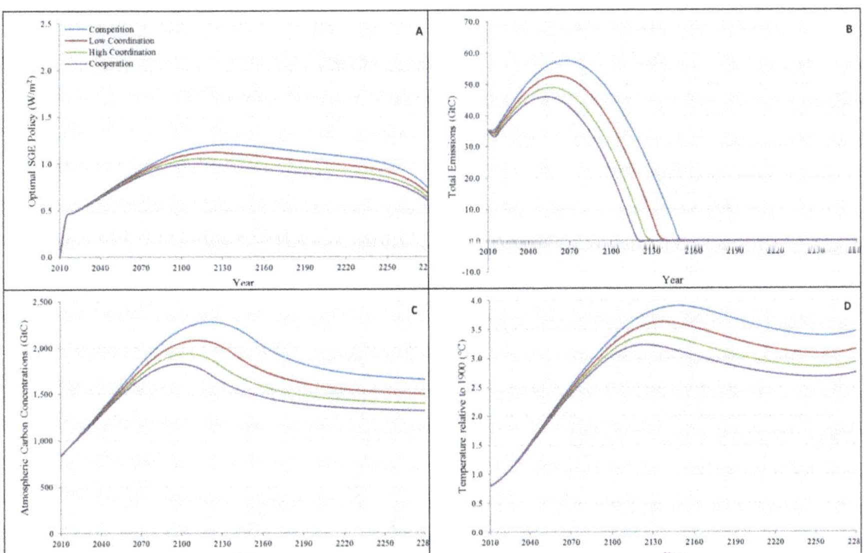

Fig. 2 Climate policies and outcomes for the model with damages from global SGE deployment. Each panel shows four scenarios of cooperation, high coordination, low coordination, and competition. Panel **A** shows the optimal SGE policy. Panel **B** represents the total emissions. Panel **C** shows the total carbon concentration in the atmosphere and panel **D** is the temperature change relative to 1900

Fig. 1, when the 100% abatement point is reached, the level of SGE gradually decreases.

The results from panel ***A*** and panel ***B*** of Fig. 2 show the free-driving and free-riding effects in the context of climate change policy, respectively. For abatement action, moving away from a cooperative regime to a competitive regime reduces the regional incentives for adopting a more stringent climate policy. This is because all of the costs of abatement are local, while the benefits are global, leading to the classic free-rider problem. In contrast, individual regions in the competitive regime find it more attractive to act unilaterally and increase their contribution of SGE deployment compared to the cooperative regime. This is because, unlike for abatement and unlike for SGE under the previous assumption of local damages, here the damages from SGE are global rather than local. Therefore, individual regions have an incentive to increase their SGE level above what is optimal (under the cooperative regime). This is the free-driver problem.[11]

As a result of free riding in abatement, atmospheric concentration is higher in the competition case (panel ***C***). While in the cooperation case, carbon concentration reaches only about 1800 *GtC* by 2100, it peaks at about 2200 *GtC* in the competition case. After each case reaches the 100% abatement rate, atmospheric concentration starts declining and it stabilizes around 1300 *GtC* in the cooperation case and 1650 *GtC* in the competition case. Free-riding in abatement and free-driving in SGE have offsetting effects on temperature: lower abatement from free-riding raises temperature while higher SGE from free-driving lowers temperature. Panel ***D*** shows that the free-riding effect of abatement dominates the free-driving effect of SGE; temperature is higher in the competition case than in the cooperation case, despite the higher SGE use in that case. Temperature starts out with a gradual increase to about 3.2°C in the cooperation case, while it reaches just under 4.0°C in the competition case.

As in Fig. 1, the middle two lines in all panels of Fig. 2 show two intermediate cases with high and low degrees of coordination between the two regions.

6 Conclusion

We investigate the potential use of solar geoengineering as a policy tool to achieve a lower global temperature under different levels of international coordination. Our theoretical and numerical models suggest that (1) geoengineering, if deployed, can play an important role in the climate policy portfolio, (2) low cooperative regimes with local SGE damages result in an under-provision of abatement and SGE actions (free riding), and finally (3) low cooperative regimes with global SGE damages

[11] Martin L. Weitzman, *A Voting Architecture for the Governance of Free-Driver Externalities, with Application to Geoengineering*, 117(4) SCANDINAVIAN J. ECON. 1049, 1068 (2015).

result in an under-provision of abatement (free riding) but over-provision of SGE (free driving).

These results are important in that they highlight the need for careful examination of costs and impacts of SGE options before committing to any international accord to regulate their deployment. In setting international regulations over the future deployment of SGE, decision makers should take into account the possibility of free-riding and free-driving effects that may emerge in any level of cooperation among individual regions.

References

1. Heutel, G., Moreno-Cruz, J.B., Ricke, K.: Climate engineering economics. Annual Rev Resource Econ. **8**, 99–118 (2016a)
2. Heutel, G., Moreno-Cruz, J., Shayegh, S.: Climate tipping points and solar geoengineering. J Economic Behav & Organization. **132**, 19–45 (2016b)
3. Heutel, G., Moreno-Cruz, J., Shayegh, S.: Solar geoengineering, uncertainty, and the price of carbon. J. Environ Economic & Management. **87**, 24–41 (2018)
4. Klepper, G., Rickels, W.: Climate engineering: Economic considerations and research challenges. Rev. Envt'l Econ. & Policy. **8**(2), 270–289 (2014)
5. Latham, J., et al.: Climate engineering: exploring nuances and consequences of deliberately altering the Earth's energy budget. Philos. Transact. A Math. Phys. Eng. Sci. **372**, 2031 (2014)
6. Moreno-Cruz, J.B.: Mitigation and the geoengineering threat. Resourc Energy Economic. **41**, 248–263 (2015a)
7. Moreno-Cruz, J.B.: Mitigation and the geoengineering threat. Resourc Energy Economic. **41**, 248–263 (2015b)
8. Moreno-Cruz, J.B., Smulders, S.: Revisiting the economics of climate change: the role of geoengineering. Res. Econom. **71**(2), 212–224 (2017)
9. Nordhaus, W.: Estimates of the social cost of carbon: concepts and results from the DICE-2013R model and alternative approaches. J. Assoc. Environ. Resour. Econ. **1**, 273–312 (2014a)
10. Nordhaus, W.: Estimates of the social cost of carbon: concepts and results from the DICE-2013R model and alternative approaches. J. Assoc. Environ. Resour. Econ. **1**, 273–312 (2014b)
11. Nordhaus, W.D., Yang, Z.: A regional dynamic general-equilibrium model of alternative climate-change strategies. Am. Econ. Rev. **86**, 741–765 (1996)
12. Ricke, K.L., Moreno-Cruz, J.B., Caldeira, K.: Strategic incentives for climate geoengineering coalitions to exclude broad participation. Environ. Res. Lett. **8**(1), 014021 (2013)
13. Weitzman, M.L.: A Voting Architecture for the Governance of Free-Driver Externalities, with Application to Geoengineering. Scandinav J Economic. **117**(4), 1049–1068 (2015a)
14. Weitzman, M.L.: A Voting Architecture for the Governance of Free-Driver Externalities, with Application to Geoengineering. Scandinav J Economic. **117**(4), 1049–1068 (2015b)
15. Weitzman, M.L.: A Voting Architecture for the Governance of Free-Driver Externalities, with Application to Geoengineering. Scandinavian J Econ. **117**(4), 1049–1068 (2015c)

Geoengineering and the Evolution of Dueling Precautions

Kalyani Robbins

1 Introduction

2018 witnessed the release of a number of films depicting a new kind of future dystopia. Instead of the usual portrayal of psychological, technological, and political horrors of future dystopias, we have begun to see an increasing emphasis on humans surviving in a world with no natural ecosystems left intact. For example, the new *Blade Runner* film takes place in a future with no remaining trees – not one – and artificial sources of nutrition instead of agriculture. In the recent Spielberg film *Ready Player One*, people must escape into a virtual reality world because the real world is unlivable, having been stripped of all natural resources and devastated by climate change. This trend in depressing entertainment reflects both an increased awareness of the extremely serious environmental issues facing our planet (including biodiversity loss and climate change) and a complete failure to appreciate our dependence on a functioning planet, and its associated ecosystem services, in order to survive. Without ecosystem services life doesn't simply become unpleasant – it is over. Seeing healthy surviving protagonists living in a future with a completely-destroyed planet is perhaps the greatest fictional aspect of these films, and yet it is the least noticed. Because of our dependence on ecosystem services for survival, biodiversity is more than just our canary, it is the entire coal mine around us. In other words, we will not be able to save ourselves after failing to save everything else.

Given the urgency caused by this inadequate warning system, we find ourselves in the position of having to make decisions rather quickly – major decisions for which we might prefer to have the luxury of more time. To make matters worse, we will have to make these decisions, and then act on them, with far less concrete

K. Robbins (✉)
Loyola University Chicago School of Law, Chicago, IL, USA
e-mail: krobbins@fiu.edu

© Springer Nature Switzerland AG 2021

W. Burns et al. (eds.), *Climate Geoengineering: Science, Law and Governance*, AESS Interdisciplinary Environmental Studies and Sciences Series,
https://doi.org/10.1007/978-3-030-72372-9_11

information than we would like. How should global leaders determine the course of actions that have a planetary-scale existential impact with such limited time and information? This chapter explores this question through the lens of the precautionary principle, analyzing its applicability to the context and developing a decision-making formula to operationalize it. Because the climate disruption problem is already so advanced, and certain to worsen before we can adequately curb GHG emissions, this discussion will focus on the question of whether and how to engage in geoengineering projects. While risky, these efforts may become necessary to address the risk-taking we are already engaged in. As we shall see, climate geoengineering may simultaneously both be justified by tenets of the precautionary principle and simultaneously conflict with it.

The treaties that have addressed geoengineering thus far, including the Convention on Biological Diversity (CBD),[1] and the London Convention/Protocol,[2] have taken a precautionary approach to geoengineering methods that have not been scientifically determined to be safe. In the case of the 2010 COP to the CBD, this involved expressly applying the precautionary principle to geoengineering.[3] This is, at least for now, a reasonable articulation of the precautionary principle. All geoengineering approaches, whether categorized as carbon dioxide removal (CDR) or solar radiation management (SRM), pose meaningful risks for biodiversity, the target of the CBD. In an effort to aid policymakers in applying the precautionary principle to geoengineering, the COP has produced two technical reports[4] discussing the risks that each proposed geoengineering method would pose for biodiversity.

In the SRM category, the three most discussed approaches are sulfur aerosol injection, marine cloud brightening, and satellite-based systems, all of which seek to reflect solar radiation back to space before it reaches the earth's surface. These methods do nothing to reduce the concentration of GHGs in the atmosphere. Instead, SRM techniques would mask the continued increase in GHGs, which not only ignores the perils of ocean acidification,[5] but creates the risk of very sudden

---

[1] The Convention on Biological Diversity of 5 June 1992 (1760 U.N.T.S. 69).

[2] *See* Int'l Mar. Org. [IMO], *Statement of Concern Regarding Iron Fertilization of the Oceans to Sequester CO2*, P 1 IMO Ref. T5/5.01, LC-LP.1/Circ. 14 (July 13, 2007).

[3] UNEP/CBD/COP/DEC/X/33 (29 October 2010).

[4] These reports issued in 2012 and 2016 and contain substantial overlap. *See generally* Williamson, P., & Bodle, R. (2016), *Update on Climate Geoengineering in Relation to the Convention on Biological Diversity: Potential Impacts and Regulatory Framework*, Technical Series No.84, Secretariat of the Convention on Biological Diversity, Montreal, 158 pages; Secretariat of the Convention on Biological Diversity (2012), *Geoengineering in Relation to the Convention on Biological Diversity: Technical and Regulatory Matters*, Montreal, Technical Series No. 66, 152 pages. The 2016 report in fact confirms the continued validity of the key messages from the 2012 report and focuses on expanding detail, utilizing improved understanding of geoengineering techniques, and in some cases updating prior statements with minor changes. Because all material cited in this chapter is found in both reports, this pair of reports will hereinafter be referred to collectively as "CBD Geoengineering Reports."

[5] Some proponents have argued that the substantial reduction in permafrost and tree loss associated with SRM's cooling temperatures would exert a salutary role in relation to ocean acidification. *See*

warming in the event it is discontinued, which could be catastrophic to biodiversity.[6] Moreover, deployment of SRM could pose a "moral hazard" should policymakers believe that it provides them with license to reduce their commitments to emissions reductions. In the event that land-based methods are applied, alterations in surface albedo (utilizing reflective materials) would require huge areas of land, which could threaten habitats, and disrupt weather patterns by localizing the cooling impact.[7]

CDR approaches are also somewhat troubling. One appealing CDR method is bioenergy with carbon capture and storage (BECCS), in which biomass is burned for energy and the resulting emissions are captured and stored, or utilized for purposes such as production of chemicals or high-strength materials. This approach is "carbon negative," because of the carbon the planted biomass absorbs as it grows. It does, however, pose a number of risks. BECCS requires a great deal of land and water, resulting in harm to habitat, potential diversion of large amounts of land uses for growing crops, and huge demands on water supplies.[8] Another concern with the various CDR methods is storage of carbon dioxide. If carbon dioxide is stored in the deep ocean, ocean floor organisms are likely to perish, and if stored in sub-seafloor geological reservoirs, it may threaten microbial communities.[9] Even the relatively gentle approach of afforestation/reforestation poses risks, including harm to ecosystems that have developed since deforestation and other competing land uses.

IPCC synthesis reports'[10] assessments have gradually become more dire, with the most recent noting that neither mitigation nor adaptation responses alone will suffice to help the globe avert passing critical climatic thresholds. Both are necessary, at this point, in order to maintain life as we know it.[11] Imagine the future IPCC report that informs us that the *combination* of mitigation and adaptation is not enough. Imagine that we find ourselves surviving increasing impacts, but ecosystems are struggling and we fear that a massive tipping point could be imminent. What would then be the most precautionary approach, when it comes to geoengineering? The precautionary principle provides that when there is an identified threat of serious or irreversible damage to the environment exists, the absence of full scientific knowledge about said threat should not justify failure to take remedial action.

Juan B. Moreno-Cruz & David W. Keith, *Climate policy under uncertainty: a case for solar geoengineering*, Climatic Change (2013) 121: 431–444, at 433. However, with unchecked increases in atmospheric CO_2 it would still continue to rise, as these same proponents concede. *See id.* at 440.

[6] *See* CBD Geoengineering Reports, *supra* note 5.

[7] *See* CBD Geoengineering Reports, *supra* note 5.

[8] *See* Wil Burns, *BECCS and Human Rights*, in Recent Developments in Climate Justice, 47 ELR 11013, 11,013–14 (2017).

[9] *See* CBD Geoengineering Reports, *supra* note 5.

[10] These are the comprehensive reports, as opposed to narrower reports that the IPCC also issues as needed.

[11] IPCC, 2014: *Climate Change 2014: Synthesis Report. Contribution of Working Groups I, II and III to the Fifth Assessment Report of the Intergovernmental Panel on Climate Change* [Core Writing Team, R.K. Pachauri and L.A. Meyer (eds.)]. IPCC, Geneva, Switzerland, 151 pp.

In the context of climate geoengineering, two juxtaposed potential applications of the precautionary are possible, with the precautionary principle currently being invoked to prevent deployment, but invoked in the future to require geoengineering to rescue us from the perils of climate change. This chapter will consider the process for getting from here to there, incrementally, over the intervening years. It should be emphasized at the outset that this shifting application of the precautionary principle is likely to evolve on distinct timelines for different categories of geoengineering. Each approach has its own risk/reward ratio. While some factors might be somewhat consistent across methods (in particular the degree of climate change peril to the earth and its ecosystems existing at the time of analysis), geoengineering risks will vary. Given that the precautionary principle counsels against risky human intervention, and in light of the fact that *both* anthropogenic climate change and geoengineering are forms of human intervention, the question as to whether use or non-use of the latter is a violation of the principle could evolve and flip one method at a time. I call this evolutionary period "duelling precautions."

2 Application of the Precautionary Principle to Geoengineering

2.1 Basic Principles of Relevance

2.1.1 The Precautionary Principle

The precautionary principle has gained significant prominence in recent decades as a global tool for dealing with risk and uncertainty. Although there are numerous formulations of the principle,[12] there are also threads that tie these all together. Formalizing the traditional notion that one is "better safe than sorry," the precautionary principle counsels against potentially risky actions when it is unclear whether those actions can be rendered safe and non-destructive. Put another way, it supports regulation of activity that may cause harm, regardless of whether that regulation can be supported by hard evidence of potential harm.[13] Its practical effect is usually described as a burden of proof placed on the proponents of action – the burden of proving that action to be harmless, rather than a burden on regulators to

---

[12] *See* Stephen G. Wood et al., *Whither the Precautionary Principle? An American Assessment from an Administrative Law Perspective*, 54 Am. J. Comp. L. 581, 589 (2006) ("One aspect of the precautionary principle that has received considerable attention is that there are multiple formulations rather than a single, uniformly accepted formulation of the precautionary principle.").

[13] *See* J. Cameron & J. Abouchar, *The Precautionary Principle: A Fundamental Principle of Law and Policy for the Protection of the Global Environment,* 14 Boston Col. Int'l & Comp. L. Rev. 1, 2 (1991) ("The precautionary principle ensures that a substance or activity posing a threat to the environment is prevented from adversely affecting the environment, even if there is no conclusive scientific proof linking that particular substance or activity to environmental damage.").

prove that it is harmful.[14] Application of the precautionary principle often occurs in relation to economically motivated activity that may have a catastrophic and/or irreversible impact on important natural resources. For this reason, it is often used in contexts involving environmental protection.

2.1.2 Cost-Benefit and Risk-Benefit Analyses

Cost-benefit analysis weighs the costs of an action against the benefits of that action. It is highly dependent on information that is as complete as possible. It is akin to a mathematical formula that requires specific inputs to provide an output. Ideally one

---

[14] *See, e.g.*, Valerie J. Watnick, *The Lautenberg Chemical Safety Act of 2016: Cancer, Industry Pressure, and A Proactive Approach*, 43 Harv. Envtl. L. Rev. 373, 406–07 (2019) (discussing the "varying levels of precaution and burden shifting" across formulations of the precautionary principle); David A. Dana, *A Behavioral Economic Defense of the Precautionary Principle*, 97 Nw. U. L. Rev. 1315 (2003) ("In most formulations, the principle entails shifting the burden of proof to proponents of regulatory inaction in the face of health or environmental risk, although the precise standard for that burden of proof is not specified."); Ken Geiser, *Establishing a General Duty of Precaution in Environmental Protection Policies in the United States: A Proposal,* in Protecting Public Health and the Environment: Implementing the Precautionary Principle (Carolyn Raffensperger & Joel Tickner eds., 1999) ("The Precautionary Principle asserts that parties should take measures to protect public health and the environment, even in the absence of clear, scientific evidence of harm. It provides for two conditions. First, in the face of scientific uncertainties, parties should refrain from actions that might harm the environment, and, second, that the burden of proof for assuring the safety of an action falls on those who propose it."); Kirsten H. Engel, *State Environmental Standard-Setting: Is There A "Race" and Is It "To the Bottom"?*, 48 Hastings L.J. 271, 394 n.285 (1997) ("This is the point of the 'precautionary principle' which avers that activities should be subject to regulation before harm is demonstrated and thus shifts the burden of proving the 'harmlessness' of a challenged activity to the persons or entities who wish to engage in the activity."); David Favre, *Debate within the CITES Community: What Direction for the Future?*, 33 N.R.J. 875, 894 (1993) ("In effect this concept reflects a reallocation of the burden of proof for environmental issues. Rather than requiring that those wishing to stop the action show in advance the harm of an action, application of the precautionary principle suggests that an action should not be undertaken if it poses a risk, if not a certainty, of harm. In effect this places the burden of proof on those wishing to proceed with an action to prove lack of environmental harm before proceeding. The principle acknowledges that much of the human activity which causes environmental harm cannot be scientifically proven to cause such harm before or even after an event.").

The Wingspread Declaration provides the strongest example of this in the context of operationalizing agreements, stating: "When an activity raises threats of harm to human health or the environment, precautionary measures should be taken even if some cause-and-effect relationships are not fully established scientifically. In this context the proponent of an activity, rather than the public, should bear the burden of proof." *See* Science & Environmental Health Network, Wingspread Statement: A Common Sense Way to Protect Public Health & the Environment, Jan. 25, 1998, at 1. The Rio Declaration, on the other hand, while still shifting the burden against the action, softens the blow significantly: "In order to protect the environment, the precautionary approach shall be widely applied by States according to their capabilities. Where there are threats of serious or irreversible damage, lack of full scientific certainty shall not be used as a reason for postponing cost-effective measures to prevent environmental degradation." Rio Declaration on Environment and Development, UN Doc. A/CONF.151/26 (vol. I); 31 ILM 874 (1992).

is working with a definite cost (the expenses of building a new factory) and well-supported expected returns (the expected profits from the new factory, based on careful analysis of industry data). Sometimes one is forced to internalize externalities, which adds to the cost, while at other times external costs are not part of the calculus. Cost-benefit analysis is conducted both in the context of private actions and in regulatory decisionmaking. Indeed, federal agencies have conducted policy-level cost-benefit analyses ever since President Reagan mandated it via an executive order. It is also arguably what each of us does hundreds of times per day, before each and every move we make, albeit often unconsciously and at lightning speed.

Risk-benefit analysis is a variant of cost-benefit analysis that allows for some gaps in information, but is otherwise quite similar. Instead of (or in addition to) the costs of an action, we look at potential harms it will cause, also known as risk. For example, suppose we want to introduce a drug that will reduce the death rate for a given affliction by 10%, but it will also result in the death of one person per hundred who takes it. The people are not fungible, of course, in that some individuals will die who would otherwise have lived. Nonetheless, both sides of the equation can be quantified (one death per 100 equals .01 deaths), so the formula winds up looking similar to cost-benefit analysis, and is often considered to be a subset of cost-benefit analysis. It is key to understand the importance of risk assessment to the risk-benefit analysis,[15] which cannot be conducted without a quantified risk (whether based on empirical evidence or probabilities).

2.1.3 Risk Versus Uncertainty

Determining whether to apply the precautionary principle or risk-benefit analysis to actions that could cause harm is not simply a matter of preference. While it is certainly true that the two methods are philosophically different, and thus favored by different interests,[16] the choice between them is somewhat constrained by the nature of the data available.

The precautionary principle is an approach to dealing with uncertainty. When we do not yet have sufficient empirical data to predict how things will play out, nor a clear grasp on probabilities, this circumstance does not relieve us from making decisions. Both action and inaction are impactful choices and we must make one. The precautionary principle favors inaction until we can determine that action would be

---

[15] *See* David Driesen, *Cost-Benefit Analysis and the Precautionary Principle: Can they be Reconciled?*, 2013 Mich. St. L. Rev. 771, 776 (2013).

[16] Cass Sunstein has argued in favor of cost-benefit analysis rather than application of the precautionary principle. *See generally* Cass R. Sunstein, Laws of Fear: Beyond the Precautionary Principle (2005). *See also* Richard A. Posner, Catastrophe: Risk and Response 140 (2004). Many others, environmentalists in particular, have advocated for the precautionary principle, arguing that cost-benefit analysis fails to adequately account for non-economic interests. *See, e.g.*, Frank Ackerman & Lisa Heinzerling, Priceless: On Knowing the Price of Everything and the Value of Nothing (2004); Robert V. Percival, *Who's Afraid of the Precautionary Principle?*, 23 Pace Envtl. L. Rev. 21 (2005–2006).

safe. It is concerning to some that it tends to ignore risks tied to inaction,[17] as there are times when the benefits of a proposed action are not merely economic. While we can choose a more precautionary approach to known risk, the principle is strongly tied to uncertainty, as favouring inaction over taking known risks arguably pushes well into the realm of cost-benefit analysis with a subjectively high weighting of the known risk.

Risk-benefit analysis, which is a variant of cost-benefit analysis, is not designed to deal with uncertainty. Indeed, it is data-hungry. Risk analysis, which requires some quantification of data,[18] is an essential prerequisite to conducting a meaningful risk-benefit analysis.[19] Thus, when we have a good idea of the risks and probabilities, this data guides us into somewhat of a balancing approach. Sometimes there are risk probabilities on both sides of the equation, such as with new drugs. For purposes of this chapter it is thus important to understand the difference between risk probabilities and uncertainty, as we now turn to the rare circumstance of uncertainty on both sides of a decision whether to take a particular action.

2.2 *Geoengineering and Precaution*

2.2.1 Incompatibility with Traditional Application of the Principle

Operationalizing the precautionary principal presents significant challenges in the geoengineering context.[20] The precautionary principle traditionally pits apples against oranges. On one side of the balance, there is usually meaningful uncertainty regarding the probability of irreversible and/or catastrophic harm associated with deployment of some geoengineering options. The principle gives great weight to this side, both because of its potential severity, and in the interest of keeping all options on the table a bit longer.[21] On the other side, there is typically a known opportunity cost in the form of forgone economic activity. Opponents of the precautionary principle have at times accused proponents of disregarding this cost, but arguably it simply seeks to take advantage of this side's capacity to survive postponement.

---

[17] *See* Frank B. Cross, *Paradoxical Perils of the Precautionary Principle*, 53 Wash. & Lee L. Rev. 851, 862–63 (1996).

[18] Daniel A. Farber, *Uncertainty*, 99 Geo. L.J. 901, 909 (2011) ("Risk analysis requires that risks be quantified.").

[19] Driesen, *supra* note 11 at 776–77.

[20] Given the wide variation in techniques and risks, combined with the many interpretations of the precautionary principle (including the one considered here), it is challenging even to suggest that a universal approach to operationalizing the principle is even possible in the context of geoengineering. *See* Kevin Elliott, *Geoengineering and the Precautionary Principle*, 24 Int'l J. Applied Philosophy 237, 238 (2010).

[21] Cass R. Sunstein, *Irreversible and Catastrophic*, 91 Cornell L. Rev. 841, 845–46 (2006) (describing the idea that application of the precautionary principle is like purchasing an option).

People are generally more comfortable with limiting new benefits than with experiencing losses of what we already have – what *is* seems more valuable than what *might be*, all else being equal, as a result of the endowment effect.[22] In scientifically testing for risks, Talbot Page long ago noted the preference for false positives (erroneously finding something to be riskier than it is, resulting in unnecessary opportunity costs) over false negatives (erroneously finding something less risky than it is, resulting in greater harms to human health or environment).[23] Human beings are innately risk-averse, even though we may vary in degree.

Some of the most effective critics of the precautionary principle have avoided focusing on the *economic* opportunity costs. These scholars point instead to the non-economic risks of inaction, such as the known deaths per year from an illness that might be cured by a new drug.[24] If precautionary regulation keeps that drug off the market for another 2 years of research, we know that X people will die. This certainly takes some of the emotional victory away from the application of the principle, and yet it does not in fact alter its function. Indeed, we are still talking about opportunity costs – largely known opportunity costs – versus an unknown but possibly irreversible and catastrophic impact. *Lives not saved versus lives affirmatively taken*. Even when we escape the problem of economic versus human or environmental harm, we still tend to favor inaction over action when applying the precautionary principle. We accept our existing problems (temporarily, pending a reduction in the uncertainty that triggered the principle) rather than risk new and potentially more extreme ones.[25]

What we have yet to see in the literature on the precautionary principle is an analysis of how to apply it when we are not comparing apples with oranges. What happens when *human intervention* that creates *unquantifiable risk* of irreversible and catastrophic consequences is on both sides of the balance? This does not usually occur; typically we are deciding whether or not to take an action (or, more precisely, whether or not to regulate to restrain that action), so naturally there is always a no-action alternative, representing a mere opportunity cost. The question this chapter raises is what to do when we are without a genuine no-action alternative.

The decision whether to engage in geoengineering efforts to mitigate climate change presents a unique scenario for the application of the precautionary principle.

---

[22] "The much studied 'endowment effect' stands for the principal that people tend to value goods more when they own them than when they do not. ... A consequence of the endowment effect is the 'offer-asking gap,' which is the empirically observed phenomenon that people will often demand a higher price to sell a good that they possess than they would pay for the same good if they did not possess it at present." Russell Korobkin, *The Endowment Effect and Legal Analysis*, 97 Nw. U. L. Rev. 1227, 1228 (2003).

[23] Talbot Page, *A Generic View of Toxic Chemicals and Similar Risks*, 7 Ecology L.Q. 207 (1978).

[24] *See, e.g.*, Frank B. Cross, *Paradoxical Perils of the Precautionary Principle*, 53 Wash. & Lee L. Rev. 851 (1996); cite Farber, etc.

[25] *But see* David A. Dana, *A Behavioral Economic Defense of the Precautionary Principle*, 97 Nw. U. L. Rev. 1315, 1327–28 (2003) (arguing that our cognitive biases actually direct us the other way, in favor of avoiding the short-term opportunity cost even at risk of an uncertain later harm, so the precautionary principle actually serves to counteract that impulse and force us to do the right thing).

It pits action versus action, apples versus apples. Geoengineering methods come with planetary-scale risks of varying degrees. These risks are largely uncertain in probability and potentially irreversible and catastrophic in terms of the deployment of some geoengineering options at a large scale. The precautionary principle (if in use at all)[26] clearly applies to such action. Indeed, the more realistic the potential for geoengineering action becomes, the louder will be the chorus insisting upon application of the precautionary principle to geoengineering.[27]

The other side of the equation – the theoretical no-action alternative – is less clear. One might attempt to fit it into the traditional precautionary paradigm by defining it as the opportunity cost of failing to more rapidly mitigate climate change, arguing that it is similar to the opportunity cost of not saving lives with a new drug. This, however, is not a match. We are not actively taking the lives we might otherwise save with the drug – the drug is not the urgent antidote for a poison we are simultaneously administering.[28] Disease, however undesirable, is a baseline problem. It exists without any action on our part. Conversely, human activity is actively causing and rapidly aggravating the climate change problem. We are already tinkering with the planet in violation of the precautionary principle. The baseline condition is gone. There is only action and its consequences.

How shall we apply the principle when there is extremely risky human action on both sides of the equation? The precautionary principle "imposes a burden of proof on those who create potential risks, and it requires regulation of activities even if it cannot be shown that those activities are likely to produce significant harms."[29] This burden would theoretically apply against all "climate engineers," whether those who might intentionally address climate change via one of the SRM or CDR methods or those already adding GHGs into the atmosphere. However, it doesn't work very well in the latter case because the actors are too diffuse and efforts to collectively govern them have not been successful. Their failure to meet this burden of proof has not resulted in stopping the behavior.

The fact that we have failed to adequately apply the precautionary principle to GHG emitters does not mean that this human behavior has no role to play in the application of the principle to geoengineering decision-making. We can collect all climate engineering into one basket, whether that engineering is incidental, as in the case of GHG emissions, or purposeful, as with geoengineering. All of it is human

[26] The Rio Declaration, drafted by the United Nations Conference on Environment and Development (UNCED) in Rio de Janeiro in 1992 and signed by 172 countries, contains a set of principles for sustainable development. Rio Declaration on Environment and Development, UN Doc. A/CONF.151/26 (vol. I); 31 ILM 874 (1992). It sets forth the precautionary principle in its Principle 15, which would apply to a member country's decisionmaking regarding geoengineering, though it should be noted that the Rio Declaration is a nonbinding agreement.

[27] *See* Daniel Bodansky, *May We Engineer the Climate?*, 33 Climatic Change 309, 312 (1996).

[28] Imagine if we were to ask ourselves whether to serve up poison with or without its antidote, where not serving up poison in the first place is missing from the menu.

[29] Cass R. Sunstein, *Beyond the Precautionary Principle*, 151 U. Pa. L. Rev. 1003 (2003).

action that tinkers with the earth's climate system – none of it is entitled to baseline status.

This shift in our thinking of the status quo is highly relevant to the application of the precautionary principle, in light of its heavy status-quo preference. Under this approach, the interference already underway would not be elevated to status-quo position – put another way, the baseline would not be shifted for this analysis. We are comparing one risky action (climate change associated with rising GHG emissions) with another (climate change via GHG emissions plus some offsetting geoengineering method). We are not comparing a baseline of no-action with one of action.[30] For this analysis the baseline remains that of pre-industrial climate, and all human action is considered against that baseline.

Once we realize that the traditional methods of applying the precautionary principle do not fit well within the climate-altered model, we are left with the question of how to functionally apply precautionary analysis when uncertainty and human action fall on both sides of the equation. What will happen if the planet warms beyond our targets? Will a given geoengineering method actually cause the harm we fear it might? How severely? If we choose an SRM method are we very likely to encounter a disastrous termination effect? The precautionary principle accepts economic (and other opportunity cost) sacrifice in the face of uncertainty, preferring known costs and maintenance of the status quo, but how to weigh uncertainty against uncertainty and change against change?

This arguably requires a balancing approach in the same family as cost-benefit analysis or risk-benefit analysis, in spite of the fact that traditionally the precautionary principle rejects such approaches. However, its basis for that rejection – uncertainty surrounding potentially irreversible and catastrophic risk on just one side of the balance – is lost when both sides of the balance have this problem, both caused by ongoing or planned human action. This circumstance does not result in formal application of cost-benefit analysis, but could create something new that functions in a similar manner, but draws on more precautionary values: a *precaution-precaution analysis*.

2.2.2 Precaution-Precaution Analysis

As a practical matter, how does one compare one uncertain and hard-to-quantify risk with another? To begin with, it is key to distinguish this from risk-risk analysis. With two *known* risks, one can multiply probabilities against harms and compare results. Imagine a well-researched drug with known impacts. Without the drug, 1 in

---

[30] When there is no chance at maintenance of baseline conditions, we must weigh the relative extent of uncertainty and danger, even where there is not human action on both sides of the decision. *See* Richard Posner, Catastrophe: Risk and Response (2004). Where it is a scenario of human action versus human action, as discussed in this chapter, that increases the duty to engage in this analysis. It's not just a question of *whether* to take action when an asteroid is headed to the earth.

10 people will die of the target disease; with the drug, half of those deaths will be averted, but 1 in 100 will experience deadly side effects. Known risks can be easily compared with one another[31]; opposing uncertainties require a more complicated analysis, especially considering that even uncertainty itself is a matter of degrees.[32]

With precaution-precaution analysis we must decide between two scenarios[33] that both would have traditionally been banned by the precautionary principle. This is a rare and unenviable circumstance, as it will only arise in cases where it is too late to select the no-action alternative normally favored by the precautionary principle. It requires us to weigh two paths against each other where both are potentially disastrous. It will not be pretty, and could appear reckless, but to overcome this impression it is key to understand that those engaging in the process may not be the same people who got us into this mess. The decisionmakers applying a precaution-precaution analysis are dealt a terrible hand, but must still play it.[34]

There is no perfect way to compare two dangerous paths and choose one, but in an effort to avoid perfection-seeking paralysis we can develop some guiding principles for such undesirable circumstances. If we view all precaution as equal we cannot deal with such circumstances. The only way to compare two scenarios requiring precaution, and to select the one less deserving of that precaution (because we have to choose) is to formulate *degrees of precaution*.[35] Drawing on some of the bases for precaution generally, we can list factors to look for in measuring the degree of precaution to be applied:

---

[31] Although the math in this scenario is easy, this is not to suggest that risk-risk analysis is easy or objective. It cannot be based purely on the math without application of subjective philosophical principles, as human lives are not fungible. People have grappled for centuries with the moral riddle involving the choice of killing some people to save more. In this drug example, fewer people will die, but individuals who would have lived will be killed, which is a tough pill to swallow.

[32] *See* Daniel A. Farber, *Probabilities Behaving Badly: Complexity Theory and Environmental Uncertainty*, 37 U.C. Davis L. Rev. 145, 171 (2003) (weighing two uncertainties is inherently uneven, given that we will likely know a bit more about one than the other, such as with the societal impact of cutting carbon emissions versus the impact of future climate change).

[33] This discussion regards the binary choice of whether or not to engage in a particular (and risky) geoengineering action. Broader climate policy contemplates many possible approaches to addressing the catastrophe, including deep decarbonization, in which we move quickly and dramatically to reduce emissions.

[34] Indeed, once we find ourselves facing this type of precaution-precaution analysis, we will inevitably be forced to accept some harm to the environment. *See* Rickels, W.; Klepper, G.; Dovern, J.; Betz, G.; Brachatzek, N.; Cacean, S.; Güssow, K.; Heintzenberg J.; Hiller, S.; Hoose, C.; Leisner, T.; Oschlies, A.; Platt, U.; Proelß, A.; Renn, O.; Schäfer, S.; Zürn M. (2011): *Large-Scale Intentional Interventions into the Climate System? Assessing the Climate Engineering Debate. Scoping report conducted on behalf of the German Federal Ministry of Education and Research* (BMBF), Kiel Earth Institute, Kiel., at 102.

[35] Although we are attempting to quantify uncertainty in order to weigh it (because we have to), it is important to distinguish this from risk analysis, which seeks more reliable/mathematical data. Risk analysis drives us away from uncertainty, while precautionary analysis works with unavoidable uncertainty and ranks it based on considerations that are difficult to quantify.

1. Extent of potential harm
   a. Worst-case scenario;
   b. Most probable scenarios (even without precise probabilities it is often possible to determine one outcome to be more probable than another);
   c. Number and variability of potential harms
2. Extent of uncertainty;
   a. Applied separately to each potential harm determined above, as some risks may be ascertainable and others not as much;
   b. To what extent are we opening the door to entirely unforeseen harms, as opposed to harms we foresee but cannot assess the probability of occurring
3. Extent of avoidability (how great is the potential to control the feedbacks via adaptive management of the human actions);
4. Extent of reversibility
   a. Applied separately to each potential harm, as some harms may be more susceptible to repair than others

These factors are largely qualitative, and require some translation into quantitative values in order to be compared selectively.[36] To apply a precaution-precaution analysis we need metaphorical blocks that can be placed on a scale. We can apply weights to each factor and scores when each is applied, but doing so is inherently subjective and thus left to the leaders involved. Consider this a framework that can be utilized quite differently by people with differing subjective values.[37] Ultimately the precautionary principle, to the extent applicable in such an apples to apples situation, will counsel against the weightier side of the balance designed here. Precaution is on both sides, but one side has more of the precautionary criteria than the other and wins the preference of the precautionary principle. This sets the stage for duelling precautions.

---

[36] At least this is needed where we have a close call because both sides of the decision hold the potential to be catastrophic and irreversible, but at least one pair of scholars has already proposed that engaging in minor geoengineering research actions would be more precautionary than failing to do so. Jesse L. Reynolds & Floor Fleurke, *Climate Engineering Research: A Pre-cautionary Response to Climate Change?*, 7 Carbon & Climate L. Rev. 101 (2013). This idea works without a very challenging analysis because the authors have restricted the scope of the geoengineering action.

[37] A guiding framework with some flexibility is all that is desirable, as countries are moving toward developing their own norms and priorities, but need a method of operationalizing the principle toward those goals. *See generally* Elizabeth Tedsen & Gesa Homann, *Implementing the Precautionary Principle for Climate Engineering*, 7 Carbon & Climate L. Rev. 90 (2013).

3 The Evolution of Duelling Precautions

In the context of determining whether to proceed with geoengineering, the question becomes how to weigh the uncertain risks on both sides of the comparison, that is, between barrelling forward in the status quote, inevitably resulting in more serious climate change impacts, or including geoengineering options in the effort to avoid climate catastrophe. If the weighting given to the geoengineering action is heavier, then precaution counsels against it, but if the weighting shifts to the perils of a warming planet, precaution tells us we can no longer plummet into that abyss while sitting on our hands.

Because different methods of geoengineering – whether CDR methods such as BECCS, ocean fertilization, and direct air capture, or SRM approaches like sulphur aerosol injection or cloud brightening – come with different degrees of potential harms[38] and rewards, even if these are uncertain, they weigh differently in the balance against the risk of continued warming. Each approach places a different weight on the scale. The risk on the other side of the scale – the climate change imperilment side – increases over time. It will have one weight today, a greater weight in 20 years, and still greater in 50 years. As a result, the precautionary principle's preference will flip for each geoengineering approach at a different point in time, potentially informing regulatory choices in dealing with the various options.

Although the climate change impacts side of the balance will increase over time, neither side is static. As our understanding of the proposed geoengineering techniques improves, the blocks each has on the scale will get heavier or lighter. We will also develop new geoengineering ideas with varying precautionary weights. As a result, the balance on the scale will tip in favor of each geoengineering option at different times, if ever, and en route to these tipping points each will be in a continuous duel with the anthropogenic climate change option (which, again, is not a no-action alternative). Each time an approach wins this duel, it will instantly change the weight on both sides for all remaining geoengineering methods. Although this precaution-precaution analysis, which is unavoidable and requires unsavory quantification, looks a lot like cost-benefit analysis, it is entirely different. The factors and priorities are entirely precautionary.

4 Conclusion

We are already tinkering with the climate. The question isn't *whether* to do so or not, which would look like a traditional precautionary principle question. The answer would be no. The question is *how* to do so once it is too late to turn back, and

[38] For a description of the dangers of each major geoengineering proposal, see William C.G. Burns, *The Paris Agreement and Climate Geoengineering Governance: The Need for a Human Rights-Based Component*, CIGI Paper No. 111, pp. 12–17 (2016).

the precautionary principle would have us choose the least risky approach, to the maximum extent determinable in light of uncertainty of impacts. Whether the *how* includes geoengineering methods depends on this analysis, which will evolve over time as the approach of continuing to worsen climate change becomes riskier than the inclusion of each geoengineering approach, one by one in order of weighting.

References

1. Bodansky, D.: May we engineer the climate? Climatic Change. **33**, 309–312 (1996)
2. Cameron, J., Abouchar, J.: The precautionary principle: A fundamental principle of law and policy for the protection of the global environment. Boston Col. Int'l & Comp. L. Rev. **14**, 1–2 (1991)
3. Cross, F.B.: Paradoxical perils of the precautionary principle. Wash. & Lee L. Rev. **53**, 851–862 (1996a)
4. Cross, F.B.: Paradoxical perils of the precautionary principle. Wash. & Lee L. Rev. **53**, 851 (1996b)
5. Dana, D.A.: A behavioral economic defense of the precautionary principle. Nw. U. L. Rev. **97**, 1315 (2003a)
6. Dana, D.A.: A behavioral economic defense of the precautionary principle. Nw. U. L. Rev. **97**, 1315–1327 (2003b)
7. Elliott, K.: Geoengineering and the precautionary principle. Int'l J. Applied Philosophy. **24**, 237–238 (2010)
8. Engel, K.H.: State environmental standard-setting: Is there a "race" and is it "to the bottom"? Hastings L. J. **48**, 271–394 (1997)
9. Farber, D.A.: Probabilities behaving badly: Complexity theory and environmental uncertainty. U.C. Davis L Rev. **37**, 145–171 (2003)
10. Farber, D.A.: Uncertainty. Geo. L.J. **99**, 901–909 (2011)
11. Favre, D.: Debate within the CITES community: What direction for the future? N.R.J. **33**, 875–894 (1993)
12. Korobkin, R.: The endowment effect and legal analysis. Nw. U. L. Rev. **97**, 1227–1228 (2003)
13. Moreno-Cruz, J.B., Keith, D.W.: Climate policy under uncertainty: a case for solar geoengineering. Climatic Change. **121**, 431–444 (2013)
14. Page, T.: A generic view of toxic chemicals and similar risks. Ecology L.Q. **7**, 207 (1978)
15. Reynolds, J.L., Fleurke, F.: Climate engineering research: A pre-cautionary response to climate change? Carbon & Climate L. Rev. **7**, 101 (2013)
16. Sunstein, C.R.: Beyond the precautionary principle. U. Pa. L. Rev. **151**, 1003 (2003)
17. Sunstein, C.R.: Irreversible and catastrophic. Cornell L. Rev. **91**, 841–845 (2006)
18. Tedsen, E., Homann, G.: Implementing the precautionary principle for climate engineering. Carbon & Climate L. Rev. **7**, 90 (2013)
19. Watnick, V.J.: The Lautenberg Chemical Safety Act of 2016: Cancer, industry pressure, and a proactive approach. Harv. Envtl. L. Rev. **43**, 373–406 (2019)
20. Wood, S.G., et al.: Whither the precautionary principle? An American assessment from an administrative law perspective. Am. J. Comp. L. **54**, 581–589 (2006)

Printed in the USA
CPSIA information can be obtained
at www.ICGtesting.com
CBHW071115170424
7072CB00003B/264

9 783030 723712